Star-Formation Rates of Galaxies

Star-formation is one of the key processes that shape the current state and evolution of galaxies. This volume provides a comprehensive presentation of the different methods used to measure the intensity of recent or ongoing star-forming activity in galaxies, discussing their advantages and complications in detail. It includes a thorough overview of the theoretical underpinnings of star-formation rate indicators, including topics such as stellar evolution and stellar spectra, the stellar initial mass function, and the physical conditions in the interstellar medium. The authors bring together in one place detailed and comparative discussions of traditional and new star-formation rate indicators, star-formation rate measurements in different spatial scales, and comparisons of star-formation rate indicators probing different stellar populations, along with the corresponding theoretical background. This is a useful reference for students and researchers working in the field of extragalactic astrophysics and studying star-formation in local and higher-redshift galaxies.

ANDREAS ZEZAS is Professor at the University of Crete. He studies the X-ray emission from galaxies and its connection with their current and past star-forming activity. He has co-authored over 200 refereed publications and has been awarded an ERC Consolidator grant.

VÉRONIQUE BUAT is Professor of Astrophysics at Aix-Marseille University and Senior Member of the Academic Institute of France (IUF). She works on large multi-wavelength galaxy surveys and develops models to study star formation and interstellar obscuration from the local to the distant universe.

CAMBRIDGE ASTROPHYSICS SERIES

Series editors:
Andrew King, Douglas Lin, Stephen Maran, Jim Pringle, Martin Ward and Robert Kennicutt

Titles available in the series

28. Cataclysmic Variable Stars
    by Brian Warner
29. The Magellanic Clouds
    by Bengt E. Westerlund
30. Globular Cluster Systems
    by Keith M. Ashman and Stephen E. Zepf
33. The Origin and Evolution of Planetary Nebulae
    by Sun Kwok
34. Solar and Stellar Magnetic Activity
    by Carolus J. Schrijver and Cornelis Zwaan
35. The Galaxies of the Local Group
    by Sidney van den Bergh
36. Stellar Rotation
    by Jean-Louis Tassoul
37. Extreme Ultraviolet Astronomy
    by Martin A. Barstow and Jay B. Holberg
39. Compact Stellar X-ray Sources
    edited by Walter H. G. Lewin and Michiel van der Klis
40. Evolutionary Processes in Binary and Multiple Stars
    by Peter Eggleton
41. The Physics of the Cosmic Microwave Background
    by Pavel D. Naselsky, Dmitry I. Novikov and Igor D. Novikov
42. Molecular Collisions in the Interstellar Medium, 2nd Edition
    by David Flower
43. Classical Novae, 2nd Edition
    edited by M. F. Bode and A. Evans
44. Ultraviolet and X-ray Spectroscopy of the Solar Atmosphere
    by Kenneth J. H. Phillips, Uri Feldman and Enrico Landi
45. From Luminous Hot Stars to Starburst Galaxies
    by Peter S. Conti, Paul A. Crowther and Claus Leitherer
46. Sunspots and Starspots
    by John H. Thomas and Nigel O. Weiss
47. Accretion Processes in Star Formation, 2nd Edition
    by Lee Hartmann
48. Pulsar Astronomy, 4th Edition
    by Andrew Lyne and Francis Graham-Smith
49. Astrophysical Jets and Beams
    by Michael D. Smith
50. Maser Sources in Astrophysics
    by Malcolm Gray
51. Gamma-ray Bursts
    edited by Chryssa Kouveliotou, Ralph A. M. J. Wijers and Stan Woosley
52. Physics and Chemistry of Circumstellar Dust Shells
    by Hans-Peter Gail and Erwin Sedlmayr
53. Cosmic Magnetic Fields
    by Philipp P. Kronberg
54. The Impact of Binary Stars on Stellar Evolution
    by Giacomo Beccari and Henri M. J. Boffin
55. Star-Formation Rates of Galaxies
    edited by Andreas Zezas and Véronique Buat

Star-Formation Rates of Galaxies

Edited by

ANDREAS ZEZAS
University of Crete

VÉRONIQUE BUAT
Aix-Marseille University

CAMBRIDGE
UNIVERSITY PRESS

CAMBRIDGE
UNIVERSITY PRESS

University Printing House, Cambridge CB2 8BS, United Kingdom

One Liberty Plaza, 20th Floor, New York, NY 10006, USA

477 Williamstown Road, Port Melbourne, VIC 3207, Australia

314–321, 3rd Floor, Plot 3, Splendor Forum, Jasola District Centre, New Delhi - 110025, India

79 Anson Road, #06–04/06, Singapore 079906

Cambridge University Press is part of the University of Cambridge.

It furthers the University's mission by disseminating knowledge in the pursuit of education, learning, and research at the highest international levels of excellence.

www.cambridge.org
Information on this title: www.cambridge.org/9781107184169
DOI: 10.1017/9781316875445

First published 2021

A catalogue record for this publication is available from the British Library.

ISBN 978-1-107-18416-9 Hardback

Additional resources for this publication at www.cambridge.org/galaxies.

Contents

Figures

Colour versions of many figures can be found in the 'Resources' tab for this book on the publisher's website, www.cambridge.org/galaxies.

Tables

Contributors

Samuel Boissier
Laboratoire d'Astrophysique de Marseille (LAM), CNRS, Aix-Marseille University, France

Médéric Boquien
Astronomy Centre (CITEVA), University of Antofagasta, Chile

Denis Burgarella
Laboratoire d'Astrophysique de Marseille (LAM), CNRS, Aix-Marseille University, France

Daniela Calzetti
Department of Astronomy, University of Massachusetts, Amherst, USA

Andrew Cole
School of Natural Sciences, University of Tasmania, Hobart, Australia

Daniel Dale
Department of Physics and Astronomy, University of Wyoming, Laramie, USA

J. J. Eldridge
Department of Physics, University of Auckland, New Zealand

John. S. Gallagher III
Department of Astronomy, University of Wisconsin–Madison, USA

Karl D. Gordon
Space Telescope Science Institute, Baltimore, USA
Sterrenkundig Observatorium, Ghent University, Ghent, Belgium

Brent Groves
International Centre for Radio Astronomy Research, University of Western Australia, 7 Fairway, Crawley, WA 6009, Australia

Tereza Jerabkova
ESTEC/SCI-S
Keplerlaan 1
2200 AG Noordwijk
Netherlands

Pavel Kroupa
Helmholtz Institute for Radiation and Nuclear Physics, Bonn, Germany
Faculty of Mathematics and Physics, Astronomical Institute, Charles University, Prague, Czech Republic

Cristina Popescu
Jeremiah Horrocks Institute, University of Central Lancashire, Preston, UK
Astronomical Institute of the Romanian Academy, Bucharest, Romania

Giulia Rodighiero
Department of Physics and Astronomy "G. Galilei", University of Padova, Italy

Elena Sabbi
Space Telescope Science Institute, Baltimore, USA

E. R. Stanway
Department of Physics, University of Warwick, Coventry, UK

Andreas Zezas
Department of Physics, University of Crete, Heraklion, Greece
Institute of Astrophysics, Foundation for Research and Technology–Hellas, Heraklion, Greece

Preface

Star formation is one of the main mechanisms of energy production in the universe and one of the key processes that are linked to the evolution of galaxies. Over the past two decades we have witnessed an explosion of data from local and distant galaxies across the entire electromagnetic spectrum. These observations gave us an unprecedented picture of the star-forming activity in galaxies, the parameters it depends on (e.g. gas content, physical conditions in the interstellar medium, dynamical state of galaxies), and its evolution over cosmic time. The common denominator in all these studies is the use of diverse techniques for quantifying the recent star-forming activity in the different environments.

Indeed, the availability of a wealth of data in combination with advances in stellar astrophysics, astrophysics of the interstellar medium and radiative transfer modelling, and numerical simulations has led to the development of a variety of methods for measuring the intensity of star-forming activity using proxies such as direct detection of stars or their remnants, direct measurement of their stellar light, and measurements of the reprocessed stellar emission by the interstellar medium.

The purpose of this book is to provide an up-to-date and comprehensive review of the methods used to measure the intensity of star-forming activity in galaxies (their star-formation rates). However, a presentation of these relevant methods would be incomplete without discussing their astrophysical foundation, and the different factors that affect their precision and accuracy.

Therefore, in Part I of this book we present a detailed account of the stellar Initial Mass Function, stellar populations and their evolution, and absorption of stellar radiation by the interstellar medium. Special care is taken to discuss how these factors influence our measurements of star-formation rates. In Part II of this volume, we present the different methods for measuring star-formation rates: resolved stellar populations, broad-band photometry, emission lines, spectral energy distributions, and emerging indicators such as high-energy emission and gravitational-wave sources. Special care is taken to discuss the advantages and limitations of different indicators, as well as their cross-calibration in galaxy-wide and sub-galactic scales.

Although the subject of this volume is rather technical (but relevant to most aspects of extragalactic astrophysics), we tried to give an overview of the latest advances in the field, while providing the relevant introductory material. The book is written at the advanced undergraduate/starting graduate level, expecting from the reader familiarity with astrophysics terminology, and at least basic knowledge of stellar evolution and astrophysics of the interstellar medium and galaxies. We hope that this volume will be a useful resource for graduate students and researchers who would like to learn more about how we can measure one of the most characteristic properties of galaxies, and the factors affecting these measurements.

We would like to thank all the contributors in this volume for their excellent presentation of the different topics relevant to the measurement of star-forming activity in galaxies. Also, we would like to thank the Cambridge University Press editorial staff for their help in the preparation of the manuscript and their patience during the lengthy editing process.

Part I

Background

1
Introduction

SAMUEL BOISSIER AND GIULIA RODIGHIERO

1.1 Star Formation in the Context of Galaxy Evolution

The formation of stars is of paramount importance in the context of galaxy evolution. Stars emit light. This light may change the physical condition of the gas in the universe (e.g. ionization of the neutral gas), or it may reach our telescopes, allowing us to study the history of galaxies, to understand the large-scale structure of the universe, or to probe the invisible dark matter filling the universe. Stellar light is also our main source of information concerning galaxies. The most massive stars evolve in the blink of a cosmic eye and distribute in their surroundings a variety of chemical elements (some synthesized in their core during their evolution) and a huge amount of energy that has the potential to remove material from galaxies, and prevent the formation of new stars or, on the contrary, promote it by inducing shocks or compression waves. Other elements are made in intermediate-mass stars that release them at a later time after their birth. Stars of small mass also play a very important role by trapping material for timescales larger than the age of the universe! Given that stars play such a fundamental role in the evolution of galaxies, some of the key questions in the field of galaxy evolution and cosmology are related to the history of their formation in galaxies. Some questions concern individual galaxies: How does star formation and its history depend on the environment or on the mass of galaxies? Others are related to cosmic scales: When do stars form globally in the history of the universe? When and how does starlight contribute to the reionization phase in the early universe?

A major goal of this book is to provide the reader with clear explanations and up-to-date definitions and references concerning the star-formation rate (SFR) in galaxies, its physical implications, together with an overall presentation of the theoretical and observational background.

Despite its importance, the details of the physics of star formation are still eluding, because of the interplay of several fundamental processes (such as gravity, turbulence, cooling, radiation, magnetic fields). The actual formation of individual stars is the subject of many studies (see e.g. the reviews by McKee and Ostriker, 2007; Hennebelle and Commerçon, 2014). Part of this complexity comes from the fact that these processes occur at completely different scales (from nuclear reactions starting in the very core of a newly formed star to the swing of majestic giant spiral arms extending on several tens of kpc that can locally enhance the gas density and help the collapse of a molecular cloud; and even on the largest scales with the accretion of gas along cosmic filaments, filling the gas reservoir for future star formation). This book is not focusing on the small-scale physics of star formation. Instead, it is written for students or researchers working in the field of galaxy evolution, for whom what matters is the collective behavior of star formation rather than the formation of individual stars. For those astronomers looking at galactic scales, star formation is a global process, playing the fundamental role of transforming interstellar gas into stars and metals. Star formation on galactic scales is the subject of many studies both in the Milky Way and in other galaxies (see e.g. the review by Kennicutt and Evans, 2012). It is crucial to be able to determine the SFR with good reliability, regardless of the details of the small-scale physics. Thus the SFR – being one of the most important properties of a galaxy (as a whole), a region of a galaxy, or the universe on the cosmic scale – is of paramount importance to understand how it can be determined, and the degree of realibility or uncertainty of these measurements. After basic definitions (see Section 1.2), we will briefly introduce the various ways to measure the SFR (see Section 1.3). Other chapters in this book will be devoted to the complementarity of different SFR indicators, as well as how they can introduce selection biases leading to wrong or incomplete evolutionary interpretations. Finally, we will mention some important current results concerning the SFR laws (see Section 1.4) and the SFR history (see Section 1.5). Both fields are very active and will remain so in the years to come. They both rely on proper SFR measurements.

1.2 Definitions

The SFR is by definition the mass that is turned into stars per unit time. The unit of choice is usually solar masses per year ($M_\odot\ yr^{-1}$). This is a convenient unit being the order of magnitude of the SFR in the galaxy most important to mankind: the Milky Way.

Very often, we wish to determine the SFR locally in galaxies: the SFR in a given region, or at a given radius. It is thus useful to define the SFR density, either in volume or surface density, respectively ρ_{SFR} (in $M_\odot\ pc^{-3}\ Gyr^{-1}$ and $M_\odot\ kpc^{-3}\ yr^{-1}$, which are numerically identical) and Σ_{SFR} (usually in $M_\odot\ kpc^{-2}\ yr^{-1}$ or in $M_\odot\ pc^{-2}\ Gyr^{-1}$). The volume–density definition is useful as the interstellar medium has a 3D distribution and star formation occurs in localized regions within this structure. On the other hand, surface densities have the advantage of being easy to relate to the surface photometry that we can readily measure in external galaxies. Also, several physical processes depend on the local surface density within a galaxy disk. For instance, the hydrostatic equilibrium involves the dispersion of the gas on one hand and the local gravity on the other, the latter being set by the surface density of baryons in the disk (in stellar or gas form).

Finally, cosmologists who want to determine the cosmic history of star formation also use a density ρ_{SFR} that represents a "cosmic" density, i.e. the average density over very large scales. The SFR density unit in this case is usually $M_{\odot}yr^{-1}\,Gpc^{-3}$ (e.g. Madau and Dickinson, 2014) adapted to the very large scales over which such an average may make sense.

A last introductory point concerns the distribution of the masses of stars themselves when they are formed. This Initial Mass Function (IMF) is, of course, extremely important. Indeed, the type of chemical elements that will be produced by a generation of stars or its stellar-light output as a function of time heavily depends on the IMF, since mass is the main parameter setting the properties of stars. Naturally, its influence will be discussed in many of the chapters of this book. The IMF is defined as a function of mass $\xi(m)$ describing the number of stars per mass interval dm, i.e. $dN = \xi(m)dm$. The reader should be aware of the fact that the IMF is sometimes defined per logarithmic mass interval. It is trivial to switch from one definition to the other. Finally, the IMF is often normalized in the following way:

$$\int_{M_l}^{M_u} m\xi(m)dm = 1, \tag{1.1}$$

where M_l and M_u are the lower and upper mass limits (around 0.1 and 100 $M_{\odot}$). With this normalization, the SFR describes the amount of material going into stars, and the IMF the statistical distribution in stellar masses of this material. The IMF is discussed in detail in Chapter 2.

1.3 Measuring Star-Formation Rates

Under the above definition, the SFR is instantaneous. It describes the formation of stars at a given time. This is usually what is computed in models. However, the rest of the book will demonstrate how difficult it is (and probably impossible in most cases) to determine an instantaneous SFR from the observational point of view. Indeed, most SFR tracers are, in fact, associated to a timescale on which they are sensitive. In the following text, we briefly review the observations that can be used as SFR indicators.

All methods aim at probing the emission from recently formed stars, avoiding, as much as possible, contamination from older stellar populations or other sources of emission. The only direct method available to quantify the SFR is by counting young stars or events tracing recent star formation (such as supernova remnants), but this method is currently applicable only to our Milky Way and very few nearby galaxies as it requires resolving objects on very small scales. For the vast majority of galaxies, the SFR can be derived from the integrated light, by applying calibrations that have been defined by assuming a certain IMF and a given star-formation history (SFH). Several indicators have been applied in the literature, spanning a wide range of photometric and spectral observations from ultraviolet (UV) to radio. Each one of these indicators suffers from its own drawbacks and is sensitive to emission from stars with slightly different stellar masses. The consequence is that each indicator samples a slightly different star-formation timescale.

The classical approach is to look for emission lines related to the production of young stars. Massive young stars produce a large amount of energetic photons ($\lambda = 912$Å) that ionize the surrounding gas and produce hydrogen recombination lines by cascade (Osterbrock and Ferland, 2006). The brightest of these lines, the Hα emission, is widely used as an SFR indicator, mainly because it falls in the optical regime for local galaxies, but in principle all the hydrogen recombination lines can be used as an SFR tracers (see e.g. Kennicutt, 1998a). Since only extremely massive stars ($M_{star} \simeq 20\,M_{\odot}$) produce an amount of high-energy radiation sufficient to ionize the hydrogen in the interstellar medium (ISM), the Hα emission is sensitive to the recent SFH, probing timescales $\tau_{SF} \sim 20$ Myr, and thus providing an almost instantaneous measurement of the SFR.

The integrated spectrum of a galaxy in the wavelength range 1200Å $< \lambda <$ 2800Å is dominated by the emission of young stars ($M_{star} \simeq 10\,M_{\odot}$) so that the UV emission is proportional to the SFR. Calibrations to compute the SFR from the galaxy UV emission have been derived using spectro-photometric models that assume continuous SFH on a timescale of 10^8 yr. This technique can be applied to star-forming galaxies in a wide redshift range, as it requires only photometric data that are relatively easy to obtain. However, UV emission is extremely sensitive to the form of the IMF, dust absorption, and, to a lesser extent, the galaxy metal content (as metals efficiently absorb radiation in the UV part of the spectrum). Also, UV emission can be heavily contaminated by the emission of an active galactic nucleus (AGN).

A significant fraction of the optical/UV luminosity of a galaxy can be absorbed by the interstellar dust and reemitted in the thermal IR at wavelengths of $10\mu m < \lambda < 1000\mu m$. Since most of the UV/optical emission comes from star formation, the infrared luminosity can be used as an SFR tracer. In this case, the calibration has to assume, in addition to an IMF and an SFH, a dust geometry and a dust optical depth. The most widely used calibration of Kennicutt (1998a), for example, assumes that an optically thick, compact dust component is heated by young stars in a continuous burst of $\sim$100 Myr. The presence of an AGN or an older stellar population may contribute to dust heating. Far infrared (FIR) observations provide a very suitable tool for deriving the SFR without suffering dust biases. Therefore, FIR indicators are crucial when deriving the star-formation activity of dust-rich star-forming galaxies. When broad wavelength coverage is available, indicators may be combined to derive the total SFR, for example, by adding the bolometric FIR emission to the observed UV light (Papovich et al., 2007) or the Hα emission (Kennicutt et al., 2009).

State-of-the-art techniques to recover the SFR of galaxies from spectro-photometric observations covering the electromagnetic spectrum (ideally from the far-UV up to the sub-millimeter and radio bands) include the treatment of energy balance between the radiation absorbed in the UV-to-optical regime and that reemitted at longer wavelengths in the FIR. The codes most commonly used by the community include `CIGALE`[1]

[1] Code Investigating GALaxy Emission; https://cigale.lam.fr

(Burgarella et al., 2005; Boquien et al., 2019) and `Magphys`[2] (da Cunha et al., 2008). This approach minimizes the degeneracy related to the amount of dust extinction against age and metallicity of the stellar populations, when limited to UV/optical observations. A further level of complexity is introduced by models that include the solution of the radiative-transfer equation, thus providing a physical description of the ISM and the geometry of the star birth places (e.g. `GRASIL`; Silva et al., 1998).

1.4 Star-Formation 'Laws'

1.4.1 Context

Considering the fundamental role played by stars in the evolution of galaxies and in our ability to observe the universe, it is clear that it is of paramount importance to understand how stars form on galactic scales, and to be able to describe it in physics terms. We ideally need to find large-scale "laws" telling us what should be the SFR in a galaxy (or a part of galaxy), knowing other physical conditions such as gas and dust density, temperature, velocity field, or the galaxy's local environment. Such laws are absolutely needed when we turn to simulations, to be used as recipes. In galactic chemical-evolution models, where the evolution of basic quantities (abundances, gas fraction, stellar masses and ages, star-formation rate) is followed in a simple formal way, a necessary step is to decide what the SFR is at a given time before computing all other quantities at the next epoch. In semi-analytic models (SAMs), the accretion history is given by a dark-matter halo-merger tree, but the baryon physics similarly rely on some assumptions concerning the SFR. Even in full hydrodynamical simulations following not only the dark matter but also the gas and stars, it is impossible to fully resolve the scales at which individual stars form. Thus, so-called "sub-grid" physics have to be implemented to simulate star formation.

We thus need "star-formation laws" to understand what affects the SFR on galactic scales, and, in a pragmatic way, to implement in models. In an ideal world, we would like these laws to emerge from first principles. A lot of theoretical ideas have been proposed (see e.g. a compilation in Boissier, 2013). They will not all be reviewed here, but it is worth mentioning a few of the very general ideas that have been discussed.

One important issue in this field is the existence (or not) of a threshold for star formation. The work of Toomre (1964) has presented the instability of a galactic disk against axi-symmetric (or large-scale shell) perturbations. He introduced the concept of a threshold radius beyond which the local density is too small to induce collapse with respect to the orbits and dispersion velocity. There are many variations of the "Toomre parameter" Q, taking into account, for instance, the contribution of gas and stars, or based on the

[2] Multi-wavelength Analysis of Galaxy Physical Properties; http://astronomy.swinburne.edu.au/~ecunha/ecunha/MAGPHYS.html

shear due to differential rotation. The concept has been largely discussed and compared to profiles with various SFR tracers (e.g. Martin and Kennicutt, 2001; Boissier et al., 2007; Goddard et al., 2010). Another possible reason for the existence of a threshold is the balance between a warm and cold gas phase (in which stars form), which is weakly influenced by the metallicity, the gas fraction, and the flux of ionizing photons (Schaye, 2004).

The other issue is to understand what determines the amount of star formation given the physical conditions in a galaxy (or a galaxy region). Many propositions have been discussed (e.g. Leroy et al., 2008; Boissier, 2013, and references therein). One simple way to present this question is to write (e.g. Larson, 1992; Wang and Silk, 1994):

$$\Sigma_{\rm SFR} = \epsilon \frac{\Sigma_{\rm GAS}}{\tau}, \tag{1.2}$$

where ϵ is an efficiency and τ the timescale to form stars from gas. The most basic assumption (Madore, 1977) is that τ is set by the free-fall time $\left(\tau \propto \rho_{\rm GAS}^{-0.5}\right)$. For a constant scale height, this gives that $\Sigma_{\rm SFR}$ is proportional to $\Sigma_{\rm GAS}^{1.5}$. A wide variety of ideas can be found in the literature to decide which physical processes set up the timescale τ, or if other factors affect ϵ. These ideas include the possibility to consider hydrostatic equilibrium, gravitational collapse, self regulation, the influence of spiral arms sweeping up material at the rotation frequency, and cloud-cloud collisions. As a result, we can deduce that the $\Sigma_{\rm SFR}$ may be proportional not only to $\Sigma_{\rm GAS}^{\rm N}$, with N varying between 1 and 2 (Wyse, 1986; Larson, 1992; Bigiel et al., 2008; Leroy et al., 2008) but also to the stellar surface density to some power (Abramova and Zasov, 2008; Blitz and Rosolowsky, 2006; Corbelli, 2003), or to the angular rotation velocity Ω (Wyse, 1986; Wyse and Silk, 1989; Larson, 1992; Tan, 2000). These dependencies are quite degenerate, so that it is not clear which effects are really at play or are dominant.

In the next sections, we show a few empirical works providing the state of the art on such star-formation laws that play an important role in the community.

1.4.2 Simple Relations between Gas and Star-Formation Rate

The first study trying to relate the SFR (in fact the number of young stars) and the gas density in the Milky Way dates back to Schmidt (1959). Many empirical works have followed up to today, increasing the size of the samples, improving the resolution, using different SFR and gas tracers. We will only mention a few of the most famous studies to illustrate this part. The work of Kennicutt (1998b) is especially notable because it combine circum-nuclear starbursts, centers of galaxies, and disk averages to construct a relation between the total gas density and the SFR density over five orders of magnitude. It found an index of N = 1.4 for the $\Sigma_{\rm SFR}$– $\Sigma_{\rm GAS}$ relation, close to the expectation obtained with the free-fall time scale (1.5). As can be seen in the left part of Fig. 1.1, this relation nevertheless presents some dispersion. Especially, using only the "normal" galaxies, a

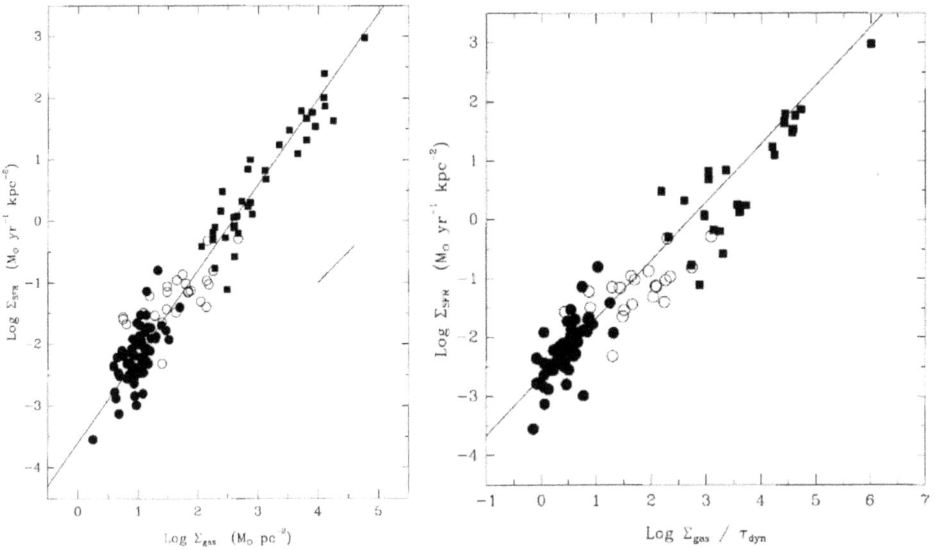

Figure 1.1 *Left*: The classical Schmidt law (slope 1.4) obtained by combining entire galaxies (circles), centers of galaxies (open circles), and circum-nuclear starbursts (squares).
Right: The dynamical Schmidt law for the same data. Figures from the seminal paper of Kennicutt (1998b), ©AAS. Reproduced with permission.

steeper index would have been found (as it was the case in other works). While this relation is not the "ultimate law" (as sometimes used by modelers), it is nonetheless a very influential one.

In more recent years, it has became possible to obtain resolved images of galaxies in multiple wavelengths, sensitive to young star emission (e.g. ultraviolet with GALEX[3]), to dust emission (with *Spitzer Space Telescope (Spitzer)*[4] or *Herschel Space Observatory (Herschel)*[5]), to ionized-gas, and to the neutral-gas content (molecular-gas maps, usually traced by CO lines, and neutral HI gas). In the coming years, we can expect even more well-resolved data concerning the gas and the star formation in galaxies, with many integral field units being now available, and with the new large radio/millimetric observatories (NOEMA,[6] ALMA,[7] SKA[8]). A noted study in the domain of resolved star-formation law

[3] GAlaxy Evolution Explorer; www.galex.caltech.edu/index.html
[4] *Spitzer Space Telescope*; www.spitzer.caltech.edu
[5] *Herschel Space Observatory*; www.cosmos.esa.int/web/herschel/overview
[6] NOrthern Extended Millimeter Array; http://iram-institute.org/EN/noema-project.php
[7] Attacama Large Millimiter/submillimeter Array; www.almaobservatory.org/en/home/
[8] Square Kilometer Array; www.skatelescope.org

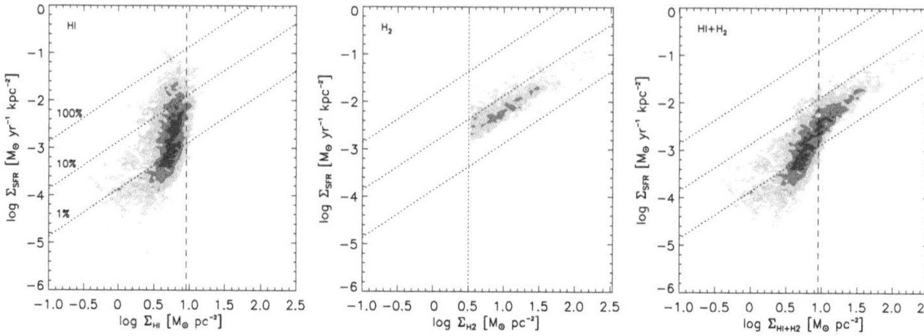

Figure 1.2 Local star-formation laws observed in few 100-parcsec pixel scales for the HI, H_2, and total gas. From Bigiel et al. (2008) ©AAS. Reproduced with permission. Color version available online.

was conducted by the THINGS[9] team which obtained exquisite HI maps of a few nearby representative galaxies. Figure 1.2 shows the star-formation law published by Bigiel et al. (2008), as observed at scales of a few hundred parsecs. A new paradigm emerged from this study with a very good correlation (with slope of unity) between the molecular gas and the SFR. In contrast, the HI saturates at a density of around 10 $M_\odot$ pc^{-2}, indicating that the gas becomes mostly molecular at high total gas density. In their results, the molecular fraction correlates with various quantities (distance from the center, stellar density, pressure, orbital timescale). Several works have also suggested that star formation is closely related to the dense gas that may be better traced by molecules such as HCN rather than CO (Gao and Solomon, 2004).

The trends in Fig. 1.2 are actually well reproduced by the simple local model of Krumholz et al. (2009), which predicts the correct amount of molecular gas and star formation once the total gas density is set. However, the spatial distribution of the total gas is not set in this model, and should result from many other physical effects (spiral arms, accretion) that could affect other star-formation laws (i.e. global or radial ones instead of the local ones).

1.4.3 Influence of Other Parameters

Kennicutt (1998b) noted that a dynamical law (right panel of Fig. 1.1) also provides a good fit to his data in nearby galaxies. In this case, the gas density is divided by the dynamical timescale of the system. Star-formation laws including a dynamical factor (e.g. rotation timescales for disks) are thus credible alternatives. Several works have also proposed

[9] The HI Nearby Galaxy Survey; www.mpia.de/THINGS/Overview.html

(and tested) the possible influence of the stellar density on star formation (e.g. Dopita and Ryder, 1994; Shi et al., 2011) that can find some theoretical support as seen above. The influence of the stellar density on the gas scale-height may result in better laws involving the volume densities rather than the surface densities. A few works have attempted to study the volume star-formation law (Abramova and Zasov, 2008; Bacchini et al., 2017). Another possibility is that several timescales must be combined to properly describe star formation itself and the inhibition to form stars that results from it afterwards (Madore, 2010; Semenov et al., 2017).

The main difficulty concerning additional parameters is to distinguish between variations with galactocentric radius, stellar density, and dynamical timescale, as all these quantities vary in a correlated way.

1.4.4 The Star-Formation Law at High Redshift

Most early works on the star-formation law were performed in the nearby universe. Its evolution with redshift is a very important issue, especially for scientists who wish to model the evolution of galaxies over cosmic times. In recent years, it became possible to obtain constraints on the gas (mostly molecular) in distant galaxies (with NOEMA or ALMA, for instance), allowing astronomers to attempt to empirically determine star-formation laws at various redshifts. In the future, we shall also have access to the HI content of distant galaxies (SKA) and pursue these studies. The first studies suggested a higher efficiency in starburst galaxies at redshifts up to $z \simeq 2.5$, by a factor of 4 or so. However, a lot of unresolved issues make this picture unclear. In particular, the possible variation of the conversion factor for the molecular gas tracer (i.e. the conversion factor between an easily traced species, typically CO, and the total molecular mass which is dominated by H_2; see e.g. Carilli and Walter, 2013). The higher efficiency could also be related to the inclusion of starbursting galaxies at high redshift, in which the mode of star formation may be more efficient, e.g. because of the compressive mode of turbulence (Renaud et al., 2014). It was also noticed that the dynamical law seems to hold at high redshift (Daddi et al., 2010; Genzel et al., 2010). Other attempts to determine the high-redshift star-formation law made use of the dust emission of ALMA (Scoville et al., 2016) or of stacked HI absorbers (Rafelski et al., 2016), and various statements on the variation of efficiency with redshift have been issued. These observations, however, involve different types of gas tracers or SFR tracers and different range of masses or density, which makes the comparison between results difficult.

Figure 1.3 shows a compilation of star-formation laws in the nearby and more distant universe, illustrating quite some diversity. More work will be needed in the future to get a complete and clear picture of the star-formation laws, on different scales, and their evolution with redshift. For this purpose, it will be important to use accurate determinations of the SFR, and be well aware of the limitations of each method.

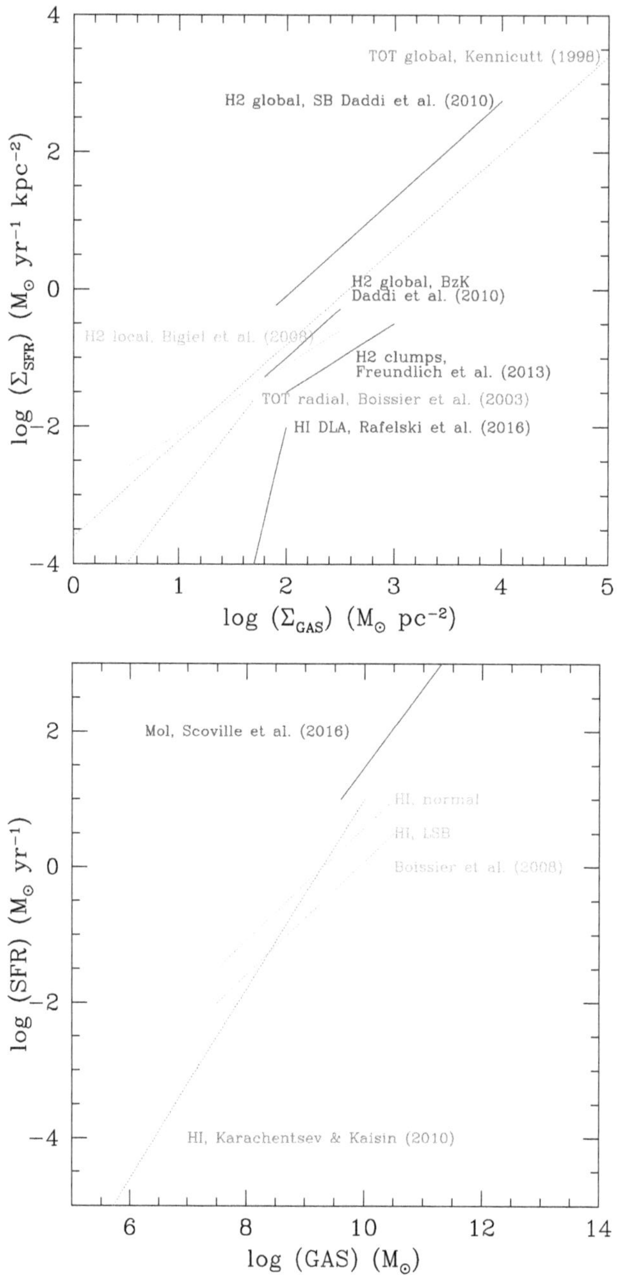

Figure 1.3 Compilation of published star-formation laws at low (dashed lines) and high (solid lines) redshifts. This is an indicative visualization where only the average trends and the range studied in each paper is shown. The gas tracer used is indicated: H_2, mol (for molecular), HI, total. In the *left panel* showing relations between surface densities, we also indicate the type of law studied: global (average over full galaxies or large regions), local (pixels) or clumps (regions), radial (azimuthal averages smoothing over orbital timescales and spiral arm frequency). The *right panel* shows global relations between the total SFR and gas mass. Color version available online.

1.5 The Star-Formation Rate History of Galaxies

A single SFR measurement in an individual galaxy provides limited information on its past, and even less on the main history of the stellar-mass assembly in various galaxy populations.

It is necessary to study the history of star formation in galaxies. In nearby galaxies, color-magnitude diagrams of individual stars or of galactic regions can give us clues about their past. Contrary to the local universe, where galaxies can be spatially resolved to study *in situ* the concomitant processes that drive the conversion of gas into young stars, if we wish to understand the role of star formation at different cosmic epochs we certainly have to deal with integrated properties of galaxies. For this reason, lessons learnt from the great laboratories in nearby galaxies should be kept in mind when extrapolating information to the whole galaxy scales.

1.5.1 Main Sequence and the Specific Star-Formation Rate

Most galaxies at high redshifts are very actively forming stars, with SFRs of the order of hundreds of $M_{\odot}\ yr^{-1}$ being quite common. In the local universe instead, galaxies with such high SFRs are very rare and are called Ultra-Luminous Infrared Galaxies (ULIRGs, with $L_{IR} > 10^{12}\ L_{\odot}$; Sanders et al., 1988). Such objects are caught in a transient, starburst event, likely driven by a merger having boosted both their SFR and their FIR luminosity. By analogy, similar high-redshift galaxies were first regarded as starburst objects until it became apparent that the data were suggesting a radically different picture. A first suspicion that a new paradigm was needed came from the discovery that over 80 per cent of a BzK K-band-selected sample of $z \sim 2$ galaxies were actually qualifying as ULIRGs (Daddi et al., 2005). Clearly, it was very unlikely that the vast majority of galaxies had all been caught in the middle of a transient event. As shown below, enhanced SFRs ought to be the norm rather than the exception at high redshifts. This was indeed demonstrated in a series of seminal papers (Daddi et al., 2007; Elbaz et al., 2007; Noeske et al., 2007), showing the existence of a tight correlation between SFR and stellar mass (M_{star}) of the form SFR $\propto f(t){M_{star}}^{1+\beta}$, where $f(t)$ is a declining function of cosmic time (an increasing function of redshift).

This relation is followed by the majority of star-forming galaxies, with a dispersion of ~0.3 dex, both at high redshifts (references above) and in the local universe (Brinchmann et al., 2004). Following Noeske et al. (2007), the correlation is called the main sequence (MS) of star-forming galaxies. The existence of the main sequence implies that most star-forming galaxies are in a quasi-steady star-formation regime, rather than stochastic. In fact, no signs of mergers have been found through dynamical measurements in many high-redshift SF galaxies (e.g. Cresci et al., 2009; Förster Schreiber et al., 2009; Law et al., 2009). But the interpretation of the main sequence goes beyond that, presenting several other important ramifications. It dictates a very rapid stellar-mass growth of galaxies at early times, paralleled by a secular growth of their SFR itself (e.g. Renzini, 2009; Peng et al., 2010), quite at odds with the widespread assumption of exponentially declining SFRs (as argued by e.g. Maraston et al., 2010; Reddy et al., 2012). Even more importantly, the slope, β, controls the relative growth of high-mass versus low-mass galaxies, thus directly impinging on the evolution of the galaxy stellar-mass function (Peng et al., 2010; Lilly et al., 2013). The absolute value of the specific SFR (sSFR $\equiv$ SFR/M_{star}) sets the clock of galaxy evolution (Peng et al., 2010), determining the growth rate of individual galaxies, hence controlling their lifetime before they are quenched. While the existence of the main sequence is generally undisputed, its slope and scatter may differ significantly from one observational study to another, depending on the sample selection and the adopted SFR and stellar-mass diagnostics. Selecting galaxies in a passband that is directly sensitive to the SFR (such as the rest-frame UV or the FIR) automatically induces a Malmquist bias in favor of low-mass galaxies with above average SFRs, thus flattening the resulting SFR–M_{star} relation. This effect is clearly seen in Herschel FIR-selected samples, where formally $\beta \sim -1$, but where only a tiny fraction of galaxies are detected at low stellar masses, i.e. those few really starbursting ones (Rodighiero et al., 2010, 2011). Rodighiero et al. (2014) achieved the reassuring conclusion that a wide variety of SFR indicators, such as the rest-frame UV continuum, the mid- and the far-IR, the 1.4 GHz radio, and the Hα luminosity all give consistent results when applied to mass-selected samples. In Fig. 1.4 we show a very recent determination of the main sequence up to $z = 3$ (from Bisigello et al., 2018) and a compilation of various estimates of the specific SFR at different cosmic epochs (from Faisst et al., 2016) showing its fast evolution with redshift, in particular at $0 < z < 1$. The trend and normalization of the sSFR are still poorly constrained at very high-z ($z > 4$), where large dust corrections and sample incompleteness turn into large uncertainties on the SFR determination.

1.5.2 Outliers of the Main Sequence of Star-Forming Galaxies: Starbursts

While the majority of galaxies occupy the locus of the main sequence, outliers are also observed with intense levels of SFR given their stellar mass. In a very recent analysis, Bisigello et al. (2018) have been able to exploit the CANDELS data (Grogin et al., 2011; Koekemoer et al., 2011) to clearly detect a separate sequence of starbursts above the main sequence that keeps evolving in normalization and slope similarly to the main sequence in the redshift range $0.5 < z < 3$ (see Fig. 1.4 above and Caputi et al., 2017, for a similar claim at $z = 4$). These two populations have been associated with different growth mechanisms (Daddi et al., 2010; Genzel et al., 2010; Elbaz et al., 2011): main-sequence galaxies are thought to grow on long timescales as a consequence of smooth gas accretion from the Intergalactic Medium (IGM), while main sequence outliers (i.e. starbursts) seem to be triggered by mergers and as a consequence they form stars with high efficiency, although this view is currently debated (e.g. Kennicutt and Evans, 2012; Narayanan et al., 2012; Santini et al., 2014; Mancuso et al., 2016). The latter, being very rare despite their high level of SFR, seem to contribute modestly to the cosmic star-formation history (Rodighiero et al., 2011; Sargent et al., 2012; Lamastra et al., 2013). However, collisions and interactions between gas-rich galaxies are thought to be pivotal stages in their formation and evolution, causing the rapid production of new stars, and possibly serving as a mechanism for fueling

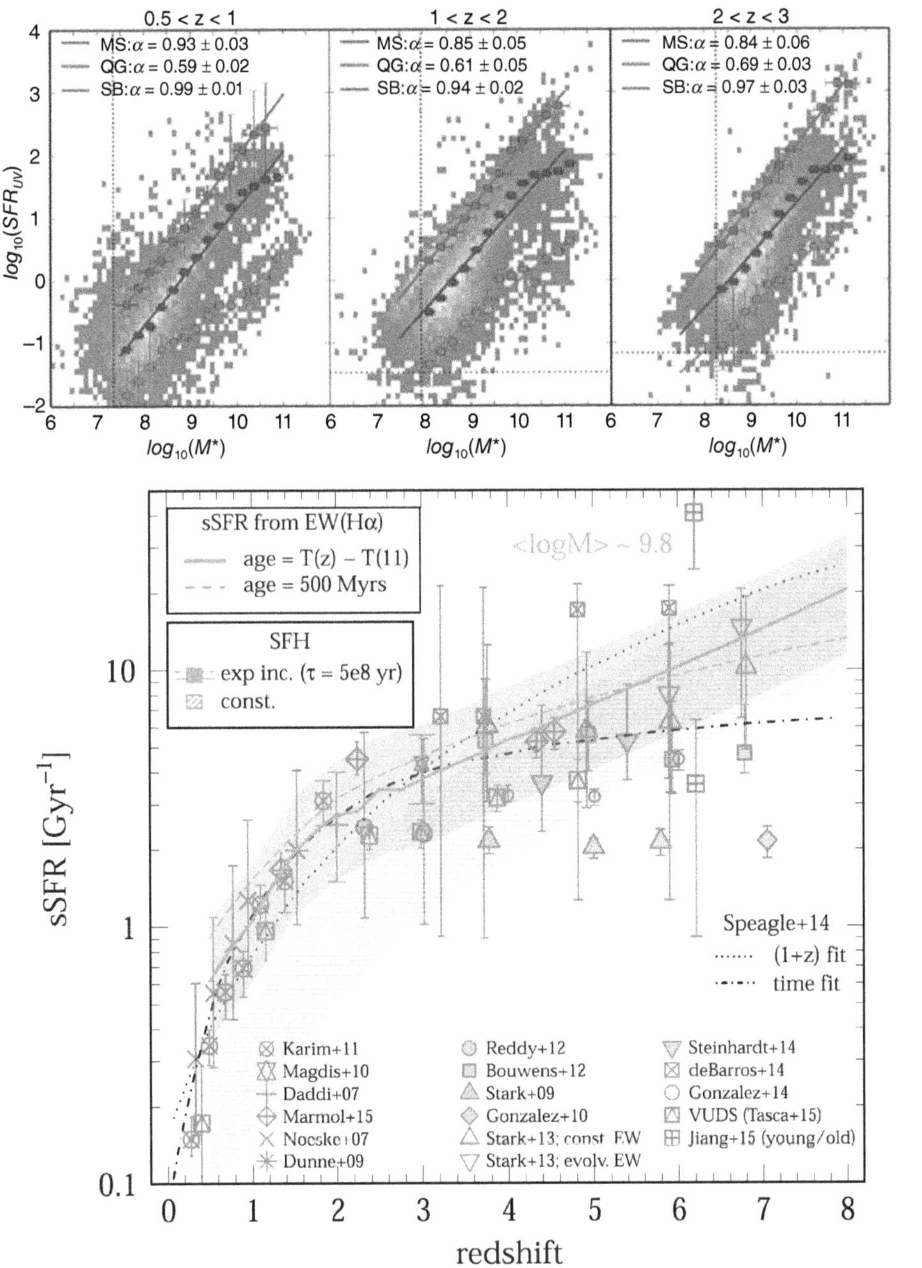

Figure 1.4 *Top panel:* SFR–M_{star} plane for three galaxy modes: quenched galaxies (bottom line), main-sequence galaxies (middle line) and starburst galaxies (top line). The slope of each sequence is reported on the top-left of each panel. In the background there are all galaxies from a mass-selected sample. The vertical black dotted lines are the 90% completeness limit in stellar mass in each redshift bin. Credit: Bisigello et al. (2018), reproduced with permission © ESO. Color version available online.
Bottom panel: Evolution of sSFR as a function of redshift, from Faisst et al. (2016) © AAS. Reproduced with permission. Color version available online.

super-massive black holes (SMBH). All in all, the results of Rodighiero et al. (2011) quite clearly show that the merger-enhanced SFR phases are relatively unimportant for the formation of stars in $z \sim 2$ galaxies. Other studies have found the same result in the local universe (e.g. Sanders and Mirabel, 1996) and recently it has been generalized to all redshifts (Schreiber et al., 2016; Bisigello et al., 2018). Still, going through this merging-driven starburst phase may be a critical phase for the transformation of star-forming galaxies into passive ellipticals (Barro et al., 2016). Cosmological simulations suggest that the high-SFR galaxies in the upper envelope of the main sequence are compact, with high gas fractions and short depletion times (blue nuggets, Zolotov et al., 2015), while the lower-SFR galaxies in the lower envelope have lower central gas densities, lower gas fractions, and longer depletion times consistent with observed gradients across the main sequence (Tacchella et al., 2016). Hence, it is believed that starbursts are likely to be crucial objects for our understanding of galaxy formation and evolution. Such discussions, however, do not always include crucial ingredients concerning the physical state of the ISM (i.e. the gas, dust, and chemical enrichment) which are further interwoven with stellar mass and SFR. These interrelationships are the imprints of both internal and external physical processes such as gas consumption, feedback, winds/outflows, and gas accretion. Such relationships are observationally supported by the fairly tight correlation between the dust and total gas content in galaxies (Magdis et al., 2012) which scales almost linearly with metallicity. The mainstream interpretation of main-sequence outlier galaxies as merging starbursts with high star-formation efficiency, as opposed to main-sequence galaxies being normal disks, has been questioned in recent papers which advocate that starbursts could instead be a sort of primordial galaxies with anomalously high total-gas masses (by a factor of several higher than their stellar masses; Scoville et al., 2014; Genzel et al., 2015; Scoville et al., 2016, 2017). In a similar vein, this alternative interpretation has also been disputed, with claims that the rise in the gas fraction up to $z \sim 2$ is much weaker than previously thought. In this case, the proposed evolutionary scenario would either require some still unknown physical mechanisms to increase the star-formation efficiency in disk galaxies to explain their high SFRs, or renouncing the idea that typical high-z main-sequence galaxies are largely unperturbed gas-rich disks. Another hotly debated topic nowadays is whether the fundamental division between main-sequence and starburst galaxies is actually warranted corresponding to bimodal distributions of galaxy properties, or if, instead, the large spread in SFRs of galaxies at a fixed stellar mass just reflects broad distributions of gas fractions or/and star-formation efficiencies (e.g. Dessauges-Zavadsky et al., 2015).

1.5.3 A Galaxy Life on the Stellar Mass–SFR Plane

The interplay of the different mechanisms at play in galaxy evolution shapes the relation between the physical parameters that we observe at different cosmic epochs, in particular, the dependence of SFR on stellar mass. Thus, it is tempting to interpret the position of a galaxy in the stellar mass–SFR plane as a pointer to its evolutionary stage (see Fig. 1.5). Different interpretations have been proposed to follow the life of sources assembling their stellar masses through different paths, with random episodes pushing them for short

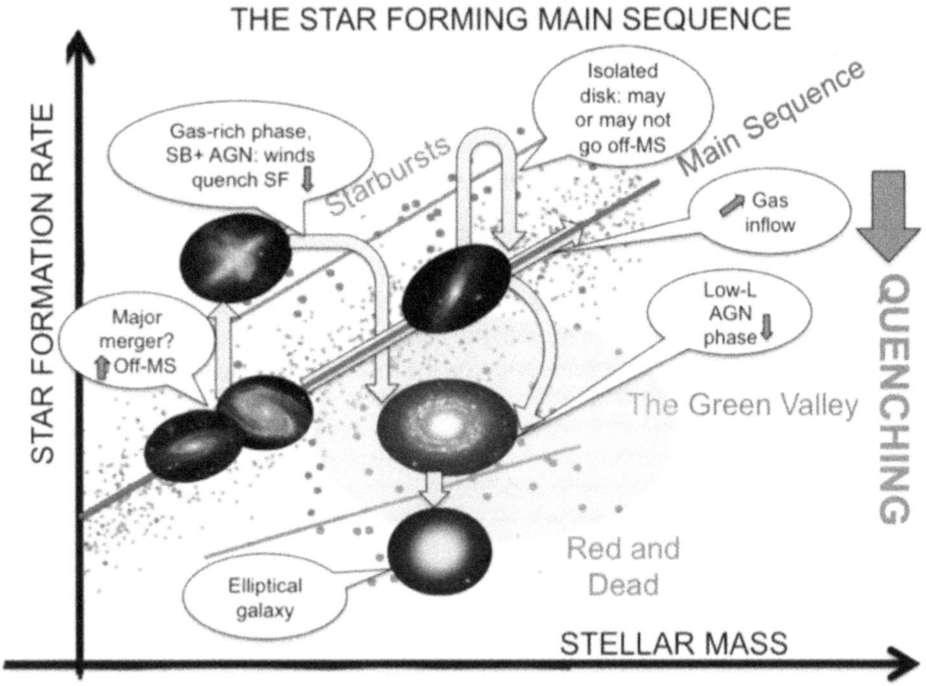

Figure 1.5 Summary of the episodes that a galaxy could experience along its evolution on the M_{star}–SFR plane. Rotationally-supported normal star-forming galaxies (on the main sequence) could merge with similar systems and increase their *starburstiness* (off main sequence). Similarly, disks through violent disk instabilities could reach similar high specific SFR. Main-sequence sources are continuously fed by pristine gas inflows that sustain their SFR on long timescales. Natural fading of SFR or AGN winds can eventually quench galaxies driving them on the elliptical-galaxies locus (below the main sequence). Credits: C. Gruppioni. Color version available online.

timescales above the main sequence, and eventually quenching the most massive sources to bury them in the graveyard of red and dead galaxies (i.e. well below the main sequence). The situation is quite complex, with various teams proposing different evolutionary tracks (e.g. Dekel and Burkert, 2014; Mancuso et al., 2016; Tacchella et al., 2016; Ciesla et al., 2017). The evolutionary schemes have to deal with the shutdown of star formation, or its drop much below the main-sequence regime (to which we refer to as quenching): this is perhaps the most dramatic change taking place during the life of galaxies. The opposite process, rejuvenation, i.e. the return to an appreciable star-formation rate, can also occur, possibly triggered by the capture of a gas-rich galaxy. But we know that over cosmic times the fraction of quenched galaxies steadily increases, approaching unity at $z = 0$ in the high-mass and high-overdensity (i.e. groups and clusters) extremes. Thus, quenching dominates over rejuvenation. This complex story is also intimately linked to the presence of nuclear activity at the very center of massive galaxies: the accretion of gas onto SMBHs is indeed one of the potential quenching mechanisms that are invoked by models

(Hopkins et al., 2008) and observations (detection of massive gas outflows in quasars at high-z Cicone et al., 2014). Indeed, understanding the physical nature of quenching has been a long-standing open issue and it is still one of the hottest topics in galaxy evolution.

1.5.4 The "Cosmic" Star-Formation Rate

Understanding the physics of star formation, and the way to recover it for unbiased samples of galaxies at different redshifts, allows the reconstruction of the history of the stellar-mass assembly of the universe. The present picture tell us that the universe has experienced a major epoch of intense star-forming activity, about 10 Gyr ago (i.e. $z \simeq 2$, see Madau and Dickinson, 2014, for a complete review on the argument and Fig. 1.6). Therefore, this

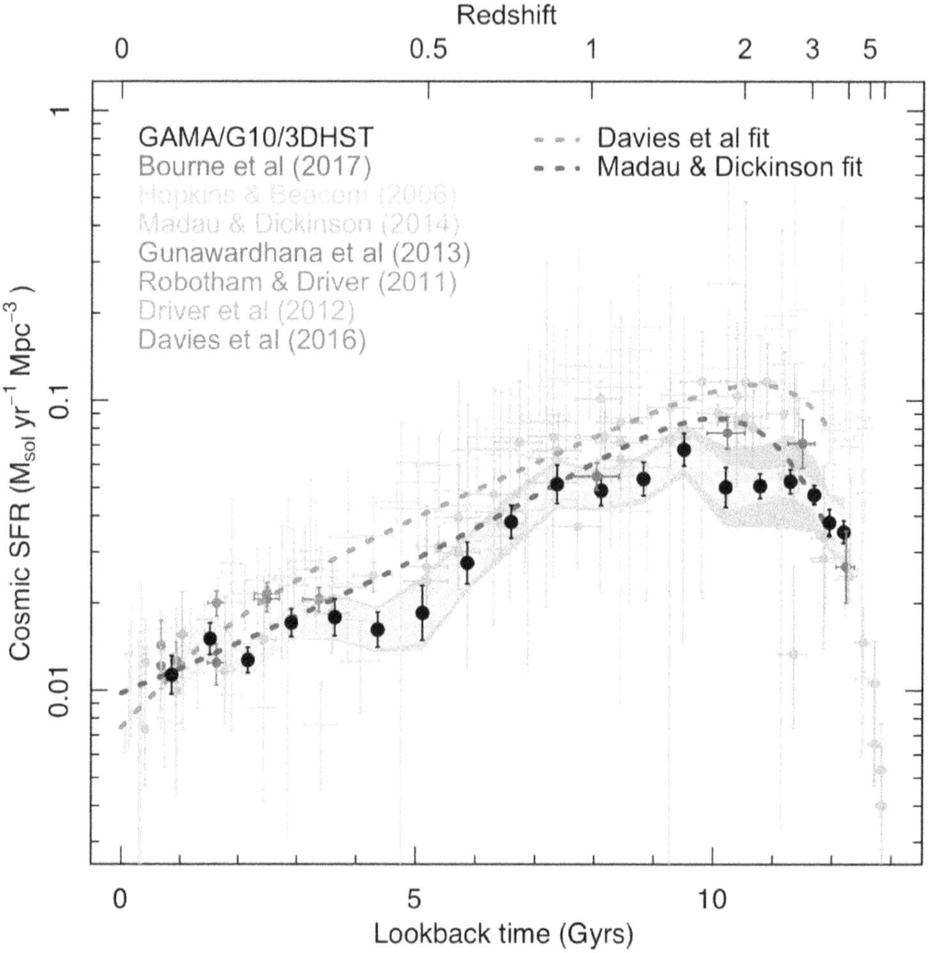

Figure 1.6 The history of cosmic star formation, from Driver et al. (2018) by permission of Oxford University Press on behalf of the Royal Astronomical Society. Color version available online.

is the best cosmic time to study the star-formation processes in large samples of galaxy populations, when large gas reservoirs were converted into stars. In the past few years, with the advent of advanced interferometric techniques working directly in the sub-millimeter and radio regimes (in particular the ALMA and the Jansky Very-Large Array[10]), a new look into the early stages of star formation has been achieved, probing the smooth rise of the cosmic SFR from $z > 10$ up to the peak at $z \simeq 2$ (e.g. Dunlop et al., 2017; Novak et al., 2017). Even the *Hubble Space Telescope*, coupled with the power of gravitational lensing in the Frontier Fields galaxy clusters program, has started to place constraints on the SFR density at $z \sim 10$ (Oesch et al., 2015), paving the way for the upcoming *James Webb Space Telescope*.

We certainly live in an exciting era when new advances in our knowledge of the SF processes in the early universe will be achieved, thus providing a complete picture of the formation and evolution of the cosmic structures within which we live today (our Galaxy).

We hope that this book will be a valuable support for students and young researchers who intend to be involved into such fascinating studies.

References

Abramova, O. V., and Zasov, A. V. 2008. The Gas Density and "Volume" Schmidt Law for Spiral Galaxies. *Astronomy Reports*, **52**(Apr.), 257–269.

Bacchini, C., Fraternali, F., Iorio, G. et al. 2017. The Role of Disc Thickness in the Schmidt-Kennicutt Law. In: V. Debattista & E. D'Onghia (eds), *The Role of Gas in Galaxy Dynamics*.

Barro, G., Kriek, M., Pérez-González, P. G. et al. 2016. Sub-kiloparsec ALMA Imaging of Compact Star-Forming Galaxies at $z \sim 2.5$: Revealing the Formation of Dense Galactic Cores in the Progenitors of Compact Quiescent Galaxies. *Astrophysical Journal*, **827**(Aug.), L32.

Bigiel, F., Leroy, A., Walter, F. et al. 2008. The Star Formation Law in Nearby Galaxies on Sub-Kpc Scales. *Astronomical Journal*, **136**(Dec.), 2846–2871.

Bisigello, L., Caputi, K. I., Grogin, N. et al. 2018. Analysis of the SFR-M* Plane at $z < 3$: Single Fitting versus Multi-Gaussian Decomposition. *Astronomy and Astrophysics*, **609**(Jan.), A82.

Blitz, L., and Rosolowsky, E. 2006. The Role of Pressure in GMC Formation II: The H_2-Pressure Relation. *Astrophysical Journal*, **650**(Oct.), 933–944.

Boissier, S. 2013. *Star Formation in Galaxies*. Page 141.

Boissier, S., Gil de Paz, A., Boselli, A. et al. 2007. Radial Variation of Attenuation and Star Formation in the Largest Late-Type Disks Observed with GALEX. *Astrophysical Journal Supplement Series*, **173**(Dec.), 524–537.

Boquien, M., Burgarella, D., Roehlly, Y. et al. 2019. CIGALE: A Python Code Investigating GALaxy Emission. *Astronomy and Astrophysics*, **622**(Feb), A103.

Brinchmann, J., Charlot, S., White, S. D. M. et al. 2004. The Physical Properties of Star-Forming Galaxies in the Low-Redshift Universe. *Monthly Notices of the Royal Astronomical Society*, **351**(July), 1151–1179.

[10] Jansky Very Large Array (JVLA); https://science.nrao.edu/facilities/vla

Burgarella, D., Buat, V., and Iglesias-Páramo, J. 2005. Star Formation and Dust Attenuation Properties in Galaxies from a Statistical Ultraviolet-to-Far-Infrared Analysis. *Monthly Notices of the Royal Astronomical Society*, **360**(4), 1413–1425.

Caputi, K. I., Deshmukh, S., Ashby, M. L. N. et al. 2017. Star Formation in Galaxies at $z \sim 4$–5 from the SMUVS Survey: A Clear Starburst/Main-Sequence Bimodality for Hα Emitters on the SFR-M* Plane. *Astrophysical Journal*, **849**(Nov.), 45.

Carilli, C. L., and Walter, F. 2013. Cool Gas in High-Redshift Galaxies. *Annual Review of Astronomy and Astrophysics*, **51**(1), 105–161.

Cicone, C., Maiolino, R., Sturm, E. et al. 2014. Massive Molecular Outflows and Evidence for AGN Feedback from CO Observations. *Astronomy and Astrophysics*, **562**(Feb.), A21.

Ciesla, L., Elbaz, D., and Fensch, J. 2017. The SFR-M_* Main Sequence Archetypal Star-Formation History and Analytical Models. *Astronomy and Astrophysics*, **608**(Dec.), A41.

Corbelli, E. 2003. Dark Matter and Visible Baryons in M33. *Monthly Notices of the Royal Astronomical Society*, **342**(June), 199–207.

Cresci, G., Hicks, E. K. S., Genzel, R. et al. 2009. The SINS Survey: Modeling the Dynamics of $z \sim 2$ Galaxies and the High-z Tully-Fisher Relation. *Astrophysical Journal*, **697**(May), 115–132.

da Cunha, E., Charlot, S., and Elbaz, D. 2008. A Simple Model to Interpret the Ultraviolet, Optical and Infrared Emission from Galaxies. *Monthly Notices of the Royal Astronomical Society*, **388**(4), 1595–1617.

Daddi, E., Dickinson, M., Chary, R. et al. 2005. The Population of BzK-Selected ULIRGs at $z \sim 2$. *Astrophysical Journal*, **631**(Sept.), L13–L16.

Daddi, E., Dickinson, M., Morrison, G. et al. 2007. Multiwavelength Study of Massive Galaxies at $z{\sim}2$. I. Star Formation and Galaxy Growth. *Astrophysical Journal*, **670**(Nov.), 156–172.

Daddi, E., Elbaz, D., Walter, F., Bournaud, F. et al. 2010. Different Star Formation Laws for Disks versus Starbursts at Low and High Redshifts. *Astrophysical Journal*, **714**(May), L118–L122.

Dekel, A., and Burkert, A. 2014. Wet Disc Contraction to Galactic Blue Nuggets and Quenching to Red Nuggets. *Monthly Notices of the Royal Astronomical Society*, **438**(Feb.), 1870–1879.

Dessauges-Zavadsky, M., Zamojski, M., Schaerer, D. et al. 2015. Molecular Gas Content in Strongly Lensed $z \sim 1.5$–3 Star-Forming Galaxies with Low Infrared Luminosities. *Astronomy and Astrophysics*, **577**(May), A50.

Dopita, M. A., and Ryder, S. D. 1994. On the Law of Star Formation in Disk Galaxies. *Astrophysical Journal*, **430**(July), 163–178.

Driver, S. P., Andrews, S. K., da Cunha, E. et al. 2018. GAMA/G10-COSMOS/3D-HST: The 0 < z <; 5 Cosmic Star Formation History, Stellar-Mass, and Dust-Mass Densities. *Monthly Notices of the Royal Astronomical Society*, **475**(3), 2891–2935.

Dunlop, J. S., McLure, R. J., Biggs, A. D. et al. 2017. A Deep ALMA Image of the Hubble Ultra Deep Field. *Monthly Notices of the Royal Astronomical Society*, **466**(Apr.), 861–883.

Elbaz, D., Daddi, E., Le Borgne, D. et al. 2007. The Reversal of the Star Formation-Density Relation in the Distant Universe. *Astronomy and Astrophysics*, **468**(June), 33–48.

Elbaz, D., Dickinson, M., Hwang, H. S. et al. 2011. GOODS-Herschel: An Infrared Main Sequence for Star-Forming Galaxies. *Astronomy and Astrophysics*, **533**(Sept.), A119.

Faisst, A. L., Capak, P., Hsieh, B. C. et al. 2016. A Coherent Study of Emission Lines from Broadband Photometry: Specific Star Formation Rates and [O iii]/Hβ Ratio at $3 > z > 6$. *Astrophysical Journal*, **821**(Apr.), 122.

Förster Schreiber, N. M., Genzel, R., Bouché, N. et al. 2009. The SINS Survey: SINFONI Integral Field Spectroscopy of z$\sim$2 Star-Forming Galaxies. *Astrophysical Journal*, **706**(Dec.), 1364–1428.

Gao, Y., and Solomon, P. M. 2004. The Star Formation Rate and Dense Molecular Gas in Galaxies. *Astrophysical Journal*, **606**(May), 271–290.

Genzel, R., Tacconi, L. J., Gracia-Carpio, J. et al. 2010. A Study of the Gas-Star Formation Relation over Cosmic Time. *Monthly Notices of the Royal Astronomical Society*, **407**(Oct.), 2091–2108.

Genzel, R., Tacconi, L. J., Lutz, D. et al. 2015. Combined CO and Dust Scaling Relations of Depletion Time and Molecular Gas Fractions with Cosmic Time, Specific Star-formation Rate, and Stellar Mass. *Astrophysical Journal*, **800**(Feb.), 20.

Goddard, Q. E., Kennicutt, R. C., and Ryan-Weber, E. V. 2010. On the Nature of Star Formation at Large Galactic Radii. *Monthly Notices of the Royal Astronomical Society*, **405**(July), 2791–2809.

Grogin, N. A., Kocevski, D. D., Faber, S. M. et al. 2011. CANDELS: The Cosmic Assembly Near-infrared Deep Extragalactic Legacy Survey. *Astrophysical Journal Supplement Series*, **197**(Dec.), 35.

Hennebelle, P., and Commerçon, B. 2014. Theories of the Massive Star Formation: A (Short) Review. Page 365 of: Stamatellos, D., Goodwin, S., and Ward-Thompson, D. (eds), *The Labyrinth of Star Formation*. Astrophysics and Space Science Proceedings, vol. 36.

Hopkins, P. F., Hernquist, L., Cox, T. J. et al. 2008. A Cosmological Framework for the Co-Evolution of Quasars, Supermassive Black Holes, and Elliptical Galaxies. I. Galaxy Mergers and Quasar Activity. *Astrophysical Journal Supplement Series*, **175**(Apr.), 356–389.

Kennicutt, Jr., R. C. 1998a. Star Formation in Galaxies along the Hubble Sequence. *Annual Review of Astronomy and Astrophysics*, **36**, 189–232.

Kennicutt, Jr., R. C. 1998b. The Global Schmidt Law in Star-Forming Galaxies. *Astrophysical Journal*, **498**(May), 541–552.

Kennicutt, Jr., R. C., and Evans, N. J. 2012. Star Formation in the Milky Way and Nearby Galaxies. *Annual Review of Astronomy and Astrophysics*, **50**(Sept.), 531–608.

Kennicutt, Jr., R. C., Hao, C.-N., Calzetti, D. et al. 2009. Dust-Corrected Star Formation Rates of Galaxies. I. Combinations of Hα and Infrared Tracers. *Astrophysical Journal*, **703**(2), 1672–1695.

Koekemoer, A. M., Faber, S. M., Ferguson, H. C. et al. 2011. CANDELS: The Cosmic Assembly Near-Infrared Deep Extragalactic Legacy Survey – The Hubble Space Telescope Observations, Imaging Data Products, and Mosaics. *Astrophysical Journal Supplement Series*, **197**(Dec.), 36.

Krumholz, M. R., McKee, C. F., and Tumlinson, J. 2009. The Star Formation Law in Atomic and Molecular Gas. *Astrophysical Journal*, **699**(July), 850–856.

Lamastra, A., Menci, N., Fiore, F. et al. 2013. The Interaction-Driven Starburst Contribution to the Cosmic Star Formation Rate Density. *Astronomy and Astrophysics*, **552**(Apr.), A44.

Larson, R. 1992. Galaxy Formation and Evolution. Page 125 of: G. Tenorio-Tagle, M. Prieto, & F. Sanchez (ed.), *Star Formation in Stellar Systems*.

Law, D. R., Steidel, C. C., Erb, D. K. et al. 2009. The Kiloparsec-Scale Kinematics of High-Redshift Star-Forming Galaxies. *Astrophysical Journal*, **697**(June), 2057–2082.

Leroy, A. K., Walter, F., Brinks, E. et al. 2008. The Star Formation Efficiency in Nearby Galaxies: Measuring Where Gas Forms Stars Effectively. *Astronomical Journal*, **136**(Dec.), 2782–2845.

Lilly, S. J., Carollo, C. M., Pipino, A. et al. 2013. Gas Regulation of Galaxies: The Evolution of the Cosmic Specific Star Formation Rate, the Metallicity-Mass-Star-Formation Rate Relation, and the Stellar Content of Halos. *Astrophysical Journal*, **772**(Aug.), 119.

Madau, P., and Dickinson, M. 2014. Cosmic Star-Formation History. *Annual Review of Astronomy and Astrophysics*, **52**(Aug.), 415–486.

Madore, B. F. 1977. Numerical Simulations of the Rate of Star Formation in External Galaxies. *Monthly Notices of the Royal Astronomical Society*, **178**(Jan.), 1–9.

Madore, B. F. 2010. Star Formation Timescales and the Schmidt Law. *Astrophysical Journal*, **716**(June), L131–L134.

Magdis, G. E., Daddi, E., Béthermin, M. et al. 2012. The Evolving Interstellar Medium of Star-forming Galaxies since z = 2 as Probed by Their Infrared Spectral Energy Distributions. *Astrophysical Journal*, **760**(Nov.), 6.

Mancuso, C., Lapi, A., Shi, J. et al. 2016. The Main Sequences of Star-Forming Galaxies and Active Galactic Nuclei at High Redshift. *Astrophysical Journal*, **833**(Dec.), 152.

Maraston, C., Pforr, J., Renzini, A. et al. 2010. Star Formation Rates and Masses of $z \sim 2$ Galaxies from Multicolour photometry. *Monthly Notices of the Royal Astronomical Society*, **407**(Sept.), 830–845.

Martin, C. L., and Kennicutt, Jr., R. C. 2001. Star Formation Thresholds in Galactic Disks. *Astrophysical Journal*, **555**(July), 301–321.

McKee, C. F., and Ostriker, E. C. 2007. Theory of Star Formation. *Annual Review of Astronomy and Astrophysics*, **45**(Sept.), 565–687.

Narayanan, D., Bothwell, M., and Davé, R. 2012. Galaxy Gas Fractions at High Redshift: The Tension between Observations and Cosmological Simulations. *Monthly Notices of the Royal Astronomical Society*, **426**(Oct.), 1178–1184.

Noeske, K. G., Weiner, B. J., Faber, S. M. et al. 2007. Star Formation in AEGIS Field Galaxies since z=1.1: The Dominance of Gradually Declining Star Formation, and the Main Sequence of Star-Forming Galaxies. *Astrophysical Journal*, **660**(May), L43–L46.

Novak, M., Smolčić, V., Delhaize, J. et al. 2017. The VLA-COSMOS 3 GHz Large Project: Cosmic Star Formation History since z 5. *Astronomy and Astrophysics*, **602**(June), A5.

Oesch, P. A., Bouwens, R. J., Illingworth, G. D. et al. 2015. First Frontier Field Constraints on the Cosmic Star Formation Rate Density at $z \sim 10$. The Impact of Lensing Shear on Completeness of High-Redshift Galaxy Samples. ApJ, **808**(July), 104.

Osterbrock, D. E., and Ferland, G. J. 2006. *Astrophysics of Gaseous Nebulae and Active Galactic Nuclei.*

Papovich, C., Rudnick, G., Le Floc'h, E. et al. 2007. Spitzer Mid- to Far-Infrared Flux Densities of Distant Galaxies. *Astrophysical Journal*, **668**(1), 45–61.

Peng, Y.-j., Lilly, S. J., Kovač, K. et al. 2010. Mass and Environment as Drivers of Galaxy Evolution in SDSS and zCOSMOS and the Origin of the Schechter Function. *Astrophysical Journal*, **721**(Sept.), 193–221.

Rafelski, M., Gardner, J. P., Fumagalli, M. et al. 2016. The Star Formation Rate Efficiency of Neutral Atomic-Dominated Hydrogen Gas in the Outskirts of Star-forming Galaxies from $z \sim 1$ to $z \sim 3$. *Astrophysical Journal*, **825**(July), 87.

Reddy, N. A., Pettini, M., Steidel, C. C. et al. 2012. The Characteristic Star Formation Histories of Galaxies at Redshifts z~2-7. *Astrophysical Journal*, **754**(July), 25.

Renaud, F., Bournaud, F., Kraljic, K. et al. 2014. Starbursts Triggered by Intergalactic Tides and Interstellar Compressive Turbulence. *Monthly Notices of the Royal Astronomical Society*, **442**(July), L33–L37.

Renzini, A. 2009. A Different Approach to Galaxy Evolution. *Monthly Notices of the Royal Astronomical Society*, **398**(Sept.), L58–L62.

Rodighiero, G., Cimatti, A., Gruppioni, C. et al. 2010. The First Herschel View of the Mass-SFR Link in High-z Galaxies. *Astronomy and Astrophysics*, **518**(July), L25.

Rodighiero, G., Daddi, E., Baronchelli, I. et al. 2011. The Lesser Role of Starbursts in Star Formation at z = 2. *Astrophysical Journal*, **739**(Oct.), L40.

Rodighiero, G., Renzini, A., Daddi, E. et al. 2014. A Multiwavelength Consensus on the Main Sequence of Star-Forming Galaxies at z $\sim$ 2. *Monthly Notices of the Royal Astronomical Society*, **443**(Sept.), 19–30.

Sanders, D. B., and Mirabel, I. F. 1996. Luminous Infrared Galaxies. *Annual Review of Astronomy and Astrophysics*, **34**, 749.

Sanders, D. B., Soifer, B. T., Elias, J. H. et al. 1988. Ultraluminous Infrared Galaxies and the Origin of Quasars. *Astrophysical Journal*, **325**(Feb.), 74–91.

Santini, P., Maiolino, R., Magnelli, B. et al. 2014. The Evolution of the Dust and Gas Content in Galaxies. *Astronomy and Astrophysics*, **562**(Feb.), A30.

Sargent, M. T., Béthermin, M., Daddi, E. et al. 2012. The Contribution of Starbursts and Normal Galaxies to Infrared Luminosity Functions at z $<$ 2. *Astrophysical Journal*, **747**(Mar.), L31.

Schaye, J. 2004. Star Formation Thresholds and Galaxy Edges: Why and Where. *Astrophysical Journal*, **609**(July), 667–682.

Schmidt, M. 1959. The Rate of Star Formation. *Astrophysical Journal*, **129**(Mar.), 243.

Schreiber, C., Elbaz, D., Pannella, M. et al. 2016. Observational Evidence of a Slow Downfall of Star Formation Efficiency in Massive Galaxies during the Past 10 Gyr. *Astronomy and Astrophysics*, **589**(May), A35.

Scoville, N., Aussel, H., Sheth, K. et al. 2014. The Evolution of Interstellar Medium Mass Probed by Dust Emission: ALMA Observations at z = 0.3–2. *Astrophysical Journal*, **783**(Mar.), 84.

Scoville, N., Lee, N., Vanden Bout, P. et al. 2017. Evolution of Interstellar Medium, Star Formation, and Accretion at High Redshift. *Astrophysical Journal*, **837**(Mar.), 150.

Scoville, N., Sheth, K., Aussel, H. et al. 2016. ISM Masses and the Star Formation Law at Z = 1 to 6: ALMA Observations of Dust Continuum in 145 Galaxies in the COSMOS Survey Field. *Astrophysical Journal*, **820**(Apr.), 83.

Semenov, V. A., Kravtsov, A. V., and Gnedin, N. Y. 2017. The Physical Origin of Long Gas Depletion Times in Galaxies. *Astrophysical Journal*, **845**(Aug.), 133.

Shi, Y., Helou, G., Yan, L. et al. 2011. Extended Schmidt Law: Role of Existing Stars in Current Star Formation. *Astrophysical Journal*, **733**(June), 87.

Silva, L., Granato, G. L., Bressan, A. et al. 1998. Modeling the Effects of Dust on Galactic Spectral Energy Distributions from the Ultraviolet to the Millimeter Band. *Astrophysical Journal*, **509**(1), 103–117.

Tacchella, S., Dekel, A., Carollo, C. M. et al. 2016. The Confinement of Star-Forming Galaxies into a Main Sequence through Episodes of Gas Compaction, Depletion and Replenishment. *Monthly Notices of the Royal Astronomical Society*, **457**(Apr.), 2790–2813.

Tan, J. C. 2000. Star Formation Rates in Disk Galaxies and Circumnuclear Starbursts from Cloud Collisions. *Astrophysical Journal*, **536**(June), 173–184.

Toomre, A. 1964. On the Gravitational Stability of a Disk of Stars. *Astrophysical Journal*, **139**(May), 1217–1238.

Wang, B., and Silk, J. 1994. Gravitational Instability and Disk Star Formation. *Astrophysical Journal*, **427**(June), 759–769.

Wyse, R. F. G. 1986. A Schmidt Law for the Molecular Gas Profiles of Disk Galaxies. *Astrophysical Journal*, **311**(Dec.), L41–L45.

Wyse, R. F. G., and Silk, J. 1989. Star Formation Rates and Abundance Gradients in Disk Galaxies. *Astrophysical Journal*, **339**(Apr.), 700–711.

Zolotov, A., Dekel, A., Mandelker, N. et al. 2015. Compaction and Quenching of High-Z Galaxies in Cosmological Simulations: Blue and Red Nuggets. *Monthly Notices of the Royal Astronomical Society*, **450**(July), 2327–2353.

2

The Initial Mass Function of Stars and the Star-Formation Rates of Galaxies

PAVEL KROUPA AND TEREZA JERABKOVA

2.1 Introduction

The measured star-formation rates (SFRs) of galaxies comprise an important constraint on galaxy evolution and also on their cosmological boundary conditions. Any available tracer of the SFR depends on the shape of the mass distribution of formed stars, i.e. on the stellar initial mass function (IMF). The luminous massive stars dominate the observed photon flux while the dim low-mass stars dominate the mass in the freshly formed population. Errors in the number ratio of the massive to low-mass stars lead to errors in SFR measurements and thus to errors concerning the gas-accretion rates and the gas-consumption timescales of galaxies. The stellar IMF has traditionally been interpreted to be a scale-invariant probability density distribution function (PDF), but it may instead be an optimal distribution function. In the PDF interpretation, the stellar IMF observed on the scales of individual star clusters is equal to the galaxy-wide IMF (gwIMF) which, by implication, would be invariant. In this chapter we discuss the fundamental properties of the IMF and the gwIMF, the nature of both and their systematic variability as indicated by measurements and theoretical expectations, and the implications for the SFRs of galaxies. The importance of the putative most-massive-star mass vs stellar-mass of the hosting embedded-cluster ($m_{\rm max} - M_{\rm ecl}$) relation and its possible establishment during the proto-stellar formation phase is emphasised.

Let dN be the number of stars in the infinitesimal initial stellar-mass interval m to $m + dm$, then $dN = \xi(m)\, dm$ where $\xi(m)$ is the stellar IMF, that is, the distribution of the initial masses of all stars formed together. It is usual to represent the IMF with a power-law function, $\xi(m) \propto m^{\ \alpha_i}$, where the power-law indices depend on the mass range considered and $\alpha_{\rm S} = 2.35$ is the Salpeter index valid for the mass range $0.4\,{\rm M}_\odot - 10\,{\rm M}_\odot$ (Salpeter, 1955).

It follows that it is in principle straight-forward to construct $\xi(m)$ because the astronomer merely needs to look up and count the stars. The difficulty in the practical execution of this is that stellar masses are only measurable directly when they are in binary orbits. For the vast majority of stars, binary orbits are not available though, and their masses can only be inferred from their luminosities, colours, and spectra relying on theoretical stellar models. Theoretical models assign these radiant properties a mass, provided an age, stellar spin, and chemical abundance is known, these properties of the star being also measured from similar data. Massive stars are extremely luminous and shine mostly in the ultraviolet with a steep mass–luminosity relation, while low-mass stars are dim and shine largely in the red and infrared with a flat mass–luminosity relation (small uncertainties in the mass translate into large uncertainties in the luminosity). Massive stars are rare and live only for a few Myr, while low mass stars dominate the population in numbers and live up to many Hubble times.

The practical construction from observational data of the IMF over all stellar masses is thus subject to very major uncertainties requiring data from different surveys to be combined. This is particularly true for the Galactic field stellar ensemble nearby to the Sun, where a complete census of the massive stars is reachable to distances of a few kpc but the low-mass stars can only be detected to within a few pc (see Section 2.3).

To infer the stellar IMF in the Galactic field the massive-star and low-mass star counts need to be combined appropriately and carefully, which requires taking care of the different spatial distribution and the star formation history (Salpeter, 1955; Miller and Scalo, 1979; Scalo, 1986; Elmegreen and Scalo, 2006). The surveys of stars in star clusters do not suffer from this complication since the stars are practically at the same distance, clusters typically being further than a few dozen pc away from us while their radii are typically 3 pc, and the stars have nearly the same ages. One problem with star clusters though is that stellar-dynamical processes remove stars from the clusters already starting at the earliest times (a few 10^5 yr) (Baumgardt and Makino, 2003; Banerjee and Kroupa, 2012; Haghi et al., 2015; Oh et al., 2015; Oh and Kroupa, 2016; Kroupa et al., 2018), binary-star evolution and dynamically induced stellar mergers also affect the inferred IMF (de Mink et al., 2014; Schneider et al., 2015; Oh and Kroupa, 2018). Also, and in all cases, in the Galactic field, star clusters, and any other stellar ensemble, the star counts always suffer from unresolved multiple systems such that the observer misses the fainter or unresolved companion. This bias is very significant for late-type stars (Kroupa et al., 1991) but is less problematic for more massive stars (Weidner et al., 2009; Kroupa and Jerabkova, 2018).

Given its important role in much of astrophysics, the IMF has been the subject of a vast research effort. Some relevant IMF-related questions that are addressed are: Can the IMF be measured? What is the shape of the IMF? What is its mathematical nature (e.g. is it a probability density distribution function)? Does the IMF vary? Does the IMF of a simple stellar population equal that of a composite population?

Reviews which address these questions can be found in Kroupa (2002b); Chabrier (2003); Bastian et al. (2010); Offner et al. (2014); and Kroupa and Jerabkova (2018). The major effort to assess the IMF for stars more massive than a few $M_{\odot}$ by Massey (2003) needs to be emphasised. The reviews by Kroupa et al. (2013) and particularly Hopkins (2018) cover the significant extragalactic evidence for a variable galaxy-wide IMF. Here we address these five most-important IMF questions in turn and place them into the context of the determination of the SFRs of galaxies. Needless to write, the answers to these questions have a rather major impact on the SFRs which are calculated given some tracer. This, in turn, affects our knowledge and understanding of the matter cycle in the Universe.

In the following two sub-sections we explain some IMF-acronyms and discuss some issues concerning how stars emerge from the interstellar medium (ISM).

2.1.1 Used IMF Terminology throughout This Text

Throughout this text we define several acronyms describing the IMF on different scales. It is useful to introduce these once a variable non-universal IMF is discussed (Hopkins, 2018; Jeřábková et al., 2018).

As in Jeřábková et al. (2018) we refer to the IMF as being the stellar initial mass function of stars formed during one star-formation event in an initially gravitationally bound region in a molecular cloud (on a 1 pc scale and within about 1 Myr in a spatially and temporarily correlated star-formation event, i.e. an embedded cluster; see Section 2.1.2). We refer to the cIMF as the composite-IMF, i.e. the sum of the IMFs over larger regions in a galaxy (e.g. stellar associations, cf. Vanbeveren 1982; Hopkins, 2018). The gwIMF is the initial stellar-mass function of *newly* formed stars in a whole galaxy, as observed, for example, with integrated-light indicators.

Throughout this text we refer to the ECMF as the embedded cluster mass function, i.e. the mass distribution of freshly formed embedded clusters (see Eq. 2.13). The same considerations as for the cIMF and the gwIMF also apply to the ECMF (cECMF and the gwECMF), but we do not discuss these here in further detail apart from mentioning that a Schechter-type form of the galaxy-wide ECMF (the gwECMF) arises naturally if the ECMF is a power-law function (Lieberz and Kroupa, 2017).

2.1.2 Turbulence, Filaments, Embedded Clusters, and Associations

The problem discussed here is how the galaxy-wide IMF, which is relevant for extragalactic research, is related to the stellar IMF which is observationally deduced in individual star-formation regions within galaxies. One ansatz is to develop a theoretical galaxy-wide description of how a turbulent ISM generates the gwIMF (Hopkins, 2012). Another ansatz is to use the IMF deduced from observations of resolved stellar populations as a boundary condition in the problem, which, if successful, has the advantage that this ansatz would be automatically consistent with observations on galaxy-wide scales as well as on molecular-cloud scales.

2.2 Can the Initial Mass Function Be Measured?

Something which does not exist can also not be measured ('The IMF Unmeasurabilty Theorem' of Kroupa et al. 2013). But the IMF can be deduced from observations as a *mathematical hilfskonstrukt* (Kroupa and Jerabkova, 2018), to allow calculations and modelling of stellar populations. But, in order to do so, statistical corrections are necessary to account for the time-evolving binary population in the embedded and exposed cluster

(Belloni et al., 2017, 2018), loss of stars through ejections (Banerjee and Kroupa, 2012; Oh et al., 2015; Oh and Kroupa, 2016), mergers, and gas expulsion, as well as the corrections for the changes in stellar mass through stellar and binary evolution (Sana et al., 2012; Schneider et al., 2015) and energy-equipartition-driven stellar-dynamical evolution (Baumgardt and Makino, 2003; Baumgardt et al., 2008).

The IMF is a mathematical hilfskonstrukt which does not have a physical representation, but which is needed for initialising stellar populations.

2.3 What Is the Shape of the Initial Mass Function?

As outlined in the reviews mentioned above, the general finding is that the IMFs reconstructed from the observed populations are consistent with the canonical IMF (see Eq. 2.1). Until recently, there has not been a significant evidence for a variation of the IMF within the star-formation events in the Milky Way (but see Section 2.5) and the notion that the IMF equals the composite Galactic-field IMF (from the solar neighbourhood ensemble of stars) was affirmed for stars with $m \lesssim 1\,M_\odot$.

THE CANONICAL IMF: (m is in units of $M_\odot$, k is the normalisation constant)

$$\xi_{\rm BD}(m) = \frac{k}{3}\left(\frac{m}{0.07}\right)^{-0.3\pm0.4} \quad , \quad 0.01 < m \lesssim 0.15,$$

$$\xi_{\rm star}(m) = k \begin{cases} \left(\frac{m}{0.07}\right)^{-1.3\pm0.3} & , \quad 0.07 < m \leq 0.5, \\ \left[\left(\frac{0.5}{0.07}\right)^{-1.3\pm0.3}\right]\left(\frac{m}{0.5}\right)^{-2.3\pm0.36} & , \quad 0.5 < m \leq 150. \end{cases} \tag{2.1}$$

We note that the canonical IMF may be either described as a two-part power-law function (see Eq. 2.1; eq. 4.55 in Kroupa et al., 2013) or a log-normal form (between the hydrogen burning mass limit, $m_{\rm L} \approx 0.07\,M_\odot$, and about $1\,M_\odot$) with a power-law extension above $\approx 1\,M_\odot$ (eq. 4.56 in Kroupa et al., 2013). The former see (Eq. 2.1) is convenient for use in computations and is adopted throughout this work. It can also be written as a multi power-law form ($\propto m^{-\alpha_i}$) with $\alpha_1 \approx 1.3 \pm 0.3$ for $0.07 \lesssim m/M_\odot < 0.5$, $\alpha_2 \approx 2.3 \pm 0.3$ for $0.5 < m/M_\odot \leq 1$ and $\alpha_3 = \alpha_2$ for $m > 1\,M_\odot$ (Jeřábková et al., 2018). The canonical log-normal+power-law form, is indistinguishable from the power-law function.

Brown dwarfs (BDs) are not discussed here. They add negligible mass and are a by-product of star formation following their own IMF (as planets follow their own IMF, cf. eq. 4.55 in Kroupa et al., 2013) and binary pairing rules different to those of stars and typically do not pair with stars (Bate et al., 2002; Thies et al., 2015). The microlensing constraints from the inner Galaxy are consistent with these results (Wegg et al., 2017). Note that in Eq. 2.1 the brown-dwarf IMF overlaps with the stellar IMF which comes about due to the fact that star-like and brown-dwarf-like formation processes overlap (Thies et al., 2015).

2.4 What Is the Mathematical Nature of the Initial Mass Function?

From a complete coeval ensemble of stars an IMF can be constructed, and thus this IMF is a description of the distribution of initial stellar masses. The physics of star formation determines the form of the IMF, but it also determines which type of distribution function the IMF is.

A turbulent-fragmentation process (see Section 2.1.2) might imply the IMF to be a probability distribution (PDF) function. A self-regulated growth process might instead imply the IMF to be an optimal distribution function (ODF), such that the distribution of stellar masses actually formed does not have Poisson scatter (Kroupa et al., 2013). Nature may well be in-between both extremes. For example, it may be probabilistic to some degree (the butterfly effect: two identical gas clouds which differ in some property by an infinitesimal amount lead to two IMFs, the distance between which is, in some metric, related to the Poisson uncertainty given by the number of stars formed). Or, it may be deterministic in the sense that the bulk properties of the cloud (its temperature, density, and chemical composition) determine the shape of the distribution function. But if the butterfly effect is minor, because the system is highly self-regulated and insensitive to small differences in initial conditions, or because any physical state of the ISM decays to a universal filamentary structure, then the ODF description may be more appropriate.

A major aspect of IMF research is thus not only to find the form of the IMF and its possible variation but also to understand which type of distribution function we are dealing with. This is important for initialising stellar populations in galaxies and the mathematical methods how to do so differ significantly: random vs optimal sampling (Kroupa et al., 2013; Schulz et al., 2015; Applebaum et al., 2019).

There are thus two main hypotheses as to the mathematical nature of the IMF: it is a PDF or an ODF. Each hypothesis can be tested against data to infer the valid nature that may lie between these two extreme descriptions.

2.4.1 The IMF as a Probability Density Function

An often-used interpretation is the IMF to be a PDF (e.g. Elmegreen 1997; Kroupa, 2001; Cerviño et al., 2013a,b): when stars form in some system, their masses appear randomly within the system, subject to being drawn from the IMF. This is the stochastic, or probabilistic, description of star formation within a galaxy. That is, the probability of finding a star of mass m in the interval m_1 to m_2 is:

$$X_{12} = \int_{m_1}^{m_2} \xi_{\mathrm{p(m)}}\, dm, \tag{2.2}$$

where $0 \leq X_{12} \leq 1$ is uniformly distributed and $X_{12} = 1$ for $m_1 \approx 0.07\,\mathrm{M}_\odot$ (the lower-mass limit, m_L) and $m_2 = m_{\mathrm{max}*}$ which is some upper limit which may be infinite or have a physical limit. Note also that the probability density distribution function is proportional to the IMF,

$$\xi_\mathrm{p}(m) \propto \xi(m) \tag{2.3}$$

such that the integral from $m_{\rm L}$ to $m_{\rm max*}$ over the former yields 1 while over the latter it yields the number of stars in the population.

If the IMF is a PDF then any observation of a coeval stellar population will be subject to Poisson-differences such that a measured IMF for population A will differ from that measured for population B, with the difference, $\xi_{\rm A}(m)\,dm - \xi_{\rm B}(m)\,dm$, being consistent with Poisson scatter for the sizes of the populations (assuming no observational errors and all individual stars to be observed). It is thus of much fundamental interest to search if such variations are observed (Elmegreen, 1999; Kroupa, 2001; Elmegreen, 2004; Dib, 2014; Ashworth et al., 2017).

A prominent test of the hypothesis that the IMF is an unconstrained (scale-free) PDF is to consider the masses of the most massive stars observed (see Box Observational Constraints I). We would expect to see stars weighing a few $1000\,{\rm M}_{\odot}$ in populous galaxies (Elmegreen, 2000). But the observed most massive stars have $m \approx 300\,{\rm M}_{\odot}$ (Crowther et al., 2010; Schneider et al., 2014). This suggests that the IMF may be a PDF but with the constraint $m_{\rm max*} \lesssim 300\,{\rm M}_{\odot}$ (more on this in Box Observational Constraints I).

It is possible to deduce that the stellar IMF varies without specifying a possible systematic variation with some physical property of the star-forming environment (Dib, 2014; Dib et al., 2017). But, by taking account of all of the relevant biases, survey differences, observational uncertainties, and also additional effects such as those arising from patchy reddening in star-forming regions and, importantly, the very large uncertainties in calculating stellar masses from the observed fluxes for pre-main sequence stars (Wuchterl and Tscharnuter, 2003), the general consensus has been reached that the IMF is invariant, at least for the stellar populations probed in the Milky Way (Kroupa, 2002a; Chabrier, 2003; Bastian et al., 2010; Kroupa et al., 2013; Offner et al., 2014). More technically, *the hypothesis that the IMF is invariant and canonical cannot be discarded with sufficient confidence*, the variation of the IMF shape being found to be too small even compared to (see Box Observational Constraints I).

The notion also emerged in the literature that massive stars can form in isolation. Finding undisputed such cases would constitute a strong argument for the IMF being a PDF and for a stochastic description of star formation (see Box Observational Constraints I). However, it is interesting that cases of presumed isolated massive star formation in the Milky Way, the Large and Small-Magellanic clouds have been shown to be untenable when improved observational data became available, and that, more distant dwarf galaxies, such as the Sextans A dwarf galaxy at about 1.3 Mpc, are now heralded as such cases in-proof (Garcia et al., 2019). If it were true that star formation is stochastic, then the observationally deduced power-law indices of the IMF, α_i, would vary according to the Poisson dispersion (even if the parent IMF were invariant, e.g. Kroupa, 2001), since some O stars may be isolated, others would be in rare clusters with a deficit of low mass stars while other populous clusters would not have any O stars.

Observational Constraints I: This box lists some key observations relevant for testing the conjecture that the IMF is a PDF.

- The most massive stars weigh $m_{\max *} \approx 150\,M_\odot$ (Weidner and Kroupa, 2004; Figer, 2005; Oey and Clarke, 2005; Koen, 2006; Maíz Apellániz et al., 2006). Crowther et al. (2010) and Schneider et al. (2014) find evidence for the existence of $m \approx 300\,M_\odot$ stars. These are consistent with $m_{\max *} \approx 150\,M_\odot$ as a result of mergers in their dense massive birth clusters (Banerjee et al., 2012), but further study is needed to ascertain if the observed number of such super-canonical stars can be obtained through mergers and binary-star evolution in dense starburst clusters (cf. Oh and Kroupa, 2018).
- The small dispersion of observationally derived IMF power-law indices (which contain all the uncertainties and come from different teams with different observational procedures), $\alpha_3 = 2.36 \pm 0.08$, above $2.5\,M_\odot$, compared to models which have no observational uncertainties but which include binaries and stellar dynamical processes and assume the IMF to be a PDF ($\alpha_3 = 2.20 \pm 0.63$), suggest the IMF to be remarkably invariant, the different populations showing differences in the IMF which are smaller than the expected Poisson scatter if the IMF were a PDF (Kroupa, 2002a).
- Most early-type stars supposed to have formed in isolation have been found to be propagating from some young cluster (Gvaramadze et al., 2012), be in hitherto unknown compact clusters (Stephens et al., 2017) or that they can be explained through the two-step ejection process (Pflamm-Altenburg and Kroupa, 2010).
- Hsu et al. (2012) observe in the Orion A cloud that the very young stellar populations formed in low-mass embedded clusters in its southern part significantly lack massive stars although, statistically, the same total number of stars (a few thousand) have formed as in the Orion Nebula Cluster in the northern part of the same cloud. If the IMF were to be a PDF, then both ensembles ought to have the same number of massive stars within the statistical uncertainties. Thus, the IMF seems to differ in the two regions and massive stars appear to correlate with the mass in stars of the embedded clusters.
- Kirk and Myers (2011, 2012) find, in their homogeneous survey of very young clusters, a pronounced correlation of the mass of the most massive star, $m_{\max}$, with the stellar mass of the embedded cluster, $M_{\rm ecl}$, for low-mass systems (the data lie near the $m_{\max} = WK(M_{\rm ecl})$ relation, see the next point).
- Stephens et al. (2017) performed a high resolution survey of previously-thought isolated massive stars in the LMC in order to test their isolated nature. They found each of them to be contained in a compact massive stellar cluster of 1–3 Myr age. The $m_{\max}$ and $M_{\rm ecl}$ values are correlated (the data lie near the $m_{\max} = WK(M_{\rm ecl})$ relation, see the next point).
- Ramírez Alegría et al. (2016) find a correlation of the most-massive star with the cluster mass in the sample of VVV-survey young (<10 Myr old) clusters.

- Weidner and Kroupa (2006) and Weidner et al. (2010, 2013c) collate data on very young clusters from the literature, selecting the clusters to be pre-supernova age ($<$4 Myr age being the only selection criterion) avoiding loss of stars through supernova explosions, and significantly limiting dynamical and stellar evolution affects. They find $m_{\rm max}$ to significantly correlate with $M_{\rm ecl}$, the dispersion of $m_{\rm max}$ values being largely consistent with the observational uncertainties such that intrinsic scatter appears to be small. They calculate that the hypothesis that the IMF is a PDF is consistent with the data at the 0.1% confidence level. The possible existence of a physical $m_{\rm max} = WK(M_{\rm ecl})$ relation is discussed (see Fig. 2.1; see also fig. 1 in Yan et al., 2017, which compare all above data by Kirk and Myers, 2011, 2012; Weidner et al., 2013c; Ramírez Alegría et al., 2016; Stephens et al., 2017). It may be a consequence of a feedback self-regulated growth process of stellar masses. Noteworthy is that the APEX/SABOCA[ab] survey of star-forming molecular cloud clumps by Lin et al. (2019) shows a similar correlation to exist between the most massive core mass and the proto-cluster clump mass (see Fig. 2.1). The flattening of this to the stellar relation above $m_{\rm max} \approx 10\,{\rm M}_{\odot}$ may be due to resolution limiting the identification of less-massive sub-clumps in the former, additional fragmentation in the former and feedback self-regulation during the transformation of the proto-stellar clump to a star (Beuther et al., 2007; Zinnecker and Yorke, 2007).
- Deviations from an assumed $m_{\rm max} = WK(M_{\rm ecl})$ relation, reported by e.g. Chené et al. (2015), can be understood naturally as arising from stellar mergers and ejections in and from embedded clusters (Oh and Kroupa, 2012, 2018).
- Observed very young embedded clusters (not older than a crossing time) have been found to be mass segregated (Bontemps et al., 2010; Kirk et al., 2014; Lane et al., 2016; Kirk et al., 2016; Plunkett et al., 2018). The ALMA[c] data by Plunkett et al. (2018) are interesting as they suggest that a proto-cluster is forming perfectly mass segregated. If this were to be the case then this suggests a high degree of regulation during the star-formation process with the individual stellar masses being a sensitive function of the core-density and thus gas reservoir (which varies radially in the embedded cluster) which, most probably, defines the accretion rate (Bonnell and Davies, 1998). Noteworthy is that this consideration appears to be consistent with the possible existence of a physical $m_{\rm max} = WK(M_{\rm ecl})$ relation noted above. N-body models of the Orion Nebula Cluster (<2.5 Myr old) also indicate it to be primordially mass segregated (Bonnell and Davies, 1998). Observationally, data on globular clusters also suggest them to have been born mass segregated (Baumgardt et al., 2008; Haghi et al., 2015). If the IMF were a PDF then the stellar mass drawn randomly from the IMF ought not to correlate with the position of the star.

[a] Atacama Pathfinder EXperiment; www.apex-telescope.org
[b] Submillimetre APEX Bolometer Camera; www.eso.org/public/teles-instr/apex/saboca/
[c] Atacama Large Millimitter/submillimeter Array; www.almaobservatory.org/en/home/

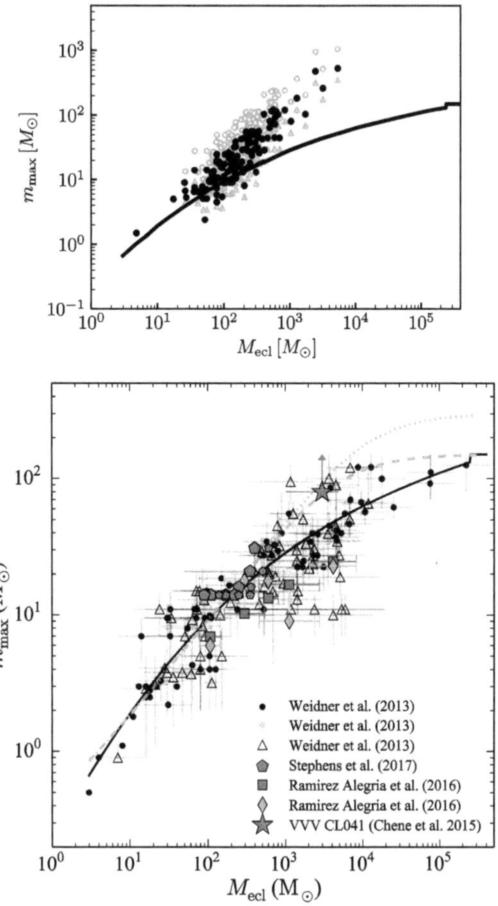

Figure 2.1 *Top panel*: Data from the Lin et al. (2019) APEX/SABOCA survey of star-forming molecular cloud clumps. The measured most-massive sub-clumps with masses $m_{\mathrm{scl}} = m_{\mathrm{max}}/\epsilon$ are plotted as grey open circles in dependence of their hosting measured proto-cluster of molecular-gas mass $M_{\mathrm{pcl}} = M_{\mathrm{ecl}}/\epsilon$ assuming each proto cluster and clump have a star-formation efficiency of $\epsilon = 1/3$. The black circles assume each $m_{\mathrm{max}} > 10\,\mathrm{M}_\odot$ value fragments into two equal mass sub-cores and the open triangles assume fragmentation into three sub-cores.
Bottom panel: The $m_{\mathrm{max}}, M_{\mathrm{ecl}}$ stellar data and the $m_{\mathrm{max}} = WK(M_{\mathrm{ecl}})$ relation (eq. 1 in Weidner et al., 2013c depicted as the black solid line, shown also in panel a). The black circles are the small-error sub-sample data in the <4 Myr old clusters used to obtain this relation. The open triangles have large uncertainties and are excluded from the fit. The red star is VVV CL041 put forward by Chené et al. (2015) as not fitting the $m_{\mathrm{max}} = WK(M_{\mathrm{ecl}})$ relation. Oh and Kroupa (2018) explain such deviant cases as being produced through mergers in the young compact embedded host cluster. The blue squares and orange diamonds are two different m_{max} estimates from the survey of <10 Myr-old clusters by Ramírez Alegría et al. (2016) and the red pentagons (large for 1 Myr age and small for 2.5 Myr age) are the data from Stephens et al. (2017). Note how the older objects (blue squares and orange diamonds) lie below the $m_{\mathrm{max}} = WK(M_{\mathrm{ecl}})$ relation, consistent with stellar evolution having removed the most massive stars. The dashed and dotted lines are the semi-analytical relation (see eqs 2.6 and 2.7 from Weidner and Kroupa, 2006) assuming $m_{\mathrm{max}*} = 150\,\mathrm{M}_\odot$ and $300\,\mathrm{M}_\odot$, respectively. From Oh and Kroupa (2018), by permission of Oxford University Press on behalf of the Royal Astronomical Society. Colour version available online.

Given the collation in Box Observational Constraints I, can the IMF still be a PDF on an embedded cluster scale? One particular observational challenge addresses this question: a physical $m_{\rm max} = WK(M_{\rm ecl})$ relation should not exist and the $m_{\rm max}, M_{\rm ecl}$ data should be consistent with randomly drawing stars from the IMF if it is a PDF. Prior to the more recent observational surveys by Kirk and Myers (2011, 2012), Ramírez Alegría et al. (2016) and Stephens et al. (2017) mentioned in Box I, Parker and Goodwin (2007); Maschberger and Clarke (2008); Cerviño et al. (2013a) addressed this problem using their own star-by-star data collations and statistical analysis negating that the relation exists. Andrews et al. (2013, 2014) also addressed this problem based on unresolved young star clusters in dwarf galaxies using broad-band photometry, spectral energy distribution (SED) modelling, and Hα flux measurements to conclude that this relation does not exist, at 1 σ significance). Being part of this team, this conclusion is echoed in the review by Krumholz (2014). For completeness, it is noted here that Weidner et al. (2014) showed these same extragalactic data to be well consistent (within 1 σ uncertainty) with the existence of a physical $m_{\rm max} = WK(M_{\rm ecl})$ relation, largely because the random sampling in connection with $m_{\rm max}$ being assumed to be a truncation limit was applied incorrectly in these and other publications (e.g. Fumagalli et al., 2011), noting also that field-star contamination in the unresolved observations plays a role in randomising any SED.

Thus, for the time being we conclude that the existence of a physical $m_{\rm max} = WK(M_{\rm ecl})$ relation needs to be studied further but with great care, and we emphasise that if it were to exist then the implications for the gwIMF are major (see Section 2.6.2).

2.4.2 The IMF as an Optimal Density Function

Independently of the possibility that a physical $m_{\rm max} = WK(M_{\rm ecl})$ relation may or may not exist (see Section 2.4.1), the observed degree of mass segregation in extremely young embedded clusters may also be seen as an indicator that a star's mass might not be random, but that it instead may be governed by the properties (mostly the density) of the molecular cloud core and thus the gas reservoir within which it forms and from which it accretes. The IMF may therefore be sensitively dependent on the physical conditions within the embedded cluster. An interpretation of the IMF as an ODF may thus be a useful consideration.

If the mathematical nature of the IMF is to be understood in terms of an optimal distribution function, then how is such a description to look like? The data (see Box Observational Constraints I) suggest that $m_{\rm max}$ is determined by $M_{\rm ecl}$. If the star-formation efficiency of embedded cluster-forming cloud cores,

$$\epsilon = \frac{M_{\rm ecl}}{M_{\rm ecl} + M_{\rm gas}}, \tag{2.4}$$

is typically in the range 0.1–0.3 (Megeath et al., 2016) then this dependency may arise if stars form through self-regulated accretion (on a timescale of 10^5 yr; cf. Wuchterl and Tscharnuter, 2003; Duarte-Cabral et al., 2013; Kristensen and Dunham, 2018) and a link between $m_{\max}$ and the total embedded-cluster forming cloud core can be made. It becomes a density dependency if the birth half-mass radii, r_h, of the embedded clusters are known[1]

$$r_h/\mathrm{pc} = 0.1\ (M_{\rm ecl}/\mathrm{M}_\odot)^{0.13}\,. \tag{2.5}$$

Concerning optimal sampling from the IMF, only the mass of the stellar population formed together in one star-formation event, i.e. an embedded cluster, is then relevant. A relation between $m_{\max}$ and $M_{\rm ecl}$ is obtained from the following two equations:

$$1 = \int_{m_{\max}}^{m_{\max *}} \xi(m)\, dm, \tag{2.6}$$

with

$$M_{\rm ecl} - m_{\max}(M_{\rm ecl}) = \int_{m_{\rm L}}^{m_{\max}} m\, \xi(m)\, dm, \tag{2.7}$$

which allow solution for the two unknowns, $m_{\max}$ and the normalisation constant k, for the population which has the physical upper mass limit $m_{\max *}$ (see Box Observational Constraints I). Both equations are also valid also if the IMF is a PDF and in this case $m_{\max}$ will be the average most massive star in an ensemble of embedded clusters (Kroupa and Weidner, 2003; Oey and Clarke, 2005). But the small dispersion of observed $m_{\max}$ values does not support this interpretation, as stated above in Box I.

If, on the other hand, the IMF is an ODF, then Eqs 2.6 and 2.7 lead directly to a good representation of the empirical $m_{\max}, M_{\rm ecl}$ data (dotted and dashed lines in Fig. 2.1; cf. Weidner et al., 2013c). In order to ensure a small or negligible scatter of the $m_{\max}$ values, ordered sampling has been invented (Weidner and Kroupa, 2006), according to which $M_{\rm ecl}$ is iteratively solved for with increasing number of stars to minimise the difference to the observed $m_{\max} = WK(M_{\rm ecl})$ function. A more elegant method to efficiently sample stars from the IMF such that this function is obeyed is given by optimal sampling (Kroupa et al., 2013; Schulz et al., 2015; Yan et al., 2017). In optimal sampling (for an improved algorithm, see Schulz et al., 2015), starting with a given $M_{\rm ecl}$ and thus $m_{\max}$, a sequence of stellar masses is generated step by step such that each equal $M_{\rm ecl}$ has exactly the same sequence of stars. This latter procedure is possible for whichever function $m_{\max} = WK(M_{\rm ecl})$ is adopted. One possibility is given by solving Eqs 2.6 and 2.7, but physically motivated functions may also be used (e.g. Weidner et al., 2010 find evidence for a possible flattening of $WK(M_{\rm ecl})$ which may be due to feedback regulation in the formation of $m_{\max} \gtrsim 25\,\mathrm{M}_\odot$ stars).

An optimally sampled stellar population has no Poisson scatter and thus fulfils two important requirements: It is consistent with the observed small dispersion of α_3 values noted above in Box I; and it is consistent with the small dispersion of observed

[1] This half-mass-radius–embedded-stellar-cluster-mass relation is derived by Marks and Kroupa (2012) from the observed binary population in open and very young clusters and assuming the truncation of the binary-binding energy distribution at low values to be a measure of the densest evolution phase of the cluster. It constitutes a theoretical approximation to the initial state of an embedded cluster for N-body modelling.

m_{max} values.[2] Dynamically stable perfectly mass-segregated embedded clusters can then be constructed readily using the methods introduced in Bonn by Šubr et al. (2008) and Baumgardt et al. (2008) in order to fulfil the observational constraint that very young embedded clusters are mass segregated. These methods assume the stellar masses to be radially arranged according to their binding energy.

Summarising, the observational evidence suggests that the IMF may be an ODF rather than a PDF. However, more research is needed to further test this assertion.

2.4.3 Emergence of Apparent Stochasticity

On stating the above, we note the following: Once a perfect (or highly constrained) system of stars is generated (or born) it will evolve towards what appears to be a stochastic version of itself. This follows from the second law of thermodynamics and comes about naturally because stars are ejected, binaries dissolve, and stars merge and age. The N-body computations of Oh et al. (2015) and Oh and Kroupa (2016) show strong dynamical activity with the very young binary-rich clusters spurting out stars of all masses within fractions of a Myr.[3]

A molecular cloud which forms all its stars in a population of embedded clusters with different masses over a few Myr will thus always lead to a distribution of isolated stars, expanded and already dissolved older clusters and a few deeply embedded still forming embedded clusters with a complex velocity field which also depends on whether the whole cloud was contracting or expanding (see Section 2.1.2). Observationally, this activity leaves, in association with observational measurement uncertainties, a population of stars which appears non-optimal and even with an age, spin and chemical spread (Kroupa, 1995a,b; Weidner et al., 2009; Marks et al., 2011; Oh and Kroupa, 2012; Oh et al., 2015; Oh and Kroupa, 2016, 2018).

The problem, to which degree an apparent probabilistic quality emerges from an optimally mono (age, metallicity, spin) case may be an interesting research project worthy of study.

2.5 Does the Initial Mass Function Vary?

Having discussed above the question of whether the IMF is a PDF or an ODF, another problem to deal with is whether the shape of the distribution function varies systematically with the physical conditions of star formation.

[2] Approaches which adopt the $m_{max} = WK(M_{ecl})$ relation but sample stars randomly from the IMF (i.e. which interpret it to be a constrained PDF) would lead to, on average, too small m_{max} values for each M_{ecl} such that the respective authors are led to erroneously conclude the WK-relation to be excluded, given unresolved data (e.g. Andrews et al., 2013, 2014). These analysis-shortcomings are elaborated on by Weidner et al. (2014).

[3] Movies based on these publications are available on youtube: 'Dynamical Ejection of Massive Stars from a Young Star Cluster' by Seungkyung Oh.

As discussed in Section 2.3, the consensus has been that the IMF is an invariant PDF, possibly with an upper stellar mass limit near $150\,M_\odot$ to $300\,M_\odot$. This was taken to be the case in most studies requiring the stellar populations in star clusters and galaxies. But, as discussed at some length in Kroupa et al. (2013), this consensus may deem unnatural because different physical conditions should lead to different distributions of stellar masses. The most elementary and thus fundamental and solid arguments based on the Jeans fragmentation scale (Larson, 1998) on one hand, and on self regulation on the other (Adams and Fatuzzo, 1996; Matzner and McKee, 2000; Federrath et al., 2014; Federrath, 2015), lead to the same expectation of how the IMF should change with temperature and metallicity of the gas cloud from which the stars form. Additional physical processes such as the coagulation of contracting pre-stellar cores in very high-density star-forming regions (Dib et al., 2007; see also Zinnecker and Yorke, 2007 and Krumholz, 2015 for reviews on the formation of massive stars) and the heating of the molecular gas though an elevated flux of supernova-generated cosmic rays in starburst regions (Papadopoulos and Thi, 2013) are likely to play a role in establishing a systematic variation of the IMF.

All these broad physical arguments lead to the following theoretical expectation: The IMF should become systematically top-heavy with increasing density, increasing temperature, and decreasing metallicity. The empirical boundary condition is that, for the star-formation conditions evident in the present-day Milky Way, the IMF should be close to canonical. This variation may involve α_3 decreasing (the IMF becoming more top-heavy) for decreasing metallicity and increasing density, and for α_1 and α_2 to increase (the IMF becoming bottom-heavy) for increasing metallicity, but a robust prediction as to how the shape of the IMF should vary with conditions does not exist to date.

Perhaps the first suggestion, based on direct star counts, as to a variation of the IMF with metallicity has been described in Kroupa (2001, 2002b). Here it was pointed out that, taken at face value, the available resolved stellar populations indicate a systematic flattening of α_1 and of α_2 with decreasing metallicity valid for $[\mathrm{Fe/H}] \gtrsim -2$. If such a systematic effect is present, then for $m < 1\,M_\odot$ $(i = 1, 2)$ and to first order,

$$\alpha_i \approx 1.3 + \Delta\alpha \log_{10}(Z/Z_\odot), \tag{2.8}$$

with $\Delta\alpha \approx 0.5$ (Kroupa, 2001) (here writing the dependent on metallicity, Z, rather than the iron abundance). The evidence was limited to late-type stars because most of the populations at disposal were old. Concerning massive stars, in a great observational effort, Massey (2003) demonstrated a high degree of invariance of α_3 on density and metallicity for young populations of stars in the Milky Way and the Large and Small Magellanic Clouds. More recently, due to improved data reaching into more extreme star-forming environments, explicit evidence for a systematic variation of α_3 with physical conditions may have begun to emerge on star-cluster scales (see Box Observational Constraints II below).

OBSERVATIONAL CONSTRAINTS II:

- The high dynamical mass to light ratios of many ultra compact dwarf (UCD) galaxies, which have present-day stellar masses larger than $10^6\,M_\odot$, can be understood if they formed under high density, ρ, with α_3 becoming smaller (more top-heavy) with increasing UCD mass (thus increasing ρ), such that a larger fraction of their initial mass ends up in stellar remnants (Dabringhausen et al., 2009; Jeřábková et al., 2017).
- The high X-ray luminosity of many UCDs can be understood if they have a larger fraction of low-mass X-ray binaries which need a larger fraction of neutron stars and stellar black holes. The data suggest α_3 to become smaller with increasing UCD mass and thus ρ (Dabringhausen et al., 2012).
- The above two points lead to similar constraints on how α_3 varies with ρ. This is an important consistency check. The systematic variation of α_3 with star-formation density implies significant expansion of the UCDs with time due to mass loss from expelled residual gas and from stellar evolution (Dabringhausen et al., 2010). This work demonstrates that the solutions for α_3 also lead to stellar-dynamical solutions to the observed UCDs (they do not dissolve due to too much mass loss).
- Combining the information on an observed systematic deficit of low-mass stars in present-day stellar mass functions in observed globular clusters (GCs) with their present-day radii and metallicities, with stellar-dynamical models which are initially mass segregated and allow for expulsion of residual gas and thus expansion of the young GCs and with the above constraints on $\alpha_3(\rho)$, yields the description of how α_3 depends on ρ and Z given by Eq. 2.9 below from Marks et al. (2012) (see also Kroupa et al., 2013).
- The observed star clusters in M31 are consistent with the dependency formulated as Eq. 2.9 (Zonoozi et al., 2016; Haghi et al., 2017).
- The young (≈ 1 Myr), compact (half-mass radius $r_h \approx 1$ pc) $M_{st} \approx 10^5\,M_\odot$ starburst cluster R136 in the 30 Dor star-forming region in the Large Magellanic Cloud has been ejecting its massive stars efficiently. Adding these back statistically into the star-counts in the cluster yields a top-heavy IMF (Banerjee and Kroupa, 2012).
- The stellar census in the massive star-forming region 30 Dor in the Large Magellanic Cloud has been found to have a top-heavy IMF (Schneider et al., 2018), confirming the Banerjee and Kroupa (2012) prediction.
- Deep adaptive-optics imaging of the Magellanic-Bridge Cluster NGC 796 reveals a top-heavy IMF (Kalari et al., 2018).
- Using multi-epoch *Hubble Space Telescope*[a] observations of the Arches cluster ($2-4$ Myr old, $M_{ecl} \approx 4-6\times 10^4\,M_\odot$) near the Galactic Centre, Hosek et al. (2019) find the IMF to be top-heavy.

[a] *Hubble Space Telescope (HST)*, www.stsci.edu/hst

Resolved stellar populations largely show an invariant IMF see (Eq. 2.1), but for star-formation rate densities SFRD $\gtrsim 0.1\,\mathrm{M_\odot\,yr^{-1}\,pc^{-3}}$ in embedded clusters, the IMF appears to become top-heavy (Marks et al., 2012). The dependence of α_3 on cluster-forming cloud density, ρ, (stars plus gas) and Z can be parametrised as

$$\begin{aligned} \alpha_3 &= \alpha_2, & m > 1\,\mathrm{M_\odot} \quad & x' < -0.87, \\ \alpha_3 &= -0.41 \times x' + 1.94, & m > 1\,\mathrm{M_\odot} \quad & x' \geq -0.87, \end{aligned} \tag{2.9}$$

$$x' = -0.14 \log_{10}(Z/Z_\odot) + 0.99 \log_{10}\left(\rho/\left(10^6\,\mathrm{M_\odot\,pc^{-3}}\right)\right). \tag{2.10}$$

A possible variation of the IMF for late-type stars with Z is suggested by Eq. 2.8. In Eq. 2.9, the density (stars plus gas at a mathematical time, t_{math}, when all the stars have been born and the remaining gas has not yet been expelled) of the star-forming molecular cloud clump is

$$\rho = \frac{3\,(M_{\mathrm{ecl}}/2)}{4\pi\epsilon\, r_{\mathrm{h}}^3}, \tag{2.11}$$

where the star-formation efficiency, ϵ, is Eq. 2.4 and r_{h} is Eq. 2.5 being the half-mass radius of the clump. Note that this clump, or rather embedded cluster, is a mathematical hilfskonstrukt which in reality does not exist because the alluded to mathematical time does not exist. In reality the stars of the embedded cluster form over about 1 Myr and the gas is blown out over a time which may be approximated by the sound speed of ionised gas ($\gtrsim 10\ \mathrm{km\,s^{-1}}$, e.g. the extremely young Treasure Chest cluster, Smith et al., 2005, and the massive starburst clusters in the Antennae galaxies, Whitmore et al., 1999; Zhang et al., 2001) and r_{h}. The stars form in filaments rather than spherical Plummer models. Nevertheless, by dynamical equivalence, any two dynamical structures (e.g. the Plummer model and a filament) lead to comparable dynamical processing of the stars formed within them if the two structures are dynamically equivalent (Kroupa, 1995a; Belloni et al., 2018). By conservation of angular momentum, stars need to form in multiple systems, whereby most form as binaries (Goodwin and Kroupa, 2005) with well defined orbital parameter distributions and internal (eigen) evolutionary processes (Kroupa, 1995b; Belloni et al., 2017). But, at the extreme density needed for the top-heavy IMF regime, our knowledge of the processes within the forming cluster remain poor. Formally, in this regime, the crossing time is shorter than the formation time of a single proto-star ($\approx 10^5$ yr).

In conclusion, high-quality star-count data appear to suggest a systematic variation of the IMF. Equations 2.9 and 2.8 are possible quantifications of this variation with density and metallicity of the star-forming gas cloud. The above observational findings are compared in Fig. 2.2. This variation is consistent with the qualitative expectations from star-formation theory noted at the beginning of this section. Extending the analysis to unresolved young clusters in other galaxies will prove to be an important consistency test whether this formulation of the variation is a good approximation (e.g. Ashworth et al., 2017).

If true, this IMF variation has important implications for galaxy-wide IMFs, and also impinges on the interpretation of the IMF as a PDF. If the IMF varies as suggested here and

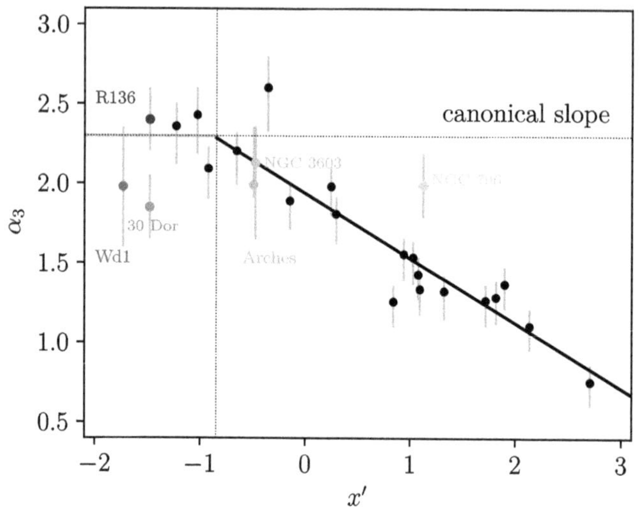

Figure 2.2 The variation of the upper end of the stellar IMF with metallicity and cloud density (solid line, Eq. 2.9) as deduced from deep observations of MW GCs using a principal-component-type analysis (Marks et al., 2012). The recent observational determination of top-heavy IMFs in two metal-poor very young clusters (30Dor: Schneider et al., 2018, NGC796: Kalari et al., 2018) and the recent Arches data (Hosek et al., 2019) are included with the data shown here. Colour version available online.

if it is to be a PDF, then it can only be a constrained PDF, because, before beginning the stochastic drawing process, the functional form of the IMF needs to be calculated based on the properties of the embedded cluster, which, however, itself is supposedly drawn stochastically from an embedded cluster mass function. It remains to be seen, perhaps more from a philosophical point of view, if, and how, a PDF interpretation of the IMF is satisfyingly compatible with a systematic variation of the IMF given by the physical conditions at star formation, since one would, on the one hand, be resorting to a stochastic process, but on the other hand, also to a process driven by physical conditions.

2.6 Is the Initial Mass Function of a Simple Stellar Population Equal to That of a Composite Population?

Consider a single star-forming region and a region containing many star-forming regions, such as a whole galaxy. We can then define the IMF of the larger region to be the composite IMF, cIMF. Logically, the cIMF is the sum of all IMFs assembled throughout the region within some time δt. Considering a whole galaxy, the cIMF becomes equal to the galaxy-wide IMF (gwIMF, Kroupa and Weidner, 2003).

From Eqs 2.2 and 2.3, for any pair m_1, m_2, X_{12} (calculated assuming $\xi_{\rm p}(m)$ is the PDF corresponding to the IMF) will be equal to another X_{12} (calculated assuming $\xi_{\rm p}(m)$ is the PDF corresponding to the cIMF) if, and only if, both PDFs are equal because m_1 and m_2

are arbitrary. This is the fundamental composite IMF-PDF theorem (see also Kroupa and Jerabkova, 2018): if the IMF is a PDF then the IMF of all stars formed in a galaxy within a short time interval δt, the galaxy-wide IMF (gwIMF) is equal to the IMF.

This theorem has been applied in most studies of star-formation in galaxies. In particular, the Kennicutt SFR Hα tracer rests on assuming this theorem holds (Kennicutt, 1998; Jeřábková et al., 2018). It assumes that any sum of IMFs will yield the same PDF which is true if all individual IMFs are fully sampled over all possible stellar masses. But is this formulation applicable to real galaxies?

Solving this problem is of paramount importance for measuring SFRs. If the IMF is invariant and equal to the gwIMF then the tracer can be calibrated simply and becomes invariant to galaxy type and mass. For example, the number of re-combination photons per unit time, i.e. the Hα flux, which is proportional to the number per unit time of ionising photons emitted by a fresh stellar population, depends on the number of the ionising (i.e. massive, m $\gtrsim 10\,M_\odot$) stars in the young population. If the gwIMF is invariant and equal to the IMF and both are fully sampled up to m_{max*} then the Hα flux is a direct measure of the total mass of stars born per unit time (Kennicutt, 1998). For this calibration to remain valid the IMF needs to be an invariant PDF. In star-by-star models of galaxies this interpretation is equivalent to a stochastic description of star formation without constraints.

2.6.1 Clustered Star Formation, but Star Formation Is Stochastic

If stars form in embedded clusters (e.g. Hopkins, 2013) which are randomly distributed throughout a galaxy and which follow a distribution of masses, i.e. an ECMF (see Eq. 2.13), then two possibilities arise:

1. The cluster mass in stars, M_{ecl}, plays no role and stars appear randomly filling a cluster to a pre-specified maximum number of stars, N_{ecl}. This yields a distribution of clusters with different numbers of stars, and thus to an ECMF. Within each cluster the IMF is sampled randomly without constraints, apart from the condition that stars have masses $m \leq m_{max*}$. In this case, the composite IMF-PDF theorem (see Section 2.6) holds, making gwIMF equal in form to the IMF and any tracer of star formation rate (e.g. the Kennicutt SFR Hα tracer, Kennicutt, 1998) will yield a correct value of the SFR upon measurement of the tracer (e.g. the Hα flux) subject to stochastic variations. This interpretation implies that a 'cluster' may consist of one massive star only. Thus, isolated massive stars would pose an important argument for this purely stochastic approach to galaxy evolution. This possibility implies that a $m_{max} - N_{ecl}$ relation does not exist. A $m_{max} - M_{ecl}$ relation emerges but has a large scatter consistent with random drawing from the IMF (Maschberger and Clarke, 2008). An issue with this possibility is that the primary variable, N_{ecl}, needs to be interpreted as a physical parameter.

   If nature were to follow this mathematical recipe, then the measured SFRs, using the Kennicutt SFR Hα tracer for example, will appropriately assess the true SFRs. Galaxies with very small SFRs will show a dispersion of SFRs which increases with decreasing SFR as a result of stochasticity.

2. Considering M_{ecl} to be the primary variable, and choosing it randomly from the ECMF, the embedded cluster is filled with stars randomly from the IMF until the mass of the stellar population matches M_{ecl}. This is constrained random sampling (Weidner and Kroupa, 2006; Weidner et al., 2010, 2013c). If the ECMF contains clusters with $M_{ecl} < m_{max*}$ then in these clusters the stellar population will lack the most massive stars. Such clusters are undersampling the IMF at large stellar masses. The deficit arises because a cluster of mass $M_{ecl} < m_{max*}$ cannot contain a star weighing more than the cluster. The whole population of all embedded clusters will thus have a systematic deficit of the most massive stars, i.e. the gwIMF will be under-sampled at high stellar masses and will not equal the IMF. This possibility implies a $m_{max} - M_{ecl}$ relation which has a large scatter since there may exist a cluster of mass M_{ecl} being composed off one star of mass $m = M_{ecl}$. The existence of isolated massive stars is of paramount importance for this possibility to be valid.

   This implies that any tracer of ionising stars, calibrated assuming the first possibility above, will systematically underestimate the true SFR of the region or galaxy in cases when under-sampled clusters are involved. This case has been extensively studied in the Stochastically Lighting Up Galaxies (SLUG[4]) approach (da Silva et al., 2012, 2014).

   If nature were to follow this mathematical recipe, then the measured SFRs, using the Kennicutt SFR Hα tracer, for example, will deviate systematically towards smaller than SFR_{true} values at low SFRs. Galaxies with very small SFRs will show a dispersion of SFRs which increases with decreasing SFR as a result of stochasticity (fig. 3 in da Silva et al., 2014).

2.6.2 Clustered Star Formation, but Star Formation Is Optimal

The above two stochastic approaches are consistent with the notion that galaxies have a stochastic history which comes about from the need of a very large number of mergers to build-up Milky-Way-class disk galaxies in the standard Lambda Cold Dark Matter (LCDM) cosmological model. The observed simplicity of galaxies (Disney et al., 2008) and lack of evidence for an intrinsic dispersion in the baryonic-Tully-Fischer relation (McGaugh, 2005, 2012), in the mass-discrepancy relation (McGaugh, 2004), and in the radial-acceleration relation (McGaugh et al., 2016; Lelli et al., 2017), as well as the existence of a tight and well defined main sequence of galaxies over different redshifts (Speagle et al., 2014), indicate that the dynamical structure and the star-formation behaviour of galaxies may follow precise rules and that the expected stochasticity may be absent (Disney et al., 2008).

[4] Stochastically Lighting Up Galaxies; www.slugsps.com/home

Observations have shown that stars form in molecular clouds (see Section 2.1.2). This suggests that, if the IMF were to be a PDF, then this PDF should be subject to constraints: Stars do not form at arbitrary positions within a galaxy, they do so only where the conditions allow their formation. Since the conditions and properties of molecular clouds change with location in a galaxy (notably, inner region vs the far-outer region, Fukui and Kawamura, 2010; Heyer and Dame, 2015) the possibility might be given that the constraints subjecting the PDF also depend on location within the galaxy and within molecular clouds (see Box Observational Constraints III).

OBSERVATIONAL CONSTRAINTS III:

- Stars do not form randomly throughout molecular clouds but in embedded clusters (see Section 2.1.2). The observed distributed population is explainable through the dynamical activity of the forming embedded clusters (see Section 2.4.3).
- The mixture of embedded clusters determines the cIMF of a region in the molecular cloud (Hsu et al., 2012). According to the discussion in Section 2.4.1, stars form in filaments which combine to embedded clusters which are mass segregated (see Box Observational Constraints I).
- The distribution of young star-cluster masses shows a radial gradient with the most-massive cluster within a radial annulus being smaller at larger galactocentric distance in the disk-galaxy M33, for example (Pflamm-Altenburg et al., 2013). This is a result of the exponentially declining surface mass-density of gas in a disk galaxy.
- A decreasing cluster-mass with galactocentric-distance relation is also found in the interacting luminous infrared galaxy Arp 299 system as a result of the gas-density distribution (Randriamanakoto et al., 2018).
- The formation of the embedded most-massive stellar sources shows a Galactocentric radially decreasing trend in the Milky Way (Urquhart et al., 2014) reminiscent of the above M33 result.
- Late-type galaxies show a pronounced correlation between their SFR and their most-massive very-young cluster, the $M_{\rm ecl,max} - SFR$ relation (Weidner et al., 2004). The dispersion of the data is consistent with observational uncertainty, and Randriamanakoto et al. (2013) point out that their own data imply that the dispersion is not consistent with stochastic scatter by being too small.

Is there thus an alternative to the above two stochastic approaches (see Section 2.6.1) for describing stellar populations in galaxies? Is it possible to derive a theory which has no intrinsic scatter, taking the observed simplicity of galaxies as a motivation? Even in such a theory, an observable dispersion would arise naturally (see Section 2.4.3) and some intrinsic dispersion can always be added if needed. It might be educational to develop a theory which has no intrinsic scatter to test how applicable this perhaps extreme description may be. If successful, we will have uncovered the laws of nature which describe how the interstellar medium in a galaxy transforms into a new stellar population and at which rate it does so.

One possibility is to begin with the rules (we may also refer to them as axioms, cf. Recchi and Kroupa, 2015; Yan et al., 2017) deduced from observations of star formation within nearby molecular clouds and from very young stellar populations in embedded and older star clusters (e.g. Eqs 2.8 and 2.9, Observational Constraints I–III), assume these rules are valid independently of cosmological epoch and calculate how they imply what galaxies with different properties ought to look like in their star-formation behaviour. This approach has the advantage that it fulfils (by construction) all observational constraints (as outlined in Observational Constraints I–III) and that it is, in particular, consistent with the stellar-mass functions in the observed star clusters and the Galaxy. These observations suggest that star formation may be significantly regulated, possibly following precise and clear rules.

In addition to the two above stochastic approaches (see Section 2.6.1), we may therefore entertain a model in which star formation is entirely optimal on every scale.

Once we know how to compute such a model, it is clear that it becomes completely predictive. Some scatter in observable properties then enters via physical processes (e.g. galaxy–galaxy encounters which change the mass distribution and SFR over some time), other physical processes (see Section 2.4.3), and measurement errors. The question is now how to construct such a deterministic, optimal model?

We thus begin with 'The composite IMF statement': If each embedded cluster[5] spawns a population of stars (this takes $\lesssim 1$ Myr) with combined mass $M_{\rm ecl}$ per cluster, then the gwIMF becomes an integral over all such embedded clusters within the galaxy, yielding the integrated galaxy-wide initial mass function (IGIMF, eq. 2.12; Kroupa and Weidner, 2003; Weidner and Kroupa, 2005; Yan et al., 2017; Jeřábková et al., 2018). A clue that this may be an interesting avenue to investigate is provided by the galactic-field gwIMF being steeper than the canonical IMF for m > 1 $M_\odot$ (Scalo, 1986; Kroupa et al., 1993; Rybizki and Just, 2015; Mor et al., 2017, 2018): This difference might be related to the gwIMF being a sum of the IMFs in the star-forming units (Kroupa and Weidner, 2003; Zonoozi et al., 2019). The IGIMF is a particular mathematical formulation of the gwIMF. The embedded clusters need to be sampled (e.g. optimally, Schulz et al., 2015) from the ECMF. This means that over some time interval, δt, the ECMF is optimally assembled, whereby this time interval needs to be investigated with care.[6]

[5] An embedded cluster can be constructed to be optimal by being mass segregated initially and by following the $m_{\rm max} - M_{\rm ecl}$ relation (Pavlík et al., 2019). Initially mass-segregated models in dynamical equilibrium, such that the mass segregation would persist in collisionless systems, can be constructed using the methods developed in Bonn, namely by energy-stratifying the binary systems in the clusters (Baumgardt et al., 2008; Šubr et al., 2008). The embedded clusters can be optimally distributed throughout the galaxy by associating the most massive cluster at any galactocentric radius with the local gas surface density (Pflamm-Altenburg and Kroupa, 2008).

[6] Note that here the optimal model is being formulated. It is possible to relax the inherent non-dispersion nature and to retain the IGIMF formulation (see Eq. 2.12) but to treat the sampling of embedded star cluster masses and of stars within the embedded clusters stochastically. This was the original notion followed by Kroupa and Weidner (2003) and forms the basis of the SLUG approach (da Silva et al., 2014).

The IGIMF is an integral over all star-formation events (embedded clusters, see Section 2.1.2) in a given star-formation 'epoch' $t, t + \delta t$,

$$\xi_{\mathrm{IGIMF}}(m;t) = \int_{M_{\mathrm{ecl,min}}}^{M_{\mathrm{ecl,max}}(SFR(t))} \xi\left(m \leq m_{\mathrm{max}}\left(M_{\mathrm{ecl}}\right)\right)\, \xi_{\mathrm{ecl}}(M_{\mathrm{ecl}})\, dM_{\mathrm{ecl}}, \tag{2.12}$$

with the normalisation conditions in Eqs 2.15 and 2.16 below and also with the conditions Eqs 2.6 and 2.7 which together yield the $m_{\mathrm{max}} - M_{\mathrm{ecl}}$ relation.

Here, $\xi(m \leq m_{\mathrm{max}})\, \xi_{\mathrm{ecl}}(M_{\mathrm{ecl}})\, dM_{\mathrm{ecl}}$ is the composite stellar IMF (i.e. the cIMF) contributed by $\xi_{\mathrm{ecl}}\, dM_{\mathrm{ecl}}$ embedded clusters with stellar mass in the interval $M_{\mathrm{ecl}}, M_{\mathrm{ecl}}+dM_{\mathrm{ecl}}$. The ECMF is often taken to be a power-law,

$$\xi_{\mathrm{ecl}}(M_{\mathrm{ecl}}) \propto M_{\mathrm{ecl}}^{-\beta}, \tag{2.13}$$

with $\beta \approx 2$ (Lada and Lada 2003; Schulz et al. 2015 and references herein). For the Milky Way and extragalactic data on non-starbursting galaxies, $\beta \approx 2.3$ may be favoured (Weidner et al., 2004; Mor et al., 2018) and β may vary with the metallicity and SFR (Yan et al., 2017).

The most-massive embedded cluster forming in a galaxy, $M_{\mathrm{ecl,max}}(SFR)$, can be assumed to come from the empirical maximum star-cluster mass *vs* global SFR of the galaxy relation,

$$M_{\mathrm{ecl,max}} = 8.5 \times 10^4 \left(\frac{\mathrm{SFR}}{\mathrm{M_\odot/yr}}\right)^{0.75}, \tag{2.14}$$

(eq. 1 in Weidner and Kroupa, 2005, as derived by Weidner et al., 2004 using observed maximum star cluster masses). A relation between $M_{\mathrm{ecl,max}}$ and SFR, which is a good description of the empirical data, can also be arrived at by resorting to optimal sampling (see Section 2.4.2). Thus, when a galaxy has, at a time, t, a $SFR(t)$ which is approximately constant over a time-span, δt, over which an optimally sampled embedded star-cluster distribution builds up with total mass $M_{\mathrm{tot}}(t)$, then there is one most massive embedded cluster with mass $M_{\mathrm{ecl,max}}$,

$$1 = \int_{M_{\mathrm{ecl,max}}(t)}^{M_{\mathrm{U}}} \xi_{\mathrm{ecl}}(M_{\mathrm{ecl}})\, dM_{\mathrm{ecl}}, \tag{2.15}$$

with M_{U} being the physical maximum star cluster that can form (for practical purposes $M_{\mathrm{U}} > 10^8\,\mathrm{M_\odot}$), and

$$SFR(t) = \frac{M_{\mathrm{tot}}(t)}{\delta t} = \frac{1}{\delta t} \int_{M_{\mathrm{ecl,min}}}^{M_{\mathrm{ecl,max}}(t)} M_{\mathrm{ecl}}\, \xi_{\mathrm{ecl}}(M_{\mathrm{ecl}})\, dM_{\mathrm{ecl}}. \tag{2.16}$$

$M_{\mathrm{ecl,min}} = 5\,\mathrm{M_\odot}$ is adopted in the standard modelling and corresponds to the smallest 'star-cluster' units observed (the few $\mathrm{M_\odot}$ heavy embedded clusters in Figure 1 in Yan et al.

2017 observed in Taurus-Auriga, e.g. Kroupa and Bouvier, 2003; Joncour et al., 2018). Note the similarity of these equations with Eqs 2.6 and 2.7. Perhaps the physical meaning of this is that within a galaxy the formation of embedded clusters may be a self-regulated growth process from the ISM, like stars in a molecular-cloud core which spawns an embedded cluster.

But what is δt? Weidner et al. (2004) define δt to be a 'star-formation epoch', within which the ECMF is sampled optimally, given a SFR. This formulation leads naturally to the observed $M_{\rm ecl,max}(SFR)$ correlation if the ECMF is invariant, $\beta \approx 2.35$, and if the 'epoch' lasts about $\delta t = 10$ Myr. Under these conditions, a galaxy forms embedded clusters with stellar masses ranging from $M_{\rm ecl,min} = 5\,{\rm M}_\odot$ to $M_{\rm ecl,max}$, the value of which increases with the SFR of the galaxy in accordance with the observed young most-massive cluster vs SFR data (see Box Observational Constraint III). Thus, the embedded-cluster mass function is optimally sampled in about 10 Myr intervals, independently of the SFR.

It is interesting to note that this timescale is consistent with the star-formation timescale in normal galactic disks measured by Egusa et al. (2004, 2009, 2017) using the offset of HII regions from the molecular clouds in spiral-wave patterns. In this view, the ISM takes about 10 Myr to transform via molecular cloud formation to a new population of young stars which optimally sample the embedded-cluster mass function (see also Section 2.1.2). Schulz et al. (2015) discuss (in their Section 3) the meaning and various observational indications for the timescale $\delta t \approx 10\,$Myr. This agreement between the independent methods to yield $\delta t \approx 10\,$Myr is encouraging since the Weidner et al. (2004) argument is independent of the arguments based on molecular-cloud lifetimes and spiral arm phase-velocities.

The IGIMF (see Eq. 2.12) can be calculated under various assumptions on how the IMF, $\xi(m)$, varies with physical conditions. Ignoring the explicit metallicity dependence but taking into account the effective density dependence, which includes an intrinsic metallicity dependence, in Eq. 2.9 above Yan et al. (2017) studies the prediction of the IGIMF theory for the variation of the shape of the gwIMF in comparison with observational constraints (see Fig. 2.4). The change of the IGIMF as a function of the SFR is shown in Fig. 2.3. The observational constraints, which indicate the gwIMF to change from a top-light form (i.e. with a deficit of massive stars) in dwarf galaxies (which have low SFRs, Lee et al., 2009) to top-heavy gwIMFs in massive late-type galaxies (which have high SFRs, Gunawardhana et al., 2011), are well covered by the IGIMF theory. It is noted here for completeness that the top-light gwIMF variation for dwarf galaxies was predicted on the basis of the IGIMF theory (Pflamm-Altenburg et al., 2007, 2009) before the data were available, with observational support from star counts being found by Watts et al. (2018) in DDO154. The calculations of the IGIMF for galaxies with high SFRs became physically relevant once the variation of the IMF (see Eqs 2.8 and 2.9) in extreme starburst clusters was quantified based on observational data (Marks et al., 2012).

A full grid of IGIMF models is provided by Jeřábková et al. (2018) in which three cases are considered: (1) the IMF is invariant, (2) the IMF varies only at the massive end (see Eq. 2.9), and (3) the IMF varies at low (see Eq. 2.8) and at high (see Eq. 2.9) masses. This latter case, IGIMF3, is considered to be the most realistic and appears to be able to

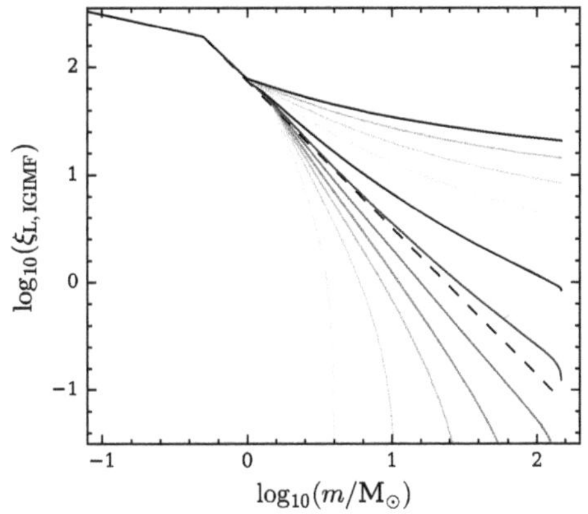

Figure 2.3 Logarithmic integrated gwIMFs (in number of stars per log-mass interval using the transformation $\xi_{\rm L} = m \ln(10)\, \xi(m)$, eq. 14 in Yan et al., 2017) for different SFRs and formed over a $\delta t = 10\,$Myr epoch. Each line is normalised to the same values at $m < 1\,M_{\odot}$. Solid curves are IGIMFs for SFR $= 10^{-5}, 10^{-4} \ldots 10^{5}\,M_{\odot}\,{\rm yr}^{-1}$ from bottom left to top right. The dashed line is the canonical IMF (see Eq. 2.1). For further details see Yan et al. (2017). Note that the explicit metal dependency of the IGIMF is not included in these calculations and that the $\alpha_3^{\rm gal}$ values plotted in Figure 2.4 are the slopes of the here shown IGIMFs in different stellar mass ranges. See Jeřábková et al. (2018) for a full grid of IGIMF models in dependency of metallicity and SFR. Colour version available online.

account for the complicated time-evolving gwIMF variation deduced to have occurred when elliptical galaxies formed as well as for the gwIMF variations deduced for late-type galaxies.

2.6.3 Some Observational Constraints

Any of the three approaches, namely pure stochastic sampling (see Section 2.6.1), constrained stochastic sampling (see Section 2.6.1) and the IGIMF formulation in terms of optimal systems (see Section 2.6.2), need to be consistent with observational constraints. Some relevant ones can be found in Box Observational Constraints IV.

OBSERVATIONAL CONSTRAINTS IV:

- Dwarf late-type galaxies, with typically small ($\approx 0.0001 - 0.1\,M_{\odot}\,{\rm yr}^{-1}$) to extremely small ($\lesssim 10^{-4}\,M_{\odot}\,{\rm yr}^{-1}$) SFRs, have been found to have a systematically increasing deficit of Hα emission relative to their UV emission with decreasing SFR. This can be readily understood as a result of an increasingly top-light gwIMF (Lee et al., 2009).
- Direct (star-count) evidence supporting this comes from the nearby dwarf galaxy DDO154 (Watts et al., 2018) and from Leo P (Jeřábková et al., 2018). Both galaxies have a deficit of massive stars, i.e. appear to have a top-light gwIMF.

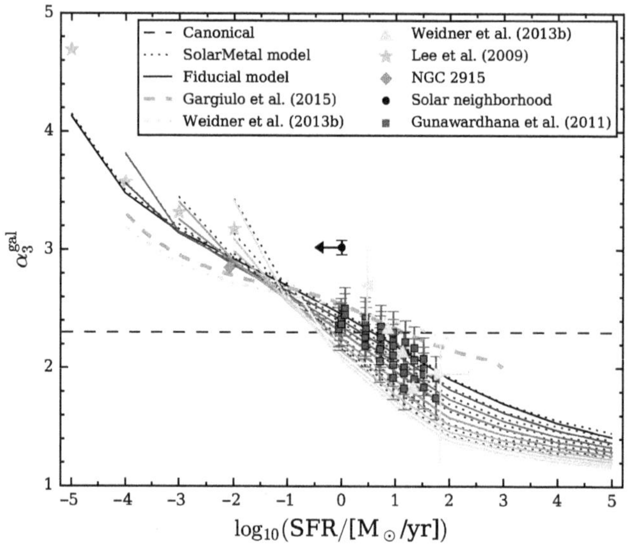

Figure 2.4 The variation of the IGIMF power-law index α_3^{gal} (for stars with $m > 1\,M_\odot$) with the galaxy-wide SFR is shown as the lines. Each line is an evaluation of the index at a particular stellar mass, showing that the IGIMF is curved since α_3^{gal} changes with m (see Fig. 2.3). The invariant canonical IMF is shown as the horizontal dashed line, values above it are top-light gwIMFs and below it are top-heavy gwIMFs. The solid and dotted lines constitute, respectively, the fiducial model (which includes an effective metallicity dependence in the density dependence of the IMF) and the solar-abundance model (for details see Yan et al., 2017). The dot-dashed and dashed coloured lines are IGIMF models from Gargiulo et al. (2015) and Weidner et al. (2013b). The symbols are various observational constraints as indicated in the key. Credit: Yan et al. (2017) reproduced with permission © ESO. Colour version available online.

- The chemical properties of the low-mass satellite galaxies of the Milky Way, for which sufficient data exist, suggest they had gwIMFs with a deficit of massive stars (Tsujimoto, 2011). This is consistent with the above two points.
- Massive late-type galaxies with typically large ($\gtrsim 1\,M_\odot\,yr^{-1}$) SFRs have been found to have a systematically more top-heavy gwIMF with increasing SFR, as deduced from their Hα flux and broad-band optical colours. This can be understood as a result of an increasingly top-heavy gwIMF (Gunawardhana et al., 2011). Evidence for this has been found before also (Hoversten and Glazebrook, 2008; Meurer et al., 2009).
- Massive elliptical galaxies have been found to have formed rapidly (within a Gyr) with SFR $\gtrsim 10^3\,M_\odot\,yr^{-1}$, as deduced from their high α-element abundances (Thomas et al., 2005; Recchi et al., 2009). Their high metallicity required them to have had top-heavy gwIMFs to generate the large mass in metals within the short time (Matteucci, 1994; Vazdekis et al., 1997; Weidner et al., 2013b,a; Ferreras et al., 2015; Martín-Navarro, 2016. This fits well into the above three points which suggest a general shift

of the gwIMF towards producing more massive stars relative to the low-mass stellar content with increasing SFR.

- The isotopes ^{13}C and ^{18}O are, respectively, released mainly by low- and intermediate-mass stars ($m < 8\,M_\odot$) and massive stars ($m > 8\,M_\odot$). Using the ALMA facility, Zhang et al. (2018) measured rotational transitions of the ^{13}CO and C^{18}O isotopologues in a number of starbursting galaxies (SFR $\gtrsim 10^3\,M_\odot\,yr^{-1}$) finding strong evidence for the galaxy-wide IMF to be top-heavy.
- Massive elliptical galaxies also show evidence that they had a significantly bottom-heavy IMF (van Dokkum and Conroy, 2010; Conroy and van Dokkum, 2012; Martín-Navarro, 2016 and references therein).
- Early-type galaxies have a trend in metallicity and α-element abundances explainable with an increasingly top-light gwIMF with decreasing galaxy mass and thus decreasing SFR during their formation (Köppen et al., 2007; Recchi et al., 2009; Recchi and Kroupa, 2015). Alternatively, element-selective outflows from star-forming galaxies may account for the observed mass–metallicity relation among galaxies. Observational surveys to establish this have failed to detect evidence for outflows (Lelli et al., 2014; Concas et al., 2017), although McQuinn et al. (2018) report observation of hot gas leaving starbursting dwarf galaxies.
- Disk galaxies have an Hα cutoff radius beyond which Hα emission is significantly reduced or absent compared to the UV emission (Boissier et al., 2007). These UV-extended disks can be understood in terms of a galactocentric-radial dependency of the ECMF (see Box Observational Constraints III) in combination with the $m_{max} - M_{ecl}$ relation (see Fig. 2.1) such that the typically low-mass embedded clusters forming in the outer regions come along with a stellar population which is deficient in ionising radiation (Pflamm-Altenburg and Kroupa, 2008).

Is it possible that instead of the optimal star formation theory developed above (see Section 2.6.2), the gwIMF remains universal and invariant and that star formation is stochastic, perhaps in a constrained manner such as in the SLUG approach (see Section 2.6.1), and that the observed correlations in and among galaxies (see Box Observational Constraints IV) are due to photon leakage, dust obscuration and redenning, and other physical effects (Calzetti, 2008, 2013)? It is likely that these are relevant, but the authors of the original research papers reporting the mentioned effects (Hoversten and Glazebrook, 2008; Meurer et al., 2009; Lee et al., 2009; Gunawardhana et al., 2011) discuss these biases at great length disfavouring them. It is, nevertheless, probably useful to study how invariant but stochastic models might be able to lead to the observed correlations. For example, a systematic deficit of massive stars in dwarf galaxies could come about if the dwarf galaxies are, as a population, going through a current lull in star-formation activity (Kennicutt and Evans, 2012). But this appears contrived and extremely unlikely (Lee et al., 2009), especially since one would need to resort to an in-step increase in the SFR of nearby massive disk galaxies in order to explain the generally observed shift of a large

Hα-deficiency becoming smaller through to an Hα-emission overabundance when increasing the stellar mass from dwarf to major disk galaxies (see Fig. 2.4).

The predictability of the IGIMF theory is an advantage though. It allows testing of the models against data (but the calculations need to be done correctly before drawing conclusions, see the end of Section 2.4.1). Also, if found to be a good representation of the observational data, we can use it to learn about star-formation at high redshift and use galaxy-wide star-formation behaviour to constrain star-formation physics on the scales of embedded clusters (Jeřábková et al., 2018).

2.7 Implications for the SFRs of Galaxies

Most measures of the SFR of galaxies rely on the photon output from massive stars ($m \gtrsim 10\,\mathrm{M}_\odot$) which are the most luminous objects. These are a good probe of the current SFR due to their short lifetimes ($\lesssim 50$ Myr before exploding as core-collapse supernovae, SNII, events). For an empirical comparison of the various SFR tracers the reader is referred to Mahajan et al. (2019).

For the purpose of this discussion of how different treatments of the gwIMF may affect SFR measurements we assume the observer detects all relevant photons (i.e. that dust obscuration, photon leakage, and other effects – see Calzetti, 2008, 2013 for a discussion – have been corrected for) and we concentrate on the Hα-based SFR measure (Kennicutt, 1989; Pflamm-Altenburg et al., 2009). It assumes that a fraction of ionising photons ionises hydrogen atoms in the nearby ISM and that these recombine. A fraction of the recombination photons is emitted as Hα photons and it has been shown that the flux of these is proportional to the ionising flux such that the Hα luminosity becomes a direct measure of the current population of massive stars. Given their short lifetimes ($\lesssim 50$ Myr), we thus obtain an estimate of the SFR if we know which mass in all young stars is associated with these ionising stars, i.e if we know the shape of the gwIMF. The Hα luminosity is the most sensitive measure of the population of ionising stars (unless direct star-counts can be performed in nearby dwarf galaxies, e.g. Watts et al., 2018). For completeness, we note that the *GALEX*[7] far-UV (FUV) flux is a measure of the stellar population which includes late-B stars and thus summarises the SFR activity over the recent 400 Myr time window (see Pflamm-Altenburg et al., 2009, who used the PEGASE[8] code; and Chapter 7 for a detailed discussion of the timescales of each SFR indicator assuming an invariant gwIMF). We do not address this measure here except to note that, in the limit where only a few ionising stars form, the FUV-flux derived SFRs are more robust and these are indeed consistent with the higher SFRs as calculated using the IGIMF1 formulation (Jeřábková et al., 2018) as shown explicitly in fig. 8 of Lee et al. (2009) who compare FUV and Hα-based SFR indicators for dwarf galaxies. The FUV flux is thus a more robust measure of SFR_{true} than the Hα flux because it assesses a much more populous stellar ensemble therewith being

[7] Galaxy Evolution Explorer; www.galex.caltech.edu/index.html
[8] Programme d'Étude des Galaxies par Synthèse Évolutive; www2.iap.fr/pegase/

less susceptible to Poisson noise, but it is more sensitive to the gwIMF of intermediate-mass stars and also offers a poorer time resolution. The rate of type-II supernovae (SNII) also provides a measure, but we do not know which fraction of massive stars implode into a black hole without producing an explosion and how this depends on metallicity and thus redshift. SNII events are too rare on a human lifetime to provide reliable global measurements except in profusely starbursting systems (e.g. as in Arp 220; Dabringhausen et al., 2012 and references therein; Jeřábková et al., 2017).

We refer to $SFR_{\mathrm{H}\alpha}$ as being the SFR measure using the Hα flux, and with SFR_{K} we mean $SFR_{\mathrm{H}\alpha}$ in the specific case of using the Kennicutt (1998) calibration which assumes a fully sampled and invariant standard IMF (the Kennicutt IMF, which is very similar to the canonical IMF, Pflamm-Altenburg et al., 2009). Therefore, if the gwIMF differs from the invariant canonical one, then $SFR_{\mathrm{K}} \neq SFR_{\mathrm{true}}$. If the gwIMF were to be invariant and a PDF then the average of SFR_{K} over a sufficiently large number of galaxies of similar baryonic mass (Speagle et al., 2014) would provide the correct measure of the SFR with increasing dispersion with decreasing $\mathrm{SFR}_{\mathrm{true}}$ (see Section 2.6.2), $SFR_{\mathrm{true}} = \overline{SFR_{\mathrm{K}}}$. We also consider SFR_{SLUG} which is the Hα-based SFR computed within the SLUG approach (see Section 2.6.1, fns 2 and 6 above). If reality were to correspond to the assumptions underlying the SLUG approach then $SFR_{\mathrm{SLUG}} = SFR_{\mathrm{true}}$, since in any particular model galaxy, the SLUG methodology knows what the model gwIMF is, such that the SFR is calculated properly. However, the SLUG approach does not allow to infer the true SFR, SFR_{true}, for an observed galaxy, since, by the stochastic aspect inherent to SLUG, the observer does not know the actual momentary gwIMF. This comes about because for a $\mathrm{SFR}_{\mathrm{true}}$ in the observed galaxy, the observer does not know whether the gwIMF is more or less top-heavy for example (due to stochastic fluctuations) as long as this gwIMF is consistent with the SFR-tracer (for example, the same Hα flux can be obtained by differently shaped gwIMFs). The SFR measure, SFR_{IGIMF}, is likewise based on the Hα flux but is calculated taking into account the number of ionising stars in the IGIMF theory (see Section 2.4.2) such that $SFR_{\mathrm{IGIMF}} = SFR_{\mathrm{true}}$ without scatter (apart from variations at the very small SFR_{true} level where a single ionising star may be born or die, see Pflamm-Altenburg et al., 2007; Jeřábková et al., 2018 for a discussion of this limit). Details of the IGIMF calculations can be found in Jeřábková et al. (2018). We define the correction factor (see eq. 17 in Jeřábková et al., 2018),

$$\Psi = \frac{SFR_{\mathrm{K\ or\ SLUG\ or\ IGIMF}}}{SFR_{\mathrm{K}}}. \tag{2.17}$$

Assuming the IMF is a PDF (see Section 2.6.1, point 1), the average over an ensemble of galaxies with the same baryonic mass, $\overline{\Psi} = 1$, for all SFR_{K} with increasing scatter as SFR_{K} decreases. The scatter in SFR_{K} will be a few orders of magnitude for $SFR_{\mathrm{true}} \lesssim 0.1\ \mathrm{M}_\odot\,\mathrm{yr}^{-1}$ as a result of randomly sampling stars from the IMF without constraints. For example, there can be galaxies which contain no ionising stars (such that $SFR_{\mathrm{K}} \approx 0\,\mathrm{M}_\odot\,\mathrm{yr}^{-1}$) despite having $SFR_{\mathrm{true}} = 1\,\mathrm{M}_\odot\,\mathrm{yr}^{-1}$. Alternatively, there can be galaxies consisting only of ionising stars and with the same SFR_{true}.

In the SLUG approach (see Section 2.6.1, point 2), $\overline{\Psi} > 1$ for $SFR_{\rm true} \lesssim 1\,{\rm M}_\odot\,{\rm yr}^{-1}$, increasing with decreasing $SFR_{\rm K}$. For decreasing $SFR_{\rm true}$, the scatter in $SFR_{\rm K}$ increases up to a few orders of magnitude. For $SFR_{\rm true} \gtrsim 1\,{\rm M}_\odot\,{\rm yr}^{-1}$, $\overline{\Psi} = 1$ (fig. 2 in da Silva et al., 2014). The reason for this comes about because at low $SFR_{\rm true}$, the galaxy is populated typically with less-massive embedded clusters such that the gwIMF is not fully sampled to the highest allowed stellar masses. As a consequence, the true SFR, $SFR_{\rm true}$, is larger than that measured from the Hα flux assuming the gwIMF is invariant and fully sampled (in this case $SFR_{\rm K}$) because the gwIMF contains fewer ionising stars per low-mass star. The scatter comes about because the IMF and ECMF are assumed to be PDFs (the IMF is a constrained PDF, the constraint being that the random drawing of stars from the IMF must add up to the mass of the pre-determined embedded cluster, $M_{\rm ecl}$). Thus, in the SLUG approach it is possible to generate a galaxy which has no ionising stars despite having $SFR_{\rm true} = 1\,{\rm M}_\odot\,{\rm yr}^{-1}$. In the SLUG approach it is possible that a galaxy with $SFR_{\rm true} = 1 \times 10^{-4}\,{\rm M}_\odot\,{\rm yr}^{-1}$ forms only a single $10\,{\rm M}_\odot$ star which needs this SFR ($10\,{\rm M}_\odot$ being assembled in 10^5 yr). In this case an observer would however conclude that $SFR_{\rm K} \gg 10^{-4}\,{\rm M}_\odot\,{\rm yr}^{-1}$ since the Hα flux would be associated with a full gwIMF (modelled via the Kennicutt IMF). This is, by the way, also true for the pure stochastic approach above, such that there Ψ varies on both sides of the value of one.

In the IGIMF approach (see Section 2.4.2) galaxies always form populations of stars and these do not contain O stars when $SFR_{\rm true} < 10^{-4}\,{\rm M}_\odot\,{\rm yr}^{-1}$ (table 3 in Weidner et al., 2013b; fig. 7 in Yan et al., 2017). This has led, in the literature, to confusion since claims were made that observed galaxies with $SFR_{\rm K} < 10^{-4}\,{\rm M}_\odot\,{\rm yr}^{-1}$ are producing O stars therewith ruling out the IGIMF theory. But these claims forgot that $\Psi \gg 1$ in these cases according to the IGIMF theory (see discussion of this in Jeřábková et al., 2018). It is evident from Figs 2.3 and 2.4 that the number of ionising stars relative to the number of low-mass stars decreases systematically with decreasing $SFR_{\rm true}$. Consequently, $\Psi > 1$ for $SFR_{\rm K} \lesssim 1\,{\rm M}_\odot\,{\rm yr}^{-1}$, increasing with decreasing $SFR_{\rm K}$. This is a similar but stronger effect than in the SLUG approach. If the IMF is invariant and canonical, then $\Psi = 1$ for $SFR_{\rm K} \gtrsim 1\,{\rm M}_\odot\,{\rm yr}^{-1}$, as in the SLUG approach, which also assumes (in the current published version and in the spirit of the IMF being a PDF, i.e. not being subject to physical limits apart from the constraint given by $M_{\rm ecl}$ when drawing stars) that the IMF is invariant. In the IGIMF theory, the IMF is however assumed to systematically vary with the physical conditions of the molecular cloud (see Eqs 2.8 and 2.9) such that $\Psi < 1$ for $SFR_{\rm K} \gtrsim 1\,{\rm M}_\odot\,{\rm yr}^{-1}$.

Kennicutt and Evans (2012) discuss the differences of these approaches and conclude that the systematic deviations observed between $SFR_{{\rm H}\alpha}$ and $SFR_{\rm UV}$ (which is a reasonably good approximation of $SFR_{\rm true}$, Pflamm-Altenburg et al., 2009) at small values of the SFR can be produced instead through temporal variations in SFRs, without having to resort to modifying the IMF. The galaxies would, however, need to be in an unexplained synchronised lull of their SFRs. Even if this were the correct conclusion, it would not be able to accommodate the systematic variation observed at high SFRs, which appear to merely be a natural continuation of the overall trend of the observationally constrained

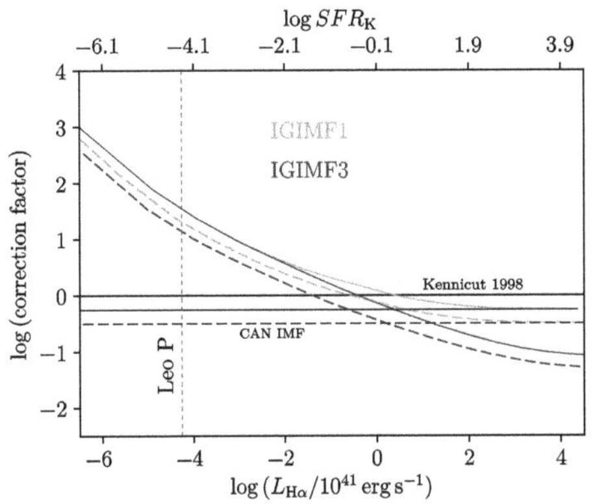

Figure 2.5 The correction factor Ψ (see Eq. 2.17, $\log x \equiv \log_{10} x$ for any x) is plotted for various cases: the horizontal solid line is $\Psi = 1$ whereby the gwIMF used by Kennicutt (1998) is assumed, the thin solid and dashed horizontal lines are for an invariant canonical IMF (see Eq. 2.1) for [Fe/H]= 0 and −2, respectively. The solid green and red curves are IGIMF models for [Fe/H]= 0 assuming, respectively, the IMF varies only through α_3 (see Eq. 2.9, 'IGIMF1') or ('IGIMF3' which is the most realistic case) at the low-mass end (see Eq. 2.8) and at the high-mass end (see Eq. 2.9). The corresponding dashed lines are for [Fe/H]= −2. Note that for [Fe/H]= 0 IGIMF1 and IGIMF3 bifurcate for $SFR \gtrsim 10^{-1.5}\, M_\odot\, yr^{-1}$ because the IMFs are identical to the canonical IMF at the low-mass end at this metallicity, while at [Fe/H]= −2 the IGIMF1 and IGIMF3 models remain separated at all SFRs because the IMFs differ at the low- and at the high-mass end at this metallicity. Thus, for example, taking the case of a dwarf galaxy with [Fe/H]= −2 and measured $SFR_K \approx 10^{-4.2}\, M_\odot\, yr^{-1}$, it would have, according to the IGIMF3 model, $SFR_{true} \approx \Psi \times 10^{-4.2}\, M_\odot\, yr^{-1}$ with $\Psi \approx 10^{1.2}$. A massive disk galaxy with [Fe/H]= 0 and $SFR_K \approx 10^{1.9}\, M_\odot\, yr^{-1}$ would have a $\Psi \approx 10^{-0.7}$ times larger SFR_{true}. If the IGIMF theory is applicable, then the Leo P dwarf galaxy has a $SFR_{true} \approx \Psi \times 10^{-4.2}\, M_\odot\, yr^{-1}$ with $\Psi \approx 10^{1.2}$ explaining why it has one or two O stars. The presence of these stars has been leading to confusion in the literature if the SFR_K value for its SFR is assumed (McQuinn et al., 2015). Credit: Jeřábková et al. (2018), reproduced with permission © ESO. Colour version available online.

gwIMF becoming increasingly top-heavy with increasing galaxy-wide SFR, beginning with a significantly top-light gwIMF in the least-massive dwarf galaxies (see Fig. 2.4).

The above discussion is quantified in Fig. 2.5 which plots Ψ for the IGIMF case (the pure PDF case, point 1 in Section 2.6.1, has $\overline{\Psi} = 1$ independently of SFR_K with a scatter which is comparable to that evident in the SLUG case) and in fig. 3 in da Silva et al. (2014) for the SLUG approach.

2.8 Conclusion

The relation of the IMF to the composite and the galaxy-wide IMF has been discussed. The mathematical treatment of the IMF (as a probability distribution function or an optimal distribution function) is closely associated with the physical processes according to which the distribution of stellar masses emerges from the interstellar medium. Three types of possibilities are being discussed and investigated in the community and these lead to different predictions on the relation between the tracer used to asses the star-formation rate of a galaxy, e.g. $SFR_{\mathrm{H}\alpha}$, and the true star-formation rate, SFR_{true}. The Hα and FUV flux are often employed to assess the SFR (over the past $\approx$50 and 400 Myr, respectively) and we used the former as an example to show the differences in the predictions. Assuming in all cases that dust obscuration, photon leakage and other effects (Calzetti, 2008, 2013) do not play a role, and that the SFR does not vary over timescales of less than about 50 Myr:

1. If the IMF is a pure PDF without constraints apart from the possible existence of an upper-mass limit to stars ($m_{\mathrm{max}*}$, see Section 2.6.1) then $SFR_{\mathrm{H}\alpha}$ is an unbiased tracer of SFR ($SFR_{\mathrm{H}\alpha} \propto SFR_{\mathrm{true}}$) and the dispersion of the SFRs, $\sigma_{\mathrm{SFR(H}\alpha)}$, increases with decreasing SFR_{true} due to stochastic variations. This dispersion can be orders of magnitude, since one can sample galaxies which have no ionising stars or galaxies which consist only of ionising stars. This model cannot reproduce the observed systematic trend of the gwIMF becoming increasingly top-light with decreasing $SFR_{\mathrm{true}} \lesssim 1\,\mathrm{M}_\odot\,\mathrm{yr}^{-1}$, nor the observed systematic trend of the gwIMF becoming increasingly top-heavy with increasing $SFR_{\mathrm{true}} \gtrsim 1\,\mathrm{M}_\odot\,\mathrm{yr}^{-1}$ (see Fig. 2.4), nor the $m_{\mathrm{max}} - M_{\mathrm{ecl}}$ data (see Fig. 2.1). If a galaxy has an observed Hα flux, then this approach does not allow the determination of SFR_{true} unless one is in the regime where $\sigma_{\mathrm{SFR(H}\alpha)}$ is sufficiently small.
2. If it is assumed that stars form in embedded clusters which are drawn stochastically from the ECMF and that the IMF is sampled stochastically within each embedded cluster with the constraint that M_{ecl} be fulfilled (the SLUG approach, see Section 2.6.1), then $SFR_{\mathrm{H}\alpha}$ increasingly underestimates SFR_{true} with decreasing $SFR_{\mathrm{true}} \lesssim 1\,\mathrm{M}_\odot\,\mathrm{yr}^{-1}$. The dispersion, $\sigma_{\mathrm{SFR(H}\alpha)}$, also increases with decreasing SFR_{true} due to stochastic variations, but to a lesser extend than under point 1, since the constraint M_{ecl} limits the possible variations. This model allows for galaxies to contain no ionising stars and also for galaxies to consist only of ionising stars and so $\sigma_{\mathrm{SFR(H}\alpha)}$ can be orders of magnitude at low SFR_{true} (da Silva et al., 2014). This model can approximately reproduce the observed systematic trend of the gwIMF becoming increasingly top-light with decreasing SFR_{true} (da Silva et al. 2014, Fig. 2.4, Eq. 2.17, Fig. 2.5), but not the $m_{\mathrm{max}} - M_{\mathrm{ecl}}$ data (see Fig. 2.1). It can also not reproduce the observed systematic trend of the gwIMF becoming increasingly top-heavy with increasing $SFR_{\mathrm{true}} \gtrsim 1\,\mathrm{M}_\odot\,\mathrm{yr}^{-1}$ (see Fig. 2.4). If a galaxy has an observed Hα flux, then this approach does not allow the determination of SFR_{true} unless one is in the regime where $\sigma_{\mathrm{SFR(H}\alpha)}$ is sufficiently small.
3. If the IGIMF theory is assumed, then $SFR_{\mathrm{H}\alpha}$ increasingly underestimates SFR_{true} with decreasing $SFR_{\mathrm{true}} \lesssim 1\,\mathrm{M}_\odot\,\mathrm{yr}^{-1}$ with a somewhat larger systematic change than under point 2. A small dispersion, $\sigma_{\mathrm{SFR(H}\alpha)}$, at small SFR is present due to the birth and death of individual most-massive stars in the galaxy. Since galaxies are interacting and have internal dynamical instabilities which affect the assembly of molecular clouds (e.g. the bar instability), a natural $\sigma_{\mathrm{SFR(H}\alpha)}$ will be observable, but this has not yet been calculated. This model can reproduce the observed systematic trend of the gwIMF becoming increasingly top-light with decreasing SFR_{true} (Lee et al., 2009, Fig. 2.4,

Eq. 2.17, Fig. 2.5) and the $m_{\rm max} - M_{\rm ecl}$ data (see Fig. 2.1). It can also reproduce the observed systematic trend of the gwIMF becoming increasingly top-heavy with increasing $SFR_{\rm true} \gtrsim 1\,{\rm M_\odot\,yr^{-1}}$ (see Fig. 2.4) due to the incorporation of the metallicity and density dependent IMF (see Eqs 2.8 and 2.9). If a galaxy has an observed Hα flux, then this approach provides a unique determination of $SFR_{\rm true}$, since the expected variation, $\sigma_{\rm SFR(H\alpha)}$, for galaxies of the same baryonic mass can (we expect) be related to the distribution of morphologies of the galaxies and there is no stochasticity.

Efficient IMF-sampling algorithms are available as downloads: for stochastic sampling see Pflamm-Altenburg and Kroupa (2006) and Kroupa et al. (2013); and for optimal sampling see Yan et al. (2017) and Kroupa et al. (2013). For the SLUG approach (relevant for points 1 and 2 in Section 2.6.1) programs are available (da Silva et al., 2012, 2014; Ashworth et al., 2017), while for the IGIMF approach also (Yan et al., 2017; Jeřábková et al., 2018).

The IGIMF theory has been computed with, and without, a variable IMF and the most realistic case is IGIMF3 (Jeřábková et al., 2018) which incorporates the full IMF dependency on density and metallicity (see Eqs 2.8 and 2.9) currently known from observational data. The two stochastic approaches discussed above (points 1 and 2 in Section 2.6.1) may be used with a variable IMF also. But, an IMF which varies systematically with physical conditions may be in tension with an interpretation of the IMF as a PDF since it implies that conditions need to be applied on the PDF. If the IMF becomes top-heavy at high SFR density (e.g. as formulated in Eq. 2.9) then galaxies with a $SFR_{\rm true} \gtrsim 1\,{\rm M_\odot\,yr^{-1}}$ will have $SFR_{\rm H\alpha}$ overestimating $SFR_{\rm true}$, just as in the IGIMF theory.

The observational constraints (see Boxes Observational Constraints I–IV) are useful for assessing which of the above possibilities are relevant for nature. The existence or non-existence of isolated massive stars and of a physical $m_{\rm max} - M_{\rm ecl}$ relation (see Fig. 2.1) and its dispersion are thus central issues in understanding SFR measurements of galaxies. For the interpretation of the IMF and thus for computing gwIMFs and galaxy-wide SFRs given some tracer, it is important to understand whether embedded clusters can be viewed as being fundamental building blocks of galaxies (Kroupa, 2005), and if stars form in a mass segregated way in these embedded clusters. If this is the case, then it is also important if the ECMF varies systematically within a galaxy, e.g. radially. To further clarify these points, which all play a role in our interpretation of the IMF and its relation to the cIMF and gwIMF, further observational work is required. But the fact that these issues are being discussed shows how rich and informative this topic is. An important constraint on any formulation of the gwIMF is that this formulation must be consistent with the IMFs deduced from resolved stellar populations (star clusters, stellar associations). The implications for our understanding of how galaxies evolve are potentially major, as discussed in this contribution. Certainly, as an example, the quantification of the main sequence of galaxies depends on how we understand the gwIMF.

Returning to elliptical galaxies (see Box Observational Constraint IV), the need for a bottom-heavy and a top-heavy gwIMF, possibly in terms of the evolution of the gwIMF, is a challenge for any theory of how elliptical galaxies formed and evolved (Weidner et al., 2013a; Ferreras et al., 2015; Jeřábková et al., 2018). It is clear that elliptical galaxies are unusual, making only a few per cent of the population of galaxies heavier than about $10^{10}\,M_{\odot}$ (Delgado-Serrano et al., 2010), and that elliptical galaxies formed rapidly and thus with $SFR_{\rm true} > 1000\,M_{\odot}\,{\rm yr}^{-1}$. Any such theory needs to be consistent with the data on star-forming galaxies. Important in this context are the constraints on the allowed dark matter (faint M dwarfs, stellar remnants) from lensing observations (Smith and Lucey, 2013; Smith, 2014) which appear to be inconsistent with the bottom-heavy gwIMF to provide a significant amount of mass. This problem is discussed also in the review on elliptical galaxies by Cappellari (2016).

References

Adams, F. C., and Fatuzzo, M. 1996. A Theory of the Initial Mass Function for Star Formation in Molecular Clouds. *Astrophysical Journal*, **464**(June), 256.

Andrews, J. E., Calzetti, D., Chandar, R. et al. 2013. An Initial Mass Function Study of the Dwarf Starburst Galaxy NGC 4214. *Astrophysical Journal*, **767**(Apr.), 51.

Andrews, J. E., Calzetti, D., Chandar, R. et al. 2014. Big Fish in Small Ponds: Massive Stars in the Low-Mass Clusters of M83. *Astrophysical Journal*, **793**(Sept.), 4.

Applebaum, E., Brooks, A. M., Quinn, T. R. et al. 2019. A Stochastically Sampled IMF Alters the Stellar Content of Simulated Dwarf Galaxies. *Monthly Notices of the Royal Astronomical Society*, Nov, 2975.

Ashworth, G., Fumagalli, M., Krumholz, M. R. et al. 2017. Exploring the IMF of Star Clusters: A Joint SLUG and LEGUS Effort. *Monthly Notices of the Royal Astronomical Society*, **469**(Aug.), 2464–2480.

Banerjee, S., and Kroupa, P. 2012. On the True Shape of the Upper End of the Stellar Initial Mass Function. The Case of R136. *Astronomy and Astrophysics*, **547**(Nov.), A23.

Banerjee, S., Kroupa, P., and Oh, S. 2012. The Emergence of Super-Canonical Stars in R136-Type Starburst Clusters. *Monthly Notices of the Royal Astronomical Society*, **426**(Oct.), 1416–1426.

Bastian, N., Covey, K. R., and Meyer, M. R. 2010. A Universal Stellar Initial Mass Function? A Critical Look at Variations. *Annual Review of Astronomy and Astrophysics*, **48**(Sept.), 339–389.

Bate, M. R., Bonnell, I. A., and Bromm, V. 2002. The Formation Mechanism of Brown Dwarfs. *Monthly Notices of the Royal Astronomical Society*, **332**(May), L65–L68.

Baumgardt, H., and Makino, J. 2003. Dynamical Evolution of Star Clusters in Tidal Fields. *Monthly Notices of the Royal Astronomical Society*, **340**(Mar.), 227–246.

Baumgardt, H., De Marchi, G., and Kroupa, P. 2008. Evidence for Primordial Mass Segregation in Globular Clusters. *Astrophysical Journal*, **685**(Sept.), 247–253.

Belloni, D., Askar, A., Giersz, M. et al. 2017. On the Initial Binary Population for Star Cluster Simulations. *Monthly Notices of the Royal Astronomical Society*, **471**(Nov.), 2812–2828.

Belloni, D., Kroupa, P., Rocha-Pinto, H. J. et al. 2018. Dynamical Equivalence, the Origin of the Galactic Field Stellar and Binary Population, and the Initial Radius-Mass Relation of Embedded Clusters. *Monthly Notices of the Royal Astronomical Society*, **474**(Mar.), 3740–3745.

Beuther, H., Churchwell, E. B., McKee, C. F. et al. 2007 (Jan.). The Formation of Massive Stars. Page 165 of: Reipurth, Bo, Jewitt, David, and Keil, Klaus (eds), *Protostars and Planets V*.

Boissier, S., Gil de Paz, A., Boselli, A. et al. 2007. Radial Variation of Attenuation and Star Formation in the Largest Late-Type Disks Observed with GALEX. *Astrophysical Journal Supplement Series*, **173**(Dec.), 524–537.

Bonnell, I. A., and Davies, M. B. 1998. Mass Segregation in Young Stellar Clusters. *Monthly Notices of the Royal Astronomical Society*, **295**(Apr.), 691.

Bontemps, S., Motte, F., Csengeri, T. et al. 2010. Fragmentation and Mass Segregation in the Massive Dense Cores of Cygnus X. *Astronomy and Astrophysics*, **524**(Dec.), A18.

Calzetti, D. 2008 (June). Star Formation Rate Determinations. Page 121 of: Knapen, J. H., Mahoney, T. J., and Vazdekis, A. (eds), *Pathways through an Eclectic Universe*. Astronomical Society of the Pacific Conference Series, vol. 390.

Calzetti, Daniela. 2013. *Star Formation Rate Indicators*. Page 419.

Cappellari, M. 2016. Structure and Kinematics of Early-Type Galaxies from Integral Field Spectroscopy. *Annual Review of Astronomy and Astrophysics*, **54**(Sept.), 597–665.

Cerviño, M., Román-Zúñiga, C., Luridiana, V. et al. 2013a. Crucial Aspects of the Initial Mass Function. I. The Statistical Correlation between the Total Mass of an Ensemble of Stars and Its Most Massive Star. *Astronomy and Astrophysics*, **553**(May), A31.

Cerviño, M., Román-Zúñiga, C., Bayo, A. et al. 2013b. Crucial Aspects of the Initial Mass Function. II. The Inference of Total Quantities from Partial Information on a Cluster. *Astronomy and Astrophysics*, **553**(May), A32.

Chabrier, G. 2003. Galactic Stellar and Substellar Initial Mass Function. *Publications of the Astronomical Society of the Pacific*, **115**(July), 763–795.

Chené, A.-N., Ramírez Alegría, S., Borissova, J. et al. 2015. Massive Open Star Clusters Using the VVV Survey. IV. WR 62-2, A New Very Massive Star in the Core of the VVV CL041 Cluster. *Astronomy and Astrophysics*, **584**(Dec.), A31.

Concas, A., Popesso, P., Brusa, et al. 2017. Light Breeze in the Local Universe. *Astronomy and Astrophysics*, **606**(Oct.), A36.

Conroy, C., and van Dokkum, P. G. 2012. The Stellar Initial Mass Function in Early-Type Galaxies From Absorption Line Spectroscopy. II. Results. *Astrophysical Journal*, **760**(Nov.), 71.

Crowther, P. A., Schnurr, O., Hirschi, R. et al. 2010. The R136 Star Cluster Hosts Several Stars Whose Individual Masses Greatly Exceed the Accepted $150M_{solar}$ Stellar Mass Limit. *Monthly Notices of the Royal Astronomical Society*, **408**(Oct.), 731–751.

da Silva, R. L., Fumagalli, M., and Krumholz, M. 2012. SLUG: Stochastically Lighting Up Galaxies. I. Methods and Validating Tests. *Astrophysical Journal*, **745**(Feb.), 145.

da Silva, R. L., Fumagalli, M., and Krumholz, M. R. 2014. SLUG—Stochastically Lighting Up Galaxies—II. Quantifying the Effects of Stochasticity on Star Formation Rate Indicators. *Monthly Notices of the Royal Astronomical Society*, **444**(Nov.), 3275–3287.

Dabringhausen, J., Fellhauer, M., and Kroupa, P. 2010. Mass Loss and Expansion of Ultra Compact Dwarf Galaxies through Gas Expulsion and Stellar Evolution for Top-Heavy Stellar Initial Mass Functions. *Monthly Notices of the Royal Astronomical Society*, **403**(Apr.), 1054–1071.

Dabringhausen, J., Kroupa, P., and Baumgardt, H. 2009. A Top-Heavy Stellar Initial Mass Function in Starbursts as an Explanation for the High Mass-to-Light Ratios of Ultra-Compact Dwarf Galaxies. *Monthly Notices of the Royal Astronomical Society*, **394**(Apr.), 1529–1543.

Dabringhausen, J., Kroupa, P., Pflamm-Altenburg, J. et al. 2012. Low-Mass X-ray Binaries Indicate a Top-Heavy Stellar Initial Mass Function in Ultracompact Dwarf Galaxies. *Astrophysical Journal*, **747**(Mar.), 72.

de Mink, S. E., Sana, H., Langer, N. et al. 2014. The Incidence of Stellar Mergers and Mass Gainers among Massive Stars. ApJ, **782**(Feb.), 7.

Delgado-Serrano, R., Hammer, F., Yang, Y. B. et al. 2010. How Was the Hubble Sequence 6 Gyr Ago? *Astronomy and Astrophysics*, **509**(Jan.), A78.

Dib, S. 2014. Testing the Universality of the IMF with Bayesian Statistics: Young Clusters. *Monthly Notices of the Royal Astronomical Society*, **444**(Oct.), 1957–1981.

Dib, S., Kim, J., and Shadmehri, M. 2007. The Origin of the Arches Stellar Cluster Mass Function. *Monthly Notices of the Royal Astronomical Society*, **381**(Oct.), L40–L44.

Dib, S., Schmeja, S., and Hony, S. 2017. Massive Stars Reveal Variations of the Stellar Initial Mass Function in the Milky Way Stellar Clusters. *Monthly Notices of the Royal Astronomical Society*, **464**(Jan.), 1738–1752.

Disney, M. J., Romano, J. D., Garcia-Appadoo, D. A. et al. 2008. Galaxies Appear Simpler than Expected. *Nature*, **455**(Oct.), 1082–1084.

Duarte-Cabral, A., Bontemps, S., Motte, F. et al. 2013. CO Outflows from High-Mass Class 0 Protostars in Cygnus-X. *Astronomy and Astrophysics*, **558**(Oct.), A125.

Egusa, F., Sofue, Y., and Nakanishi, H. 2004. Offsets between Hα and CO Arms of a Spiral Galaxy, NGC 4254: A New Method for Determining the Pattern Speed of Spiral Galaxies. *PASJ Publications of the Astronomical Society of Japan*, **56**(Dec.), L45–L48.

Egusa, F., Kohno, K., Sofue, Y. et al. 2009. Determining Star Formation Timescale and Pattern Speed in Nearby Spiral Galaxies. *Astrophysical Journal*, **697**(June), 1870–1891.

Egusa, F., Mentuch Cooper, E., Koda, J. et al. 2017. Gas and Stellar Spiral arms and Their Offsets in the Grand-Design Spiral Galaxy M51. *Monthly Notices of the Royal Astronomical Society*, **465**(Feb.), 460–471.

Elmegreen, B. G. 1997. The Initial Stellar Mass Function from Random Sampling in a Turbulent Fractal Cloud. *Astrophysical Journal*, **486**(Sept.), 944–954.

Elmegreen, B. G. 1999. The Stellar Initial Mass Function from Random Sampling in Hierarchical Clouds. II. Statistical Fluctuations and a Mass Dependence for Starbirth Positions and Times. *Astrophysical Journal*, **515**(Apr.), 323–336.

Elmegreen, B. G. 2000. Modeling a High-Mass Turn-Down in the Stellar Initial Mass Function. *Astrophysical Journal*, **539**(Aug.), 342–351.

Elmegreen, B. G. 2004. Variability in the Stellar Initial Mass Function at Low and High Mass: Three-Component IMF Models. *Monthly Notices of the Royal Astronomical Society*, **354**(Oct.), 367–374.

Elmegreen, B. G., and Scalo, J. 2006. The Effect of Star Formation History on the Inferred Stellar Initial Mass Function. *Astrophysical Journal*, **636**(Jan.), 149–157.

Federrath, C. 2015. Inefficient Star Formation through Turbulence, Magnetic Fields and Feedback. *Monthly Notices of the Royal Astronomical Society*, **450**(July), 4035–4042.

Federrath, C., Schrön, M., Banerjee, R. et al. 2014. Modeling Jet and Outflow Feedback during Star Cluster Formation. *Astrophysical Journal*, **790**(Aug.), 128.

Ferreras, I., Weidner, C., Vazdekis, A. et al. 2015. Further Evidence for a Time-Dependent Initial Mass Function in Massive Early-Type Galaxies. *Monthly Notices of the Royal Astronomical Society*, **448**(Mar.), L82–L86.

Figer, D. F. 2005. An Upper Limit to the Masses of Stars. *Nature*, **434**(Mar.), 192–194.

Fukui, Y., and Kawamura, A. 2010. Molecular Clouds in Nearby Galaxies. *Annual Review of Astronomy and Astrophysics*, **48**(Sept.), 547–580.

Fumagalli, M., da Silva, R. L., and Krumholz, M. R. 2011. Stochastic Star Formation and a (Nearly) Uniform Stellar Initial Mass Function. *Astrophysical Journal*, **741**(Nov.), L26.

Garcia, M., Herrero, A., Najarro, F. et al. 2019. Ongoing Star Formation at the Outskirts of Sextans A: Spectroscopic Detection of Early-O Type Stars. *arXiv e-prints*, Jan., arXiv:1901.02466.

Gargiulo, I. D., Cora, S. A., Padilla, N. D. et al. 2015. Chemoarchaeological Downsizing in a Hierarchical Universe: Impact of a Top-Heavy IGIMF. *Monthly Notices of the Royal Astronomical Society*, **446**(Feb.), 3820–3841.

Goodwin, S. P., and Kroupa, P. 2005. Limits on the Primordial Stellar Multiplicity. *Astronomy and Astrophysics*, **439**(Aug.), 565–569.

Gunawardhana, M. L. P., Hopkins, A. M., Sharp, R. G. et al. 2011. Galaxy and Mass Assembly (GAMA): The Star Formation Rate Dependence of the Stellar Initial Mass Function. *Monthly Notices of the Royal Astronomical Society*, **415**(Aug.), 1647–1662.

Gvaramadze, V. V., Weidner, C., Kroupa, P. et al. 2012. Field O Stars: Formed In Situ or as Runaways? *Monthly Notices of the Royal Astronomical Society*, **424**(Aug.), 3037–3049.

Haghi, H., Zonoozi, A. H., Kroupa, P. et al. 2015. Possible Smoking-Gun Evidence for Initial Mass Segregation in Re-virialized Post-Gas Expulsion Globular Clusters. *Monthly Notices of the Royal Astronomical Society*, **454**(Dec.), 3872–3885.

Haghi, H., Khalaj, P., Hasani Zonoozi, A. et al. 2017. A Possible Solution for the M/L-[Fe/H] Relation of Globular Clusters in M31. II. The Age-Metallicity Relation. *Astrophysical Journal*, **839**(Apr.), 60.

Heyer, M., and Dame, T. M. 2015. Molecular Clouds in the Milky Way. *Annual Review of Astronomy and Astrophysics*, **53**(Aug.), 583–629.

Hopkins, A. M. 2018. The Dawes Review 8: Measuring the Stellar Initial Mass Function. *Publications of the Astronomical Society of Australia*, **35**(Nov.), 39.

Hopkins, P. F. 2012. The Stellar Initial Mass Function, Core Mass Function and the Last-Crossing Distribution. *Monthly Notices of the Royal Astronomical Society*, **423**(July), 2037–2044.

Hopkins, P. F. 2013. Why Do Stars Form in Clusters? An Analytic Model for Stellar Correlation Functions. *Monthly Notices of the Royal Astronomical Society*, **428**(Jan.), 1950–1957.

Hosek, Jr., M. W., Lu, J. R., Anderson, J. et al. 2019. The Unusual Initial Mass Function of the Arches Cluster. *Astrophysical Journal*, **870**(1), 44.

Hoversten, E. A., and Glazebrook, K. 2008. Evidence for a Nonuniversal Stellar Initial Mass Function from the Integrated Properties of SDSS Galaxies. *Astrophysical Journal*, **675**(Mar.), 163–187.

Hsu, W.-H., Hartmann, L., Allen, L. et al. 2012. The Low-Mass Stellar Population in L1641: Evidence for Environmental Dependence of the Stellar Initial Mass Function. *Astrophysical Journal*, **752**(June), 59.

Jeřábková, T., Hasani Zonoozi, A., Kroupa, P. et al. 2018. Impact of Metallicity and Star Formation Rate on the Time-Dependent, Galaxy-Wide Stellar Initial Mass Function. *Astronomy and Astrophysics*, **620**(Nov.), A39.

Jeřábková, T., Kroupa, P., Dabringhausen, J. et al. 2017. The Formation of Ultra Compact Dwarf Galaxies and Massive Globular Clusters. Quasar-Like Objects to Test for a Variable Stellar Initial Mass Function. *Astronomy and Astrophysics*, **608**(Dec.), A53.

Joncour, I., Duchêne, G., Moraux, E. et al. 2018. Multiplicity and Clustering in Taurus Star Forming Region. II. From Ultra-Wide Pairs to Dense NESTs. *Astronomy and Astrophysics*, **620**(Nov.), A27.

Kalari, V. M., Carraro, G., Evans, C. J. et al. 2018. The Magellanic Bridge Cluster NGC 796: Deep Optical AO Imaging Reveals the Stellar Content and Initial Mass Function of a Massive Open Cluster. *Astrophysical Journal*, **857**(Apr.), 132.

Kennicutt, R. C., and Evans, N. J. 2012. Star Formation in the Milky Way and Nearby Galaxies. *Annual Review of Astronomy and Astrophysics*, **50**(Sept.), 531–608.

Kennicutt, Jr., R. C. 1989. The Star Formation Law in Galactic Disks. *Astrophysical Journal*, **344**(Sept.), 685–703.

Kennicutt, Jr., R. C. 1998. The Global Schmidt Law in Star-Forming Galaxies. *Astrophysical Journal*, **498**(May), 541–552.

Kirk, H., and Myers, P. C. 2011. Young Stellar Groups and Their Most Massive Stars. *Astrophysical Journal*, **727**(Feb.), 64.

Kirk, H., and Myers, P. C. 2012. Variations in the Mass Functions of Clustered and Isolated Young Stellar Objects. *Astrophysical Journal*, **745**(Feb.), 131.

Kirk, H., Offner, S. S. R., and Redmond, K. J. 2014. The Formation and Evolution of Small Star Clusters. *Monthly Notices of the Royal Astronomical Society*, **439**(Feb.), 1765–1780.

Kirk, H., Johnstone, D., Di Francesco, J. et al. 2016. The JCMT Gould Belt Survey: Dense Core Clusters in Orion B. *Astrophysical Journal*, **821**(Apr.), 98.

Koen, C. 2006. On the Upper Limit on Stellar Masses in the Large Magellanic Cloud Cluster R136. *Monthly Notices of the Royal Astronomical Society*, **365**(Jan.), 590–594.

Köppen, J., Weidner, C., and Kroupa, P. 2007. A Possible Origin of the Mass-Metallicity Relation of Galaxies. *Monthly Notices of the Royal Astronomical Society*, **375**(Feb.), 673–684.

Kristensen, L. E., and Dunham, M. M. 2018. Protostellar Half-Life: New Methodology and Estimates. *Astronomy and Astrophysics*, **618**(Oct.), A158.

Kroupa, P. 1995a. Inverse Dynamical Population Synthesis and Star Formation. *Monthly Notices of the Royal Astronomical Society*, **277**(Dec.), 1491.

Kroupa, P. 1995b. The Dynamical Properties of Stellar Systems in the Galactic Disc. *Monthly Notices of the Royal Astronomical Society*, **277**(Dec.).

Kroupa, P. 2001. On the Variation of the Initial Mass Function. *Monthly Notices of the Royal Astronomical Society*, **322**(Apr.), 231–246.

Kroupa, P. 2002a. The Initial Mass Function of Stars: Evidence for Uniformity in Variable Systems. *Science*, **295**(Jan.), 82–91.

Kroupa, P. 2002b. Thickening of Galactic Discs through Clustered Star Formation. *Monthly Notices of the Royal Astronomical Society*, **330**(Mar.), 707–718.

Kroupa, P. 2005 (Jan.). The Fundamental Building Blocks of Galaxies. Page 629 of: Turon, C., O'Flaherty, K. S., and Perryman, M. A. C. (eds), *The Three-Dimensional Universe with Gaia*. ESA Special Publication, vol. 576.

Kroupa, P., and Bouvier, J. 2003. The Dynamical Evolution of Taurus-Auriga-type Aggregates. *Monthly Notices of the Royal Astronomical Society*, **346**(Dec.), 343–353.

Kroupa, P., Jeřábková, T., Dinnbier, F. et al. 2018. Evidence for Feedback and Stellar-Dynamically Regulated Bursty Star Cluster Formation: The Case of the Orion Nebula Cluster. *Astronomy and Astrophysics*, **612**(Apr.), A74.

Kroupa, P., Weidner, C., Pflamm-Altenburg, J. et al. 2013. *The Stellar and Sub-Stellar Initial Mass Function of Simple and Composite Populations*. Page 115.

Kroupa, P., and Weidner, C. 2003. Galactic-Field Initial Mass Functions of Massive Stars. *Astrophysical Journal*, **598**(Dec.), 1076–1078.

Kroupa, P., Gilmore, G., and Tout, C. A. 1991. The Effects of Unresolved Binary Stars on the Determination of the Stellar Mass Function. *Monthly Notices of the Royal Astronomical Society*, **251**(July), 293–302.

Kroupa, P., Tout, C. A., and Gilmore, G. 1993. The Distribution of Low-Mass Stars in the Galactic Disc. *Monthly Notices of the Royal Astronomical Society*, **262**(June), 545–587.

Kroupa, P., and Jerabkova, T. 2018. The Impact of Binaries on the Stellar Initial Mass Function. *arXiv e-prints*, Jun, arXiv:1806.10605.

Krumholz, M. R. 2014. The Big Problems in Star Formation: The Star Formation Rate, Stellar Clustering, and the Initial Mass Function. *Physics Reports*, **539**(June), 49–134.

Krumholz, M. R. 2015. The Formation of Very Massive Stars. Page 43 of: Vink, J. S. (ed), *Very Massive Stars in the Local Universe*. Astrophysics and Space Science Library, vol. 412.

Lada, C. J., and Lada, E. A. 2003. Embedded Clusters in Molecular Clouds. *Annual Review of Astronomy and Astrophysics*, **41**, 57–115.

Lane, J., Kirk, H., Johnstone, D. et al. 2016. The JCMT Gould Belt Survey: Dense Core Clusters in Orion A. *Astrophysical Journal*, **833**(Dec.), 44.

Larson, R. B. 1998. Early Star Formation and the Evolution of the Stellar Initial Mass Function in Galaxies. *Monthly Notices of the Royal Astronomical Society*, **301**(Dec.), 569–581.

Lee, J. C., Gil de Paz, A., Tremonti, C. et al. 2009. Comparison of Hα and UV Star Formation Rates in the Local Volume: Systematic Discrepancies for Dwarf Galaxies. *Astrophysical Journal*, **706**(Nov.), 599–613.

Lelli, F., Verheijen, M., and Fraternali, F. 2014. Dynamics of Starbursting Dwarf Galaxies. III. A H I Study of 18 Nearby Objects. *Astronomy and Astrophysics*, **566**(June), A71.

Lelli, F., McGaugh, S. S., Schombert, J. M. et al. 2017. One Law to Rule Them All: The Radial Acceleration Relation of Galaxies. *Astrophysical Journal*, **836**(Feb.), 152.

Lieberz, P., and Kroupa, P. 2017. On the Origin of the Schechter-Like Mass Function of Young Star Clusters in Disc Galaxies. *Monthly Notices of the Royal Astronomical Society*, **465**(Mar.), 3775–3783.

Lin, Y., Csengeri, T., Wyrowski, F. et al. 2019. Fragmentation and Filaments at the Onset of Star and Cluster Formation. SABOCA 350 μm View of ATLASGAL-Selected Massive Clumps. *Astronomy and Astrophysics*, **631**(Nov.), A72.

Mahajan, S., Ashby, M. L. N., Willner, S. P. et al. 2019. The Star Formation Reference Survey—III. A Multiwavelength View of Star Formation in Nearby Galaxies. *Monthly Notices of the Royal Astronomical Society*, **482**(Jan.), 560–577.

Maíz Apellániz, J., Walborn, N. R., Morrell, N. I. et al. 2006. Pismis 24-1: The Stellar Upper Mass Limit Preserved. *ArXiv Astrophysics e-prints*, Dec.

Marks, M., and Kroupa, P. 2012. Inverse Dynamical Population Synthesis. Constraining the Initial Conditions of Young Stellar Clusters by Studying Their Binary Populations. *Astronomy and Astrophysics*, **543**(July), A8.

Marks, M., Kroupa, P., and Oh, S. 2011. An Analytical Description of the Evolution of Binary Orbital-Parameter Distributions in N-body Computations of Star Clusters. *Monthly Notices of the Royal Astronomical Society*, **417**(Nov.), 1684–1701.

Marks, M., Kroupa, P., Dabringhausen, J. et al. 2012. Evidence for Top-Heavy Stellar Initial Mass Functions with Increasing Density and Decreasing Metallicity. *Monthly Notices of the Royal Astronomical Society*, **422**(May), 2246–2254.

Martín-Navarro, I. 2016. Revisiting the Classics: Is [Mg/Fe] A Good Proxy for Galaxy Formation Time-Scales? *Monthly Notices of the Royal Astronomical Society*, **456**(Feb.), L104–L108.

Maschberger, T., and Clarke, C. J. 2008. Maximum Stellar Mass Versus Cluster Membership Number Revisited. *Monthly Notices of the Royal Astronomical Society*, **391**(Dec.), 711–717.

Massey, P. 2003. Massive Stars in the Local Group: Implications for Stellar Evolution and Star Formation. *Annual Review of Astronomy and Astrophysics*, **41**, 15–56.

Matteucci, F. 1994. Abundance Ratios in Ellipticals and Galaxy Formation. *Astronomy and Astrophysics*, **288**(Aug.), 57–64.

Matzner, C. D., and McKee, C. F. 2000. Efficiencies of Low-Mass Star and Star Cluster Formation. *Astrophysical Journal*, **545**(Dec.), 364–378.

McGaugh, S. S. 2004. The Mass Discrepancy-Acceleration Relation: Disk Mass and the Dark Matter Distribution. *Astrophysical Journal*, **609**(July), 652–666.

McGaugh, S. S. 2005. The Baryonic Tully-Fisher Relation of Galaxies with Extended Rotation Curves and the Stellar Mass of Rotating Galaxies. *Astrophysical Journal*, **632**(Oct.), 859–871.

McGaugh, S. S. 2012. The Baryonic Tully-Fisher Relation of Gas-Rich Galaxies as a Test of ΛCDM and MOND. *Astronomical Journal*, **143**(Feb.), 40.

McGaugh, S. S., Lelli, F., and Schombert, J. M. 2016. Radial Acceleration Relation in Rotationally Supported Galaxies. *Physical Review Letters*, **117**(20), 201101.

McQuinn, K. B. W., Skillman, E. D., Dolphin, A. et al. 2015. Leo P: An Unquenched Very Low-Mass Galaxy. *Astrophysical Journal*, **812**(Oct.), 158.

McQuinn, K. B. W., Skillman, E. D., Heilman, T. N. et al. 2018. Galactic Outflows, Star Formation Histories, and Time-Scales in Starburst Dwarf Galaxies from STARBIRDS. *Monthly Notices of the Royal Astronomical Society*, **477**(July), 3164–3177.

Megeath, S. T., Gutermuth, R., Muzerolle, J. et al. 2016. The Spitzer Space Telescope Survey of the Orion A and B Molecular Clouds. II. The Spatial Distribution and Demographics of Dusty Young Stellar Objects. *Astronomical Journal*, **151**(Jan.), 5.

Meurer, G. R., Wong, O. I., Kim, J. H. et al. 2009. Evidence for a Nonuniform Initial Mass Function in the Local Universe. *Astrophysical Journal*, **695**(Apr.), 765–780.

Miller, G. E., and Scalo, J. M. 1979. The Initial Mass Function and Stellar Birthrate in the Solar Neighborhood. *Astrophysical Journal Supplement Series*, **41**(Nov.), 513–547.

Mor, R., Robin, A. C., Figueras, F. et al. 2017. Constraining the Thin Disc Initial Mass Function Using Galactic Classical Cepheids. *Astronomy and Astrophysics*, **599**(Mar.), A17.

Mor, R., Robin, A. C., Figueras, F. et al. 2018. BGM FASt: Besan\c{c}on Galaxy Model for Big Data. Simultaneous Inference of the IMF, SFH and Density in the Solar Neighbourhood. *ArXiv e-prints*, Sept.

Oey, M. S., and Clarke, C. J. 2005. Statistical Confirmation of a Stellar Upper Mass Limit. *Astrophysical Journal*, **620**(Feb.), L43–L46.

Offner, S. S. R., Clark, P. C., Hennebelle, P. et al. 2014. The Origin and Universality of the Stellar Initial Mass Function. *Protostars and Planets VI*, 53–75.

Oh, S., and Kroupa, P. 2012. The Influence of Stellar Dynamical Ejections and Collisions on the Relation between the Maximum Stellar and Star Cluster Mass. *Monthly Notices of the Royal Astronomical Society*, **424**(July), 65–79.

Oh, S., and Kroupa, P. 2016. Dynamical Ejections of Massive Stars from Young Star Clusters Under Diverse Initial Conditions. *Astronomy and Astrophysics*, **590**(May), A107.

Oh, S., and Kroupa, P. 2018. Very Massive Stars in Not So Massive clusters. *Monthly Notices of the Royal Astronomical Society*, **481**(Nov.), 153–163.

Oh, S., Kroupa, P., and Pflamm-Altenburg, J. 2015. Dependency of Dynamical Ejections of O Stars on the Masses of Very Young Star Clusters. *Astrophysical Journal*, **805**(June), 92.

Papadopoulos, P. P., and Thi, W.-F. 2013. The Initial Conditions of Star Formation: Cosmic Rays as the Fundamental Regulators. Page 41 of: Torres, D. F., and Reimer, O. (eds), *Cosmic Rays in Star-Forming Environments*. Advances in Solid State Physics, vol. 34.

Parker, R. J., and Goodwin, S. P. 2007. Do O-Stars Form in Isolation? *Monthly Notices of the Royal Astronomical Society*, **380**(Sept.), 1271–1275.

Pavlík, V., Kroupa, P., and Šubr, L. 2019. Do Star Clusters Form in a Completely Mass-Segregated Way? *Astronomy and Astrophysics*, **626**(Jun), A79.

Pflamm-Altenburg, J., and Kroupa, P. 2006. A Highly Abnormal Massive Star Mass Function in the Orion Nebula Cluster and the Dynamical Decay of Trapezium Systems. *Monthly Notices of the Royal Astronomical Society*, **373**(Nov.), 295–304.

Pflamm-Altenburg, J., and Kroupa, P. 2008. Clustered Star Formation as a Natural Explanation for the Hα Cut-Off in Disk Galaxies. *Nature*, **455**(Oct.), 641–643.

Pflamm-Altenburg, J., and Kroupa, P. 2010. The Two-Step Ejection of Massive Stars and the Issue of their Formation in Isolation. *Monthly Notices of the Royal Astronomical Society*, **404**(May), 1564–1568.

Pflamm-Altenburg, J., Weidner, C., and Kroupa, P. 2007. Converting Hα Luminosities into Star Formation Rates. *Astrophysical Journal*, **671**(Dec.), 1550–1558.

Pflamm-Altenburg, J., Weidner, C., and Kroupa, P. 2009. Diverging UV and Hα Fluxes of Star-Forming Galaxies Predicted by the IGIMF Theory. *Monthly Notices of the Royal Astronomical Society*, **395**(May), 394–400.

Pflamm-Altenburg, J., González-Lópezlira, R. A., and Kroupa, P. 2013. The Galactocentric Radius Dependent Upper Mass Limit of Young Star Clusters: Stochastic Star Formation Ruled Out. *Monthly Notices of the Royal Astronomical Society*, **435**(Nov.), 2604–2609.

Plunkett, A. L., Fernández-López, M., Arce, H. G. et al. 2018. Distribution of Serpens South Protostars Revealed with ALMA. *Astronomy and Astrophysics*, **615**(July), A9.

Ramírez Alegría, S., Borissova, J., Chené, A.-N. et al. 2016. Massive Open Star Clusters Using the VVV Survey. V. Young Clusters with an OB Stellar Population. *Astronomy and Astrophysics*, **588**(Apr.), A40.

Randriamanakoto, Z., Escala, A., Väisänen, P. et al. 2013. Near-Infrared Adaptive Optics Imaging of Infrared Luminous Galaxies: The Brightest Cluster Magnitude-Star Formation Rate Relation. *Astrophysical Journal*, **775**(Oct.), L38.

Randriamanakoto, Z., Väisänen, P., Ryder, S. D. et al. 2018. Young Massive Clusters in the Interacting LIRG Arp 299: Evidence for the Dependence of Star Cluster Formation and Evolution on Environment. *Monthly Notices of the Royal Astronomical Society*, Oct.

Recchi, S., and Kroupa, P. 2015. The Chemical Evolution of Galaxies with a Variable Integrated Galactic Initial Mass Function. *Monthly Notices of the Royal Astronomical Society*, **446**(Feb.), 4168–4175.

Recchi, S., Calura, F., and Kroupa, P. 2009. The Chemical Evolution of Galaxies Within the IGIMF Theory: the [α/Fe] Ratios and Downsizing. *Astronomy and Astrophysics*, **499**(June), 711–722.

Rybizki, J., and Just, A. 2015. Towards a Fully Consistent Milky Way Disc Model - III. Constraining the Initial Mass Function. *Monthly Notices of the Royal Astronomical Society*, **447**(Mar.), 3880–3891.

Salpeter, E. E. 1955. The Luminosity Function and Stellar Evolution. *Astrophysical Journal*, **121**(Jan.), 161.

Sana, H., de Mink, S. E., de Koter, A. et al. 2012. Binary Interaction Dominates the Evolution of Massive Stars. *Science*, **337**(July), 444.

Scalo, J. M. 1986. The Stellar Initial Mass Function. *Fundamentals of Cosmic Physics*, **11**(May), 1–278.

Schneider, F. R. N., Izzard, R. G., de Mink, S. E. et al. 2014. Ages of Young Star Clusters, Massive Blue Stragglers, and the Upper Mass Limit of Stars: Analyzing Age-Dependent Stellar Mass Functions. *Astrophysical Journal*, **780**(Jan.), 117.

Schneider, F. R. N., Izzard, R. G., Langer, N. et al. 2015. Evolution of Mass Functions of Coeval Stars Through Wind Mass Loss and Binary Interactions. *Astrophysical Journal*, **805**(May), 20.

Schneider, F. R. N., Sana, H., Evans, C. J. et al. 2018. An Excess of Massive Stars in the Local 30 Doradus Starburst. *Science*, **359**(Jan.), 69–71.

Schulz, C., Pflamm-Altenburg, J., and Kroupa, P. 2015. Mass Distributions of Star Clusters for Different Star Formation Histories in a Galaxy Cluster Environment. *Astronomy and Astrophysics*, **582**(Oct.), A93.

Smith, N., Stassun, K. G., and Bally, J. 2005. Opening the Treasure Chest: A Newborn Star Cluster Emerges from Its Dust Pillar in Carina. *Astronomical Journal*, **129**(Feb.), 888–899.

Smith, R. J. 2014. Variations in the Initial Mass Function in Early-Type Galaxies: A Critical Comparison between Dynamical and Spectroscopic Results. *Monthly Notices of the Royal Astronomical Society*, **443**(Sept.), L69–L73.

Smith, R. J., and Lucey, J. R. 2013. A Giant Elliptical Galaxy with a Lightweight Initial Mass Function. *Monthly Notices of the Royal Astronomical Society*, **434**(Sept.), 1964–1977.

Speagle, J. S., Steinhardt, C. L., Capak, P. L. et al. 2014. A Highly Consistent Framework for the Evolution of the Star-Forming 'Main Sequence' from $z \sim 0$–6. *Astrophysical Journal Supplement Series*, **214**(Oct.), 15.

Stephens, I. W., Gouliermis, D., Looney, L. W. et al. 2017. Stellar Clusterings around 'Isolated' Massive YSOs in the LMC. *Astrophysical Journal*, **834**(Jan.), 94.

Thies, I., Pflamm-Altenburg, J., Kroupa, P. et al. 2015. Characterizing the Brown Dwarf Formation Channels from the Initial Mass Function and Binary-Star Dynamics. *Astrophysical Journal*, **800**(Feb.), 72.

Thomas, D., Maraston, C., Bender, R. et al. 2005. The Epochs of Early-Type Galaxy Formation as a Function of Environment. *Astrophysical Journal*, **621**(Mar.), 673–694.

Tsujimoto, T. 2011. Chemical Signature Indicating a Lack of Massive Stars in Dwarf Galaxies. *Astrophysical Journal*, **736**(Aug.), 113.

Urquhart, J. S., Figura, C. C., Moore, T. J. T. et al. 2014. The RMS Survey: Galactic Distribution of Massive Star Formation. *Monthly Notices of the Royal Astronomical Society*, **437**(Jan.), 1791–1807.

Šubr, L., Kroupa, P., and Baumgardt, H. 2008. A New Method to Create Initially Mass Segregated Star Clusters in Virial Equilibrium. *Monthly Notices of the Royal Astronomical Society*, **385**(Apr.), 1673–1680.

van Dokkum, P. G., and Conroy, C. 2010. A Substantial Population of Low-Mass Stars in Luminous Elliptical Galaxies. *Nature*, **468**(Dec.), 940–942.

Vanbeveren, D. 1982. On the Difference between the Initial Mass Function of Single Stars and of Primaries of Binaries. *Astronomy and Astrophysics*, **115**(Nov.), 65–68.

Vazdekis, A., Peletier, R. F., Beckman, J. E. et al. 1997. A New Chemo-evolutionary Population Synthesis Model for Early-Type Galaxies. II. Observations and Results. *Astrophysical Journal Supplement Series*, **111**(July), 203–232.

Watts, A. B., Meurer, G. R., Lagos, C. D. P. et al. 2018. Star Formation in the Outskirts of DDO 154: A Top-Light IMF in a Nearly Dormant Disc. *Monthly Notices of the Royal Astronomical Society*, **477**(July), 5554–5567.

Wegg, C., Gerhard, O., and Portail, M. 2017. The Initial Mass Function of the Inner Galaxy Measured from OGLE-III Microlensing Timescales. *Astrophysical Journal*, **843**(July), L5.

Weidner, C., and Kroupa, P. 2004. Evidence for a Fundamental Stellar Upper Mass Limit from Clustered Star Formation. *Monthly Notices of the Royal Astronomical Society*, **348**(Feb.), 187–191.

Weidner, C., and Kroupa, P. 2005. The Variation of Integrated Star Initial Mass Functions among Galaxies. *Astrophysical Journal*, **625**(June), 754–762.

Weidner, C., and Kroupa, P. 2006. The Maximum Stellar Mass, Star-Cluster Formation and Composite Stellar Populations. *Monthly Notices of the Royal Astronomical Society*, **365**(Feb.), 1333–1347.

Weidner, C., Kroupa, P., and Larsen, S. S. 2004. Implications for the Formation of Star Clusters from Extragalactic Star Formation Rates. *Monthly Notices of the Royal Astronomical Society*, **350**(June), 1503–1510.

Weidner, C., Kroupa, P., and Maschberger, T. 2009. The Influence of Multiple Stars on the High-Mass Stellar Initial Mass Function and Age Dating of Young Massive Star Clusters. *Monthly Notices of the Royal Astronomical Society*, **393**(Feb.), 663–680.

Weidner, C., Kroupa, P., and Bonnell, I. A. D. 2010. The Relation between the Most-Massive Star and Its Parental Star Cluster Mass. *Monthly Notices of the Royal Astronomical Society*, **401**(Jan.), 275–293.

Weidner, C., Ferreras, I., Vazdekis, A. et al. 2013a. The (Galaxy-Wide) IMF in Giant Elliptical Galaxies: From Top to Bottom. *Monthly Notices of the Royal Astronomical Society*, **435**(Nov.), 2274–2280.

Weidner, C., Kroupa, P., Pflamm-Altenburg, J. et al. 2013b. The Galaxy-Wide Initial Mass Function of Dwarf Late-Type to Massive Early-Type Galaxies. *Monthly Notices of the Royal Astronomical Society*, **436**(Dec.), 3309–3320.

Weidner, C., Kroupa, P., and Pflamm-Altenburg, J. 2013c. The m_{max}-M_{ecl} Relation, the IMF and IGIMF: Probabilistically Sampled Functions. *Monthly Notices of the Royal Astronomical Society*, **434**(Sept.), 84–101.

Weidner, C., Kroupa, P., and Pflamm-Altenburg, J. 2014. Sampling Methods for Stellar Masses and the mmax-Mecl Relation in the Starburst Dwarf Galaxy NGC 4214. *ArXiv e-prints*, Apr.

Whitmore, B. C., Zhang, Q., Leitherer, C. et al. 1999. The Luminosity Function of Young Star Clusters in 'the Antennae' Galaxies (NGC 4038-4039). *Astronomical Journal*, **118**(Oct.), 1551–1576.

Wuchterl, G., and Tscharnuter, W. M. 2003. From Clouds to Stars. Protostellar Collapse and the Evolution to the Pre-main Sequence I. Equations and Evolution in the Hertzsprung-Russell Diagram. *Astronomy and Astrophysics*, **398**(Feb.), 1081–1090.

Yan, Z., Jerabkova, T., and Kroupa, P. 2017. The Optimally Sampled Galaxy-Wide Stellar Initial Mass Function. Observational Tests and the Publicly Available GalIMF Code. *Astronomy and Astrophysics*, **607**(Nov.), A126.

Zhang, Q., Fall, S. M., and Whitmore, B. C. 2001. A Multiwavelength Study of the Young Star Clusters and Interstellar Medium in the Antennae Galaxies. *Astrophysical Journal*, **561**(Nov.), 727–750.

Zhang, Z.-Y., Romano, D., Ivison, R. J. et al. 2018. Stellar Populations Dominated by Massive Stars in Dusty Starburst Galaxies Across Cosmic Time. *Nature*, **558**(June), 260–263.

Zinnecker, H., and Yorke, H. W. 2007. Toward Understanding Massive Star Formation. *Annual Review of Astronomy and Astrophysics*, **45**(Sept.), 481–563.

Zonoozi, A. H., Haghi, H., and Kroupa, P. 2016. A Possible Solution for the M/L-[Fe/H] Relation of Globular Clusters in M3. I. A Metallicity- and Density-Dependent Top-Heavy IMF. *Astrophysical Journal*, **826**(July), 89.

Zonoozi, A. H., Mahani, H., and Kroupa, P. 2019. Was the Milky Way a Chain Galaxy? Using the IGIMF Theory to Constrain the Thin-Disc Star Formation History and Mass. *Monthly Notices of the Royal Astronomical Society*, **483**(1), 46–56.

3

Stellar Populations, Stellar Evolution, and Stellar Atmospheres

J. J. ELDRIDGE AND E. R. STANWAY

3.1 Introduction

Everything we know about galaxies and the stars that form within them comes from the photons we detect across the electromagnetic spectrum. While astronomers can now detect non-electromagnetic particles such as neutrinos, and even gravitational waves, these do not have a direct relationship to star formation. Gaining the greatest possible knowledge from the light we detect is thus key to understanding young stellar populations. To do this requires a detailed model of the physical processes producing the luminous signal we detect and quantify. However, there are multiple factors to consider: the evolution of the stars as they fuse new heavy elements in their cores to provide the energy supply; the physics of the stellar photosphere which reprocesses the energy from the core fusion reactions; the gas and dust surrounding the stars that can interact with that light – including that in the natal cloud, the circumgalactic medium, and the intergalactic medium the light interacts with before it reaches the Earth.

In this chapter we will concentrate on the details of stars and stellar populations. We will address how we can model stars and predict how they appear, and thus how we derive the star-formation rate (SFR) of observed galaxies by comparing theoretical predictions to observations. We will discuss the current understanding in this area and highlight significant recent advances that have modified this understanding. First we discuss the evolution of stars, followed by modelling of their atmospheres. Then we consider how we can combine these to create model stellar populations and eventually synthesise a predicted spectrum. Finally, we discuss other factors and caveats that must be considered in spectral synthesis, before looking towards the future of this field.

3.2 Stellar Evolution

One of the most critical steps in understanding the Universe is understanding the evolution of stars. At their simplest, stars are gravitationally confined nuclear fusion reactors. The energy released in creating ever-heavier nuclei from hydrogen at the stellar core slowly escapes to the surface, and this escaping light causes the sky to shine. There are many textbooks that cover the subject in varying levels of accessibility and detail, but the core picture remains important.

Taking our own star, the Sun, for instance, we know that it is currently fusing hydrogen atoms to helium in its radiative core and is approximately halfway through its 10-billion year main-sequence lifetime. The energy released in the core takes thousands of years to reach the convective surface of the Sun and escape to reach our planet. We understand the Sun in exquisite detail because of its proximity. Not only can we observe the surface but helioseismology and solar neutrinos allow us to peer into its interior structure. Despite this there are still uncertain factors, even in this best-studied of stars.

3.2.1 Composition

The most frequently overlooked uncertainty regarding the Sun is its composition. While there is some uncertainty about the amount of hydrogen and helium in stellar models, the largest issue is the estimation of its metal content and its relation to that in other stars. There are two problems here. The first is that what is frequently used as a measure of metallicity in studies of star-forming regions is [O/H], while what is really important for driving the stellar evolution and the effect of nuclear burning on its envelope is [Fe/H]. This is because iron provides the dominant contribution to the opacity of stellar interiors. Thus iron abundance affects internal structure, surface radii, and temperatures, as well as strongly influencing stellar mass-loss rates, both of which modify evolutionary pathways. The effect of the radii and surface temperatures is also important when modelling the nebular emission from the gas surrounding hot stars. Lower-metallicity main-sequence stars are on average hotter, meaning they can excite some nebular lines more strongly (e.g. Xiao et al., 2018). However, it is not always remembered that [O/Fe] is not constant. There is good evidence that, in the past, stars were significantly more oxygen-rich for the amount of iron within them (e.g. Steidel et al., 2016).

A second factor is that the solar composition itself is still uncertain. While many stellar-evolution codes still use the Grevesse and Sauval (1998) prescription, more recent estimates (e.g. Grevesse et al., 2007; Asplund et al., 2009; Caffau et al., 2011) mean that the range of given [Fe/H] values range from 7.45 to 7.52, while [O/H] varies from 8.66 to 8.83. The issue is primarily that it is difficult to measure oxygen abundance from the Solar atmosphere. Helioseismology suggests that the internal oxygen abundance should, in fact, be towards the upper end of the reported range. Another confusing factor is that the Sun was not born where it is today in the Galaxy and was much closer to the centre of the disc, as a result we might expect its composition to be higher (Nieva and Przybilla, 2012). If we calibrate our stellar models instead against the Solar neighbourhood, we should be using the recent cosmic abundance standard of Nieva and Przybilla (2012) derived from nearby B stars. Studies of stars across the Galactic disk should then account for how metallicity scales with Galactocentric radius.

There is no easy solution to this problem but it remains one to be aware of. Future stellar evolution and population synthesis models will need to carefully consider the best composition to adopt as a Solar standard, which will be different to that for galaxy stellar populations, particularly in terms of enhancement of oxygen and other elements relative to iron as lower metallicities (e.g. Steidel et al., 2016; Conroy et al., 2018).

3.2.2 Mass-Loss Rates

The evolution of a star is driven by a race between how quickly the nuclear fusion reactions in its core can progress and how fast mass is lost from its surface. If there was no mass loss from stars then we could not exist, as earlier generations of stars would never have enriched the Universe with metals. Without mass loss, stars like our Sun would still evolve to become white dwarfs, but every star with an initial mass above the Chandrasekhar limit would eventually experience a supernova (SN) or core-collapse event, whereas many do not. Thus mass loss from stars is important for the chemical evolution of the Universe and creating more diversity in the evolutionary outcomes.

For massive stars, above about $15\,M_{\odot}$, stellar winds can significantly reduce the mass of the star over its main-sequence lifetime, especially for the most massive stars. However, for most stars, mass loss is strongest during the post main-sequence phases of evolution. Intermediate- and low-mass stars, those below $\approx 8\,M_{\odot}$, experience their strongest mass loss during the red giant and asymptotic giant branches. Here, the low surface gravities and high luminosities drive strong stellar winds. These can be enhanced by dust production (and hence increased opacity) in the cool atmospheres and remove the hydrogen envelopes of such stars, leaving the exposed cores that eventually cool to become white dwarfs. During the final removal of the hydrogen envelope these are observed as planetary nebulae as the hot cores are able to photo-ionise the ejected material. The asymptotic giant branch (AGB) stars also have significant amounts of nucleosynthesis occurring in the burning shells surrounding the stellar core and are sites of significant production of carbon, nitrogen, and s-process elements (e.g. Herwig, 2005). Some fraction of the synthesised elements are carried to the surface as a result of third dredge-up as the burning shells evolve. The mass-loss rates of these stars are highly uncertain although recent work is beginning to better constrain what these must be (e.g. Rosenfield et al., 2014). AGB stars are very cool but luminous and produce significant infrared flux. Thus accurate modelling of these stars is vital not just for chemical evolution but also spectral synthesis of stellar populations.

For massive stars, with masses greater than $\approx 8\,M_{\odot}$, mass loss again is most important after the main sequence. We show example evolution tracks in the Hertzsprung–Russel diagram for these stars in Fig. 3.1. In this figure the stellar tracks shown by dashed lines represent standard single star, non-rotating stellar models. The other lines are discussed in the following sections. Each track indicates how a star with a given initial mass changes in luminosity and temperature as it evolves from the stellar main sequence onto the cool-but-bright giant branch and through additional phases due to winds, rotation, and binary interactions. Massive stars are luminous and evolve to become red supergiants. If mass-loss rates are high enough during this phase then the hydrogen envelope can be removed, but rather than forming white dwarfs they instead become naked-helium or Wolf-Rayet (WR) stars. These are stars that have no hydrogen but are still powered by helium-burning and

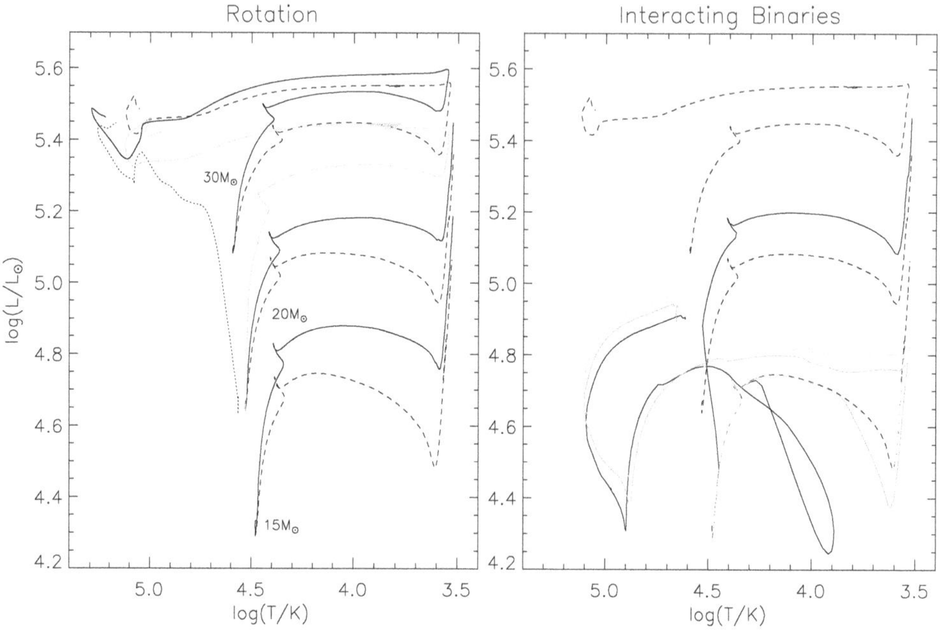

Figure 3.1 Hertzsprung-Russel diagrams showing the evolutionary tracks for different mass stars and how adding new physics changes these models. In both panels the dashed lines show the evolution of 15, 20, and 30 $M_{\odot}$ non-rotating BPASS (Binary Population and Spectral Synthesis code, see http://bpass.auckland.ac.nz) single-star models at Solar metallicity ($Z = 0.020$). In the the left panel we show tracks that have been given extra-mixing to simulate the effect of rotation. The black solid lines are for an initial rotation rate of 50% of the critical rotation velocity. The light grey track is for a 20 $M_{\odot}$ star at 90% of the critical rotation velocity. The light grey dashed line is a representative track of a 20 $M_{\odot}$ star that experiences chemically homogeneous evolution so is full mixed during the main-sequence, it has a metallicity of $Z = 0.004$. In the right panel we show sample binary tracks for the 15 $M_{\odot}$ primary star. Each has a 7.5 $M_{\odot}$ companion with the initial period increasing from the black solid line to the light grey solid line. The models shown are for $\log(P_{\rm i}/{\rm days})$ of 0.2, 0.8, 2, and 2.8 where $P_{\rm i}$ is the initial orbital period for the binary.

later nuclear stages in their cores. These stars perhaps have the most uncertain mass-loss rates (e.g. McClelland and Eldridge, 2016; Yoon, 2017). The initial mass of progenitors for these stars lies somewhere above $\approx 20\,M_{\odot}$ for single stars. In the left panel of Fig. 3.1, for example, only the $30\,M_{\odot}$ star has strong enough mass loss to move across the diagram to become a hot Wolf-Rayet star. However, the observed and predicted mass-loss rates of red supergiants is uncertain (e.g. Schröder and Cuntz, 2007; Georgy, 2012). Part of the issue is that it is commonly assumed that all stars are single and many of the uncertainties might be removed if we consider that many stars are in binaries (see Section 3.2.4).

A further complication of mass-loss rates links back to the uncertainties already discussed regarding composition. Winds of hot stars are radiation-driven, primarily due to the opacity of the many weak lines of iron. Thus the mass-loss rates are metallicity dependent and this must be included when calculating stellar models of different metallicities. Typically, the rates are scaled by a factor of $(Z/Z_{\odot})^{0.5}$, but this exponent value is poorly constrained and lies in the range from 0.5 to 1. Although it does appear that the scaling is understood for OB stars (Vink et al., 2001), there is evidence is might be steeper for Wolf-Rayet stars (e.g. Hainich et al., 2015).

3.2.3 Stellar Rotation

By the beginning of the twenty-first century, it was understood that the single-star models employed in population and spectral-synthesis codes were inadequate to perfectly reproduce observed stellar populations. There are two strong contenders for the root cause of this issue: interacting binaries; and stellar rotation. It was the latter that was taken up more widely by the stellar community due to pioneering efforts of the Geneva stellar evolution group in creating detailed models which account for the effects of rotation (Meynet and Maeder, 1997, 2000, 2005; Maeder and Meynet, 2001, 2004; Hirschi et al., 2004; Ekström et al., 2012; Georgy et al., 2012).

The centripetal forces from stellar rotation change the surface of gravitational potential which is no longer the same as the surface of constant pressure and so the star moves out of thermal equilibrium. To re-balance itself the star drives a star-wide motion referred to as an Eddington–Sweet circulation. We note that there are also other mixing processes instigated by rotation but they will not be discussed in detail here (see e.g. Heger et al., 2000, for more details). This circulation, unless suppressed by steep composition gradients, causes extra-mixing in radiative zones that would normally be assumed to be unmixed. Therefore, fresh hydrogen can be introduced into the stellar core, extending its lifetime. Nuclear burning products, especially nitrogen, can also be mixed to the surface where nitrogen enhancement can be observed (e.g. Hunter et al., 2008).

Rotation also reduces the effective equatorial gravity of a star, and so it can increase the mass-loss rate. This process is feedback-controlled since the strong winds extract angular momentum and slow the star down but extra complication arises from how angular momentum is transported within the stellar interior, for example: Can the surface slow down but the core remain rapidly rotating? The entire situation is further complicated by the addition of magnetic fields which allow for more efficient transport of angular momentum and ensure rigid body rotation. Although prescriptions accounting for these effects exist (e.g. Maeder and Meynet, 2004; Heger et al., 2005; Potter et al., 2012), it is difficult to know how correct these models are. None seem to fully explain the full distribution of stars that are nitrogen enriched and rapidly rotating (Hunter et al., 2008).

In summary, there is good observational and theoretical evidence that rotation should be included in stellar models but uncertainties remain regarding how to do so. Evolution is only very different for the most rapidly rotating stars, which leads to the question: Why do some stars rotate more rapidly than others of the same mass? Comparisons of rotating models to observational data requires all stars to be born with a $300\,\mathrm{km\,s^{-1}}$ equatorial rotation velocity: is this likely? An alternative, which explains both the diversity of rotation rates at a given mass, and may require less severe birth-rotation speeds, has arisen from the development of models including interacting binaries. Assuming reasonable initial-parameter ranges, these are able to match the same observational constraints without invoking rapid initial rotation.

3.2.4 Binary and Multiple Stars

A recent development in stellar evolution is the growing certainty that it is essential to account for interacting binary stars when studying any stellar system. While indirect evidence had been showing for some time that interacting binary stars were important for stellar studies (e.g. Podsiadlowski et al., 1992; Van Bever and Vanbeveren, 2003; Eldridge et al., 2008), observational surveys of the initial binary parameters of young stellar populations have now demonstrated just how prolific binaries are (Sana et al., 2012; Moe and Di Stefano, 2017). These studies showed that every O star can expect to be born in a binary system, and that 70% of these O stars have evolution altered by a binary interaction. Binary evolution is thus the norm; it is single-star evolution that is now the exception to the rule. It is true that the binary fraction drops to around 40% in $\sim 1\,M_{\odot}$ stars, but this is still a significant departure from the enduring single-star paradigm of stellar research.

Binary evolution provides a natural context in which to consider stellar rotation. Tidal forces between the secondary and primary stars of the binary system can modify rotation, although these are generally only strong if a star has a convective surface and is close to filling its Roche lobe. In this circumstance, tides tend to synchronise the rotation rate of the star to the orbit of the binary, exchanging angular momentum between the two. Rotation can also be modified by mass transfer, which occurs when a star fills its Roche lobe and transfers both mass and angular momentum to its companion. This extra angular momentum spins up the companion and very little material needs to be accreted before the star approaches the critical rotation rate (e.g. Cantiello et al., 2007; de Mink et al., 2013; Vanbeveren et al., 2018). In fact, de Mink et al. (2013) was able to show that the observed rotational velocity distribution of B stars could be reproduced naturally by this process, assuming a realistic population of binary stars. Thus, the effects of stellar rotation are important, but are likely strongest only within binary (or higher order multiple) systems.

Given this now overwhelming evidence of the importance of binaries, we have to ask: How else do binary stars change our understanding of stellar evolution? The answer is that they introduce new evolution pathways due to mass loss, mass gain, and angular-momentum gain. We have already discussed the effect of the mass and angular-momentum gain by Roche-lobe overflow. This tends to lead to rejuvenation of the accreting star as rapid rotation mixes fresh hydrogen into the core, as well as increasing the mass of the star and thus its luminosity and surface temperature. In a stellar population these stars would appear as blue stragglers beyond the main-sequence turn-off of a cluster, for example. Such stars also arise from mergers of binary components, which may occur if the expansion of the star cannot be stopped by mass-loss during Roche-lobe overflow, and the stars enter a common-envelope evolution phase.

A unique phase of evolution that results from binary interactions, but is not possible from stellar rotation, is the creation of helium stars from stars with initial masses in the range from 10 to 20 $M_{\odot}$. We show this evolutionary path in Fig. 3.1 for a 15 $M_{\odot}$ star (solid grey lines). Interactions remove the hydrogen envelope and the star moves across to high effective temperatures of the order of 10^5 K. Stars more massive than approximately 20 $M_{\odot}$ have stellar winds that are strong enough to to remove their hydrogen envelope and expose the helium core. These helium cores have masses greater than $\sim$5 $M_{\odot}$ and can drive strong optically thick stellar winds and so are typically observed as Wolf-Rayet stars. Helium stars less massive than this tend to be less luminous and cannot drive these winds and so are not observed to be Wolf-Rayet stars, in fact observational evidence for these sources is rare due to their relatively low luminosities and very high temperatures (Yoon et al., 2012; Yoon, 2017; Götberg et al., 2017, 2018). With typical temperatures $\sim$100000K, weak spectral features, and their proximity to an O or B star which has accreted significant mass and is thus luminous (Götberg et al., 2017, 2018), they are exceptionally challenging to detect. The only plausibly confirmed example of this class is HD45166 (Groh et al., 2008). Stars with initial masses below the 10–20 $M_{\odot}$ range tend to form white dwarfs and sdBO stars from binary interactions.

While the lack of galactic examples of 1.4–5 $M_{\odot}$ helium stars may give cause for scepticism, there are many indirect indicators that these stars exist. One example is that observed progenitor of type Ib supernovae iPTF13bvn was most likely a helium star of $\sim$3 $M_{\odot}$ (Bersten et al., 2014; Kim et al., 2015; Eldridge and Maund, 2016). Furthermore, the relative rates of type Ib/c (hydrogen deficient) to type II (hydrogen rich) supernovae can only be explained if a significant population of helium stars exist, as can the large number of non-detected type Ib/c progenitors (Eldridge et al., 2013). There is also growing evidence from the study of nebula regions that these stars also produce many ionising photons at ages beyond the typical 10 Myr cutoff for HII region ages from single stars (Xiao et al., 2019, 2018). Also, while they lead to increased ionising-photon production, they also increase the amount of hard ionising photons and this has important implications for modelling of HII regions and the epoch of re-ionisation (Stanway et al., 2016; Xiao et al., 2018). The growing indirect evidence and recent predictions about how these stars will appear suggest that it is only a matter of time before examples of these helium stars are found.

Including interacting binary stars does introduce more uncertainties into stellar models. The initial binary parameters appropriate for the local universe are finally becoming reasonably well constrained (e.g. Moe and Di Stefano, 2017). It is the interactions between the stars that are most uncertain. One major uncertainty is the efficiency of Roche-lobe overflow; the reaction of the orbit to the mass transfer is sensitive to how much mass is accreted by the companion relative to how much is lost. The less efficient the transfer, the more likely the system will be stable. The dependence of efficiency on binary parameters is not precisely known, but it appears that stars of similar masses can transfer significant amounts of material between them while dissimilar mass binaries have poor efficiency.

If Roche-lobe overflow is not stable for a particular binary and the donor star continues to grow in radius, it can eventually engulf its companion. This leads to the common-envelope evolution phase, where the companion and the core of the donor star can spiral closer together as angular momentum and orbital energy is carried away together with the expelled hydrogen envelope. Sometimes, the core and companion merge, at other times two very close stars are produced if the envelope is ejected before the final merger. As this is a dynamical process the evolution is highly uncertain, although some progress is being made (Ivanova et al., 2013; De Marco and Izzard, 2017). The final orbit of the star is strongly dependent on how efficiently orbital energy is extracted to expel the hydrogen envelope. No single efficiency parameter is able to explain every known occurrence of common-envelope evolution so some more complex prescription is likely required.

A further major uncertainty arises when the first core-collapse supernova occurs within a binary. If more than half the total mass of the binary is expelled in the supernova explosion then the system will become unbound. Many neutron stars and some black holes may, however, acquire extra momentum in the explosion. This can change the result of the supernova either by keeping a binary bound or unbinding a binary that would otherwise remain stable. Typically when modelling binaries a kick is chosen at random from some distribution, but there is growing evidence that kicks are more complex and may be closely related to the structure of the exploding star (Hobbs et al., 2005; Bray and Eldridge, 2016, 2018; Tauris et al., 2017). This then determines the future evolution of the companion star post-SN. Kicks also have serious implications for the production of a gravitational wave signal by modifying the probability and timescales for two compact remnants to form, and then merge, in a binary (Belczynski et al., 2016; Eldridge and Stanway, 2016).

In summary, while binaries introduce complexity and uncertainty into models, predictions for stellar populations accounting for them naturally reproduce the complexity we observe when studying stellar systems and this cannot be neglected.

3.3 Stellar Atmospheres

Stellar models only calculate the interior structure and surface parameters of a star, chiefly: its radius, effective temperature, luminosity, and mass. While these can be used to make a crude estimate of a star's appearance, detailed work requires a model of the emission from the stellar atmosphere at the surface. This is complicated, and the calculation of a stellar atmosphere at one age/physical state can take as long as the calculation of a single star's entire evolution. In the past this caused problems, but modern computing power has now made it possible to begin to calculate large grids of atmosphere models (e.g. Conroy et al., 2018). This has also enabled development of more complex models, leading to increased realism (e.g. Gustafsson et al., 2008; Hillier, 2011; Sander et al., 2015). For simple applications, such as interpreting broad-band photometry, a detailed atmosphere model may not be necessary – as is the case for white dwarfs, which contribute,

to first approximation, a simple black body to most optical/UV photometry. However, when spectra are obtained, it is necessary to use a theoretical model at the same resolution in their analysis: the same white dwarfs will have absorption-line features strongly dependent on subtleties in the atmosphere model and evolutionary stage. Thus, application determines the necessary detail in an atmosphere model.

The emergent spectrum of a star is dependent on its effective temperature, surface gravity, and composition. The simplest model atmospheres assume that the atmospheres are optically thin and that the atmosphere's geometry can be treated in a plane approximation. For most stars, these are good approximations. They break down for Wolf-Rayet stars, which have high luminosities that drive optically thick winds (Hillier, 2011; Sander et al., 2015) and for red supergiants whose atmospheres are sufficiently extended that spherical geometry is essential (Gustafsson et al., 2008). Broadly speaking, most stars can be divided into one of three groups which require different atmosphere model codes: hot stars, cool stars, and white dwarfs.

Hot-star atmospheres, with effective temperatures above about 10000K, are generally the simplest models to make as the surface of these stars are radiative and any stellar winds are radiatively driven. At the higher end of this temperature range, however, models for optically thick winds may also be required. These have to include the hydrodynamics of the wind and become extremely computationally demanding. Nonetheless, great strides have been taken in the last few years in improving the accuracy and prediction of these codes (Hillier, 2011; Sander et al., 2015).

Cool stars, typically, have convective envelopes and have the problem that more and more of the material on the surface is becoming neutral with fewer free electrons, while simple molecules are also formed. This complicates calculations as the electron configurations and lines of more atoms and molecules must be included to accurately reproduce the emerging spectrum. The most luminous cool stars, large red giants, and supergiants are also large in radius. The stars can be so large and cool that the opacity of the photosphere becomes strongly wavelength dependent. This means the apparent radius of the star also become wavelength dependent and the simple plane-parallel approximation which relies on identifying a physical radius must be abandoned. These problems lead to difficulties in defining the effective temperature of red supergiants, which are still to some degree debated theoretically and observationally (Levesque et al., 2006; Davies et al., 2013). The atmospheres of the coolest, asymptotic giant branch (AGB) stars are so cool that dust formation is strong, which again adds further complication to models of their spectrum, especially since the dust is likely to have different composition to that normally assumed for interstellar extinction.

At their simplest level, the atmospheres of white dwarfs, the stellar remnants that end the life of most main sequence stars, may be the simplest to model given that their high gravities lead to thin atmospheres. However, the difficulty of measuring these atmospheres, together with the uncertainties in convective mixing and diffusion lengths for these objects, means that complications do arise, particularly for the lowest masses and also for certain subtypes (e.g. Tremblay et al., 2011, 2015a,b). Generally, the most accurate models are not required for population synthesis but only when studying individual white dwarfs in detail.

So, where do the largest uncertainties and problems arise in atmosphere models? Yet again, there is the issue of composition. While many older libraries assumed solar-scaled abundances, as already mentioned, the [O/Fe] abundance has varied over the history of the Universe. Therefore, when applying low-metallicity atmospheres, their abundances must be treated with care, with the enhancement on α-element abundances considered. A second issue is how accurately the model atmospheres reproduce the parts of the electromagnetic spectrum outside the narrow optical window. This is important as future telescopes are designed to work primarily in the infrared and observational data from these facilities will require panchromatic models.

An unexpected problem is the dilemma of what model atmospheres to use for the helium stars discussed in Section 3.2.4. Without known, observed examples, we are forced to extrapolate from other atmosphere models, although pioneering work has been performed by Kim et al. (2015) and Götberg et al. (2017, 2018) in calculating atmosphere models for these stars. The sole population synthesis code to model these stars, BPASS (see Section 3.5.3), employs WR spectra that have been extrapolated to these masses. Some constraints may be obtainable from modelling the HII emission nebulae irradiated by individual stars (e.g. Simón-Díaz and Stasińska, 2008) and HII regions around young stellar populations (e.g. Byler et al., 2017; Xiao et al., 2019, 2018).

Finally, issues remain concerning how best to combine the atmosphere models to the stellar-evolution models. The simplest method is to pick the atmosphere with the closest surface gravity and temperature to that provided from the stellar model, although for stars with optically thick winds such as Wolf-Rayet stars a more complex match needs to be performed. There are potential improvements over a simple cross match between grids. One is to calculate surface-atmosphere models that are linked to the surface conditions of the stellar-evolution tracks as necessary rather than using a precomputed grid (e.g. Rix et al., 2004; Groh et al., 2014; Topping and Shull, 2015). This is most accurate but does become computationally inefficient when using a large number of stellar models for a full population synthesis, as discussed in the next section. A second approach is to match the interior structure of the atmosphere to the interior structure from the stellar model to make a fully consistent model from the star's centre to the outermost point of the atmosphere. Again, this is difficult for a population synthesis but does provide a way to check the more simple matching method commonly employed. It is most vital for the stars with optically thick winds because for these stars there is no hard boundary or edge to the star but a transition from the interior into the stellar wind (e.g. Groh et al., 2014) making the definition of surface conditions difficult.

In summary, model atmosphere spectra are now more accurate than ever and they are increasingly becoming entwined with stellar-evolution models. The demanding nature of this work leads to a fine line between what is most accurate to do and most computationally efficient.

3.4 Principles of Stellar Population and Spectral Synthesis

The identification of star-formation sensitive emission, and hence SFR indicators and their calibration from a population of stellar evolution and atmosphere models, is discussed in detail elsewhere in this book (Chapters 1, 5–8). Binary interactions add complexity to such calibration, but any procedure requires identifying the answers to certain key questions:

3.4.1 Which Stars Power the Emission?

Naturally, the answer to this question depends on the SFR indicator in question. In Fig. 3.2 we show a schematic overview of the stellar masses contributing to different indicators and their timescales. For the majority of widely used indicators (ultraviolet continuum, nebular-line emission, thermal infrared continuum), the ultimate power source for the emission is from stars with masses substantially larger than the Sun, due to their short lifetimes. These produce sufficiently high temperatures and luminosities while on the main sequence to ionise their surroundings, and the ionising photons reprocessed through gas and dust interactions dominate the emission in the indicator. Since the stars are short-lived, they indicate the presence of recent star formation.

The radio continuum and X-ray continuum are also tied to the presence of short-lived massive stars; however, in both cases, the immediate source of the emission is

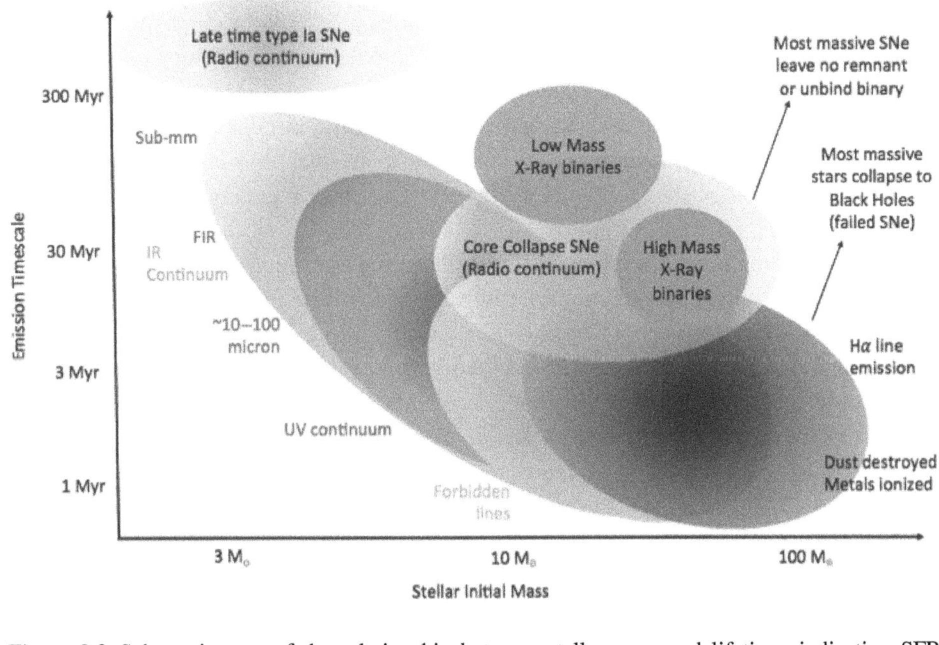

Figure 3.2 Schematic map of the relationship between stellar mass and lifetime, indicating SFR indicator contributions from different single-star populations.

more indirect. The synchrotron-emission continuum which dominates at radio frequencies $\sim 1-10\,$GHz is believed to arise by the acceleration of relativistic electrons through magnetic fields both on a local and galactic scale. The dominant source of these accelerated electrons is the energy input from core-collapse supernova explosions, while supernova remnants act as magnetic accelerators (see e.g. Condon, 1992; Tabatabaei et al., 2017). By contrast, the X-ray emission associated with star formation is believed to arise from a complicated combination of supernovae and their remnants, accretion onto massive compact objects, and (rarely) massive stars (see Chapter 11). Again, the existence of either a supernova and a massive compact object implies a massive-star progenitor, but as we have seen, the end of a stellar life depends on a wide range of evolutionary processes and parameters. As a result, radio and X-ray emission are usually calibrated empirically against other indicators, but this will introduce further uncertainties since their sources differ in both age and mass.

So, how massive is massive? The minimum initial stellar mass required to ionise its surroundings will depend both on the properties of the star and its surrounding nebular gas, and a different mass range will contribute to the emission in each line. It will also depend on the atmosphere model for a star with known properties and, in particular, its ionising spectrum. The mass required to create a black hole (and hence X-ray binary) will be dependent on both its metallicity and its evolutionary pathway including any binary interactions. The thermal response of the dust grains radiating in the infrared will depend on the grain composition and size. Nonetheless, it is possible to generalise somewhat: the ultraviolet continuum has significant contributions from young stars of a few solar masses, while hydrogen recombination lines such as Hα likely require more massive, shorter-lived stars with a typical mass $>8\,M_\odot$. The populations contributing to the thermal infrared are comparable to those which dominate the ultraviolet, since this is where dust absorption is most efficient. However, dust extinction extends through the optical where the less massive, but far more numerous, stars contribute to the flux, allowing a still larger range of stellar masses to contribute. Dust attenuation is described in greater detail in Chapter 4 of this book.

As discussed in previous sections, however, these generalisations do not account for the uncertainties and spread in evolutionary pathways which arise from binary interactions: a concern for a massive star population which, typically, has a binary fraction approaching unity (Sana et al., 2012; Moe and Di Stefano, 2017). Either stripping or accretion in a binary system can lead to the formation of very hot helium stars, or boost the mass of an initially lower mass star such that its ionising luminosity increases. Given the very blue spectra of helium stars, they can dominate in both ionising-photon production and ultraviolet continuum over the conventional main-sequence OB stars, even if rare. The less extreme case of rejuvenated, intermediate-mass companions may be more common, and while each star is substantially less luminous than an individual stripped helium star, their cumulative effect in populations of ages 10–100 Myr can be significant.

The supernova-dependent indicators – radio and X-ray continuum – are also affected by binary interactions. The accretion of material of a binary partner can lead to supernovae

occurring in stars with lower initial masses than a single star model would permit. This also allows supernovae to occur at later ages than expected from single star evolution (Zapartas et al., 2017). Binary interactions that lead to mass loss can also prevent supernovae and modify the distribution of remnant masses (and so accretion luminosities in the X-ray and elsewhere).

In short, while broad generalisation can be made, a precision theoretical SFR indicator calibration requires a full suite of binary stellar evolution and atmosphere models to identify contributing stars.

3.4.2 Is the Stellar Population Ever in a Steady State?

In a standard calibration of star-formation indicators a steady state is typically assumed (e.g. Kennicutt and Evans, 2012). The emission in a SFR indicator reaches a steady state (i.e. a constant ratio between the emission and the SFR) if the population has been forming stars at a constant rate per year for a period which exceeds the lifetime of the least massive or longest-lived star which contributes measurably to the emission source.

Given that the lifetime of a star depends on: (a) its initial mass, (b) its binary properties, and (c) its metallicity, again models are required to calculate the interval over which star formation can be considered constant. In general, the effect of incorporating binary models is to extend the lifetime of an ionising stellar population, thus extending the necessary assumption of a constant epoch from ~10–100 Myr to ~100–300 Myr. Figure 3.3 illustrates these timescales specifically for the binary case, which can be contrasted against the timescales discussed in Chapter 7.

Is any galaxy capable of sustaining a constant SFR over such an extended period? In the local universe, where SFRs are typically $<1\,M_\odot\,yr^{-1}$ and galaxy-evolution timescales are long, this may be a reasonable assumption. In the distant universe, much higher SFRs are common and these are usually inferred to be varying substantially on timescales of a few hundred Myr or less. A common prescription is that SFRs are varying exponentially (either rising or falling) with timescales of ~ 1 Gyr. There is also strong evidence for self-regulation and quenching of starbursts due to star-formation powered galactic winds driving gas out of galaxies, which suggests that a high constant SFR cannot be sustained except in a very massive galaxy. When modelling the spectral output of a galaxy the variability of star formation history is typically accounted for (Conroy, 2013), but this must be done on a case by case basis and does not produce a simple conversion factor from flux to SFR. Given the relatively young (<1 Gyr) stellar populations involved, the assumed star formation history will have a statistically significant effect on the calibration of simpler SFR indicators.

Note that for those indicators driven by supernovae or supernova remnants (e.g. radio or X-ray luminosity) there will also be a delay time between the onset of star formation and the death of the source progenitors. In the case of young or rapidly varying star formation this will lead to offsets in the rates derived in different indicators unless population synthesis models are invoked in their age-dependent calibration (see Greis et al., 2017).

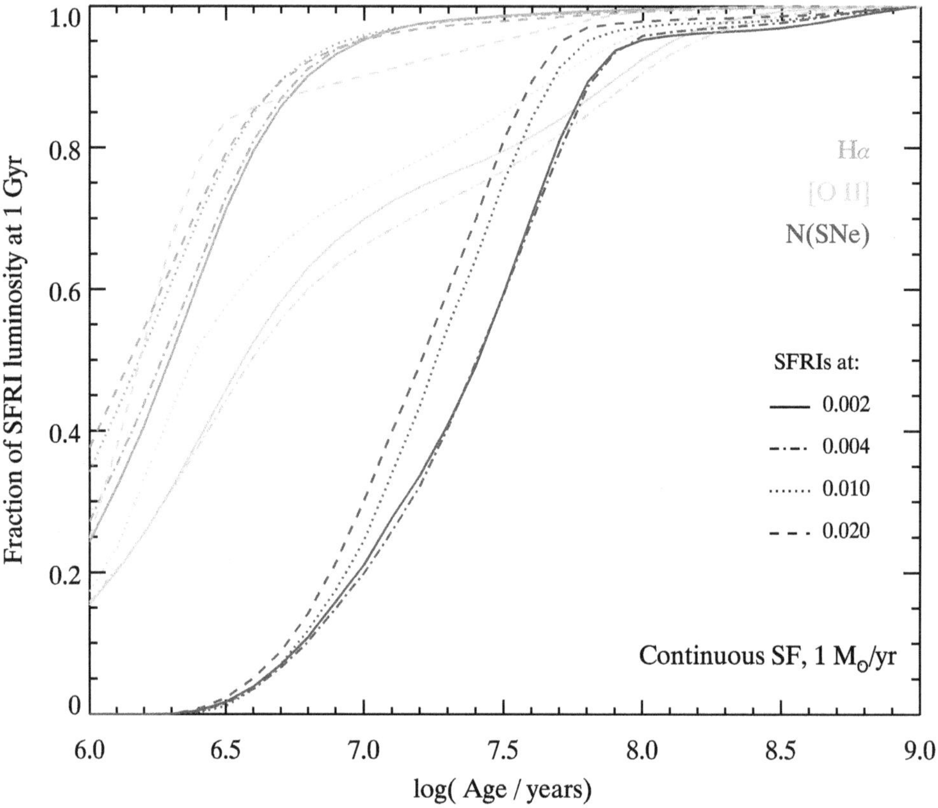

Figure 3.3 The timescales required for the flux in selected SFR indicators to reach a steady state in the binary population synthesis case, assuming a constant SFR and at four different metallicities. Hα and [OII] are derived from the combination of a BPASS stellar population model with a CLOUDY model for radiative transfer through nebular gas. N(SN) indicates the number of supernovae occurring per unit time, which is believed to correlate with synchrotron continuum at radio frequencies. All rates are scaled to unity at the value reached after 1 Gyr of continuous star formation. In the case of supernova rate, this does not represent a true steady state since type Ia events are just beginning to significantly contribute to the total rate at this age. Colour version available online.

As Cerviño et al. (2016) discuss, it may not be possible to define a single timescale for SFR estimates, or to establish the recent star-formation history from a single indicator. Combination of a number of indicators, with different mass sensitivities and timescales will likely provide a more robust estimate.

3.4.3 What Fraction of the Stellar Population is Involved?

To calculate the fraction f of the stellar mass of a population which contributes directly to the emission in the SFR indicator, the required quantity is:

$$f = \int_{M_1}^{M_2} \int_{a_1}^{a_2} \int_{q_1}^{q_2} \phi(M)\, \xi(M,a)\, \eta(M,a,q)\, p(M,a,q)\, dq\, da\, dM,$$

where $\phi(M)$ describes the initial mass function (IMF; Chapter 2) such that the number of stars in a given mass interval $dN = \phi(M)\,dM$; $\xi(M,a)$ is the distribution of initial binary separations a as a function of initial primary mass M; $\eta(M,a,q)$ is the distribution of initial binary mass-ratios q as a function of initial mass and binary separation; and $p(M,a,q)$ is the extent to which a star with given initial parameters contributes to the star-formation diagnostic. The functions must be normalised such that the integral over all permitted values yields a total of one. In the single-star evolution case, functions $\xi(M,a)$ and $\eta(M,a,q)$ collapse to delta functions with $q = 0$ and $a = \infty$, and the probability of a star contributing to the chosen indicator becomes dependent on mass only.

The IMF and its dependence, if any, on environment remains an active area of research and is discussed in detail in Chapter 2. The Salpeter (1955) simple power-law function with constant slope, $\phi(M) \propto M^{-2.35}$, with M in the range 1–100 $M_\odot$, is still favoured by the majority of studies for stars with $M > 1\,M_\odot$. However, it is now clear that this function overestimates the number of low-mass stars if extended to the hydrogen-burning limit ($0.08\,M_\odot$). To overcome this, most recent studies implement either a break to a shallower power law (e.g. Kroupa et al., 1993; Kroupa, 2001) or an exponential turn-over (e.g. Chabrier, 2003) at the low-mass end. These variations introduce a $\sim$30% change in the ionising-photon production rate from a simple stellar population and, therefore, (more or less directly) in its SFR indicators.

However this simple prescription may neglect some of the intrinsic variation and uncertainty in the IMF. Recent studies in the Large Magellanic Cloud by the VLT-FLAMES Tarantula Nebula Survey (VFTS) have identified not only a number of very massive stars (in the range 100–300 $M_\odot$) but also evidence that the massive star population may have a much flatter power-law slope (Schneider et al., 2018). Both extending the IMF and boosting the number of high-mass stars have the effect of increasing the ionising-photon production for a fixed stellar mass (see Stanway et al., 2016) which will lead to an overestimate of the SFR for populations where massive stars dominate (e.g. young or ongoing star-formation events). There is a risk of over-interpreting results derived from the single intense starburst that can be studied in detail, although other studies have also shown evidence for local, perhaps metallicity-dependent, variations in the IMF (e.g. Lagattuta et al., 2017; Vaughan et al., 2018).

By contrast, the properties of $\xi(M,a)$ and $\eta(M,a,q)$ remain extremely poorly constrained. Estimates of the multiplicity fraction beyond the immediate Solar neighbourhood are hampered by the difficulty of resolving systems either astrometrically or spectroscopically, and determining physical parameters (masses, temperatures, period, or separation) for the components of those systems which are identified. Nonetheless, progress in resolving these issues has been made in recent years as a result of the increasing availability of large-survey data. Sana et al. (2012) demonstrated that the multiplicity fraction for massive stars identified in the VFTS survey approaches unity. More recently, Moe and di Stefano (2017) compiled data from a range of sources to provide a meta-analysis of stellar multiplicity in the Milky Way. They demonstrated that the fraction of stars in a multiple system varies with both primary star mass and the binary period, with the total multiplicity fraction ranging from 40% for solar-like stars to >90% for O-type stars. The more massive stars are also more likely to be in close, short-period orbits, with an increased chance of interaction. The mass-ratio distribution was described as a broken power law, with indices which depended both on primary mass and binary period. They also found strong evidence for an excess of 'twin systems' (i.e. equal-mass binaries), particularly at the low-mass end of the distribution. This comprehensive study makes clear the importance of considering binary interactions in determining the relevant fraction of the stellar population, and is likely the best current model for these, but it does not represent a full understanding. Studies of this kind must correct for observational biases, which are difficult to assess consistently without introducing further assumptions. They are only plausible for nearby stellar populations and therefore show a metallicity bias, the effects of which are currently impossible to quantify. They also capture the properties of far-from-simple stellar populations. Unlike in models, the stars identified in these surveys do not have a single, well known age, and so further uncertainty is introduced if these observed distributions are treated as identical to the initial model distributions (i.e. without taking into account evolution of binary parameters).

The final function, $p(M,a,q)$, incorporates all the remaining uncertainties in stellar evolution, integrated over the stellar lifetimes, as discussed elsewhere in this chapter. This includes the various evolution pathways that are only possible as a result of binary interactions, the probability of supernovae disrupting multiple systems, and the relative flux of each source at relevant wavelengths.

3.4.4 What Is the Emission from Each Contributing Star?

A key input to $p(M,a,q)$ is the bridge between theoretical models and observed data: the process of spectra synthesis. While stellar-evolution models estimate the mass, radius, temperature, and surface composition of a star at each timestep, translating these to a luminosity as a function of wavelength requires matching the model output to an appropriate stellar-atmosphere model. To reproduce the observable output of an integrated stellar population, such as its line or continuum emission, the spectra produced by the atmosphere models are combined in proportions given by the population synthesis.

A number of stellar-atmosphere model sets have now been publicly available. Each represents a set of spectra for stars with known temperature, surface gravity, and surface composition. Since the required input physics is different, model-sets for post-main sequence stars such as WR stars, or remnants such as white dwarfs, are often distinct from those prepared for the bulk of the stellar population. Due to the computational expense of generating an atmosphere model, these are typically presented as a discrete grid in the available parameters. Matching these to the continuous distribution of parameters derived by evolution models requires approximation – either through selecting a model at the nearest grid-point or by interpolating between them. In the case of most stellar models this is straightforward since the spectra vary smoothly with a given parameter, but problems arise if a stellar-evolution model lies outside the selected atmosphere grid, requiring extrapolation rather than interpolation. This most commonly occurs as a result of binary interactions populating regions of the Hertzsprung-Russell diagram which are empty or sparse if they are neglected. Short-lived evolutionary phases which dominate the flux when present, but rarely occur in old stellar populations, are also often neglected in construction of model grids. Surface metallicity, and more importantly, the relative abundance of different metals, also represents a large and undersampled parameter space, over which approximation, interpolation, or extrapolation is often required.

Having established the spectral energy distribution of the stellar population from a population and spectral synthesis, further processing is required to build the full spectral energy distribution related to stars in a galaxy. A nebular gas and dust model is required to translate the ionising photon production in the stellar spectrum to observable emission-line and nebular-continuum fluxes, while a cool-dust model is required to reprocess the ultraviolet and optical photons into thermal infrared.

3.5 Existing Population Synthesis Models

A large number of population and spectral-synthesis models exist. They vary in their input physics, assumptions, and between pure population synthesis (i.e. determining the number of stars with a given mass, radius, temperature as a function of age) and a full treatment of the output spectra (combining this information with atmosphere models). Most only include single stars. Recent reviews and comparisons have been presented by Conroy (2013) and Wofford et al. (2016).

3.5.1 Pure Stellar Population Synthesis

Models which perform population synthesis are primarily used for analysis in two areas: the rates of astrophysical transients, and the interpretation of resolved stellar populations. As such they have a limited immediate impact on SFR indicators, but are used to estimate the ages of young populations by, for example, their ratios of Wolf-Rayet to O-type stars. They have come to increasing prominence in recent years since they are also able to predict not only the relative occurrence rate of different types of supernovae but also the rate of the compact object mergers detectable as gravitational-wave chirp signatures (e.g. `StarTrack`; Belczynski et al., 2008).

A notable example which incorporates binary evolution is `binary_c`[1] (Izzard et al., 2009), which is an evolution of the older `BSE`[2] models of Hurley et al. (2002). We note that nearly every current binary population-synthesis code employs these models. These are built on a rapid population framework which interpolates between a fixed grid of stellar models, based on semi-analytic approximations or physical prescriptions for the behaviour of stars. This has the benefit that individual stars do not need to be modelled in detail and

1 http://personal.ph.surrey.ac.uk/~ri0005/binary_c.html
2 Binary-Star Evolution; http://astronomy.swin.edu.au/~jhurley/

hence the input physics can be varied and new synthesis models rapidly produced. However, this approach will miss much of the detail and non-linear behaviour that is possible in binary interactions, for example, predicting different outcomes for how the envelope of a star responds to mass loss.

3.5.2 Single-Star Population and Spectral Synthesis

Stellar-population and spectral-synthesis models built on evolution tracks for isolated, single stars are in widespread use in extragalactic astronomy and are commonly used to estimate the SFR in galaxies. They are discussed in detail elsewhere in this book (see Chapters 5–8). While such models have existed since the early 1970s, they have grown in complexity and detail over the years. Model sets such as `Starburst99`[3] (Leitherer et al., 1999, 2014) focus on the evolution of young, simple stellar populations, others such as the `GALAXEV`[4] models of Bruzual and Charlot (2003) or `PÉGASE`[5] models (Fioc and Rocca-Volmerange, 1997; Le Borgne et al., 2004) have aimed to recreate the complex evolution of mature galaxies; a task which requires modelling many more stages of low-mass and post main-sequence evolution.

Even when based on the same underlying stellar isochrones, there is scope for models to differ in their implementation, relative weighting and prescriptions for later evolution stages. The Maraston models (Maraston, 2005),[6] for example, place a greater emphasis on the thermally pulsating AGB phase than previous synthesis models, arguing that its effects had been underestimated based on a new fuel consumption theorem. Recently, much interest has been focussed on evolutionary pathways for rotating massive stars which have a higher ionising-photon output (and hence modified SFR calibrations) relative to non-rotating stars. These have been implemented for subsets of stellar tracks by both `Starburst99` and the `FSPS`[7] models (Conroy et al., 2009, 2018).

Also noteworthy are the `MILES`[8] models (Vazdekis et al., 2016). These use theoretical isochrones to reconstruct the evolution of a population, but a large library of empirical stellar spectra for the spectral-synthesis step (as indeed does the high-resolution version of `PÉGASE`). This has the advantage of reproducing the detailed contributions of atomic and molecular species which may not be included by theoretical atmosphere models to the output spectrum, but is subject to uncertainties in classifying the physical conditions in the spectra relative to the isochrones and may omit rare object types or those at extreme metallicities.

[3] www.stsci.edu/science/starburst99/docs/default.htm
[4] Galaxy Spectral Evolution Library; www.bruzual.org/bc03/
[5] Programme d'Étude des Galaxies par Synthèse Évolutive; www2.iap.fr/pegase/
[6] www.icg.port.ac.uk/~maraston/Claudia's_Stellar_Population_Model.html
[7] Flexible Stellar Population Synthesis; https://github.com/cconroy20/fsps
[8] Medium-resolution Isaac Newton Telescope Library of Empirical Spectra; www.iac.es/proyecto/miles/pages/ssp-models.php

3.5.3 Binary Star Population and Spectral Synthesis

Binary population and spectral-synthesis models are relatively rare. This is largely due to the increased complexity in modelling the stellar evolution and the large number of input models (both evolution and atmosphere) required to fully sample the range of outcomes of binary evolution. Early influential work by the Brussels group (e.g. Van Bever and Vanbeveren, 2003; Vanbeveren et al., 2007) has not resulted in a publicly released set of synthesis models, and the team have recently veered away from this area.

The Yunnan group[9] have focussed on evolutionary population synthesis, but also have undertaken spectral synthesis (Belkus et al., 2003; Li and Han, 2008; Zhang et al., 2005, 2015). These models are built on the same grid of binary stellar-evolution models as BSE (discussed above) and a grid of model atmospheres which are relatively low in resolution for modern spectral work but sufficient for deriving SFR indicators. Their uptake by the wider community has been relatively limited.

The Binary Population and Spectral Synthesis (BPASS) project[10] is an effort to fill this gap with population synthesis built on a large grid of custom, detailed binary and single stellar-evolution models, combined with publicly available, large grids of high-resolution theoretical stellar atmospheres (Eldridge and Stanway, 2009, 2012; Eldridge et al., 2017; Stanway and Eldridge, 2018). These models also span a range of metallicities from 0.05% solar to twice solar. Despite being limited to a fixed grid of elemental abundances, and a limited range of input initial mass and binary parameter distributions, the BPASS models have seen increasing popularity in both the stellar and extragalactic community in recent years (e.g. Steidel et al., 2016; Henry et al., 2018; Van Dyk et al., 2018).

3.6 Further Considerations

We have discussed the main features of stellar-evolution models and their application in population synthesis of simple stellar populations. However, there are additional uncertainties which are less frequently discussed, particularly those which arise when comparing models to the more complex stellar populations present in observations.

3.6.1 Non-Stellar Components and Post-Processing

While stars dominate the energy production in typical stellar populations, it is not unusual for star light to be processed by nebular gas, dust, or other mechanisms as discussed above. A number of models exist which combine the stellar population and spectral synthesis with further post-processing by non-stellar components. In the case of dust this may be involve application of a relatively simple extinction and emission curve. For gas, a more complex radiative transfer code such as CLOUDY[11] (Ferland et al., 1998) is required.

[9] www1.ynao.ac.cn/~zhangfh/
[10] http://bpass.auckland.ac.nz
[11] https://trac.nublado.org

Where reprocessing is included in population synthesis, this is usually done in the context of template-fitting codes designed for the determination of galaxy photometric redshifts or physical parameters from photometry.

For example the `Starburst99` models allow users to add nebular reprocessing, while the `CIGALE`[12] fitting code (Burgarella et al., 2005; Boquien et al., 2016) allows combination of either `GALAXEV` or Maraston stellar population models with a star formation history and components derived from nebular reprocessing, AGN and dust emission, for comparison with observed galaxy photometry. Similarly the `MAGPHYS`[13] models (da Cunha et al., 2008) combine the `GALAXEV` stellar populations with a dust-emission curve constructed using energy-balance arguments to account for those photons attenuated from the optical and ultraviolet. The `FSPS` (Byler et al., 2017), `GALAXEV` (Gutkin et al., 2016) and `BPASS` (Xiao et al., 2019) groups have also released data-sets processed through a limited grid of nebular models.

In each case, assumptions must be made regarding the appropriate nebular and dust conditions (e.g. ionisation parameter, density, composition, grain size) for the source, or these must be constrained from the data itself, leading to larger uncertainties and degeneracies in model fits. Of particular concern for SFR measurements is the effect of the changing stellar populations at high redshift on their gas and dust environment. If a stellar population is capable of destroying dust (for example in shocks or in highly irradiated environments), or of creating it (for example in the winds of AGB stars or the environs of late-type Wolf-Rayet stars), then self-consistent reprocessing must be handled with more care.

3.6.2 Stochasticity

Emission in most SFR indicators, as already established, is sensitive to the most massive, most luminous stars in a young or ongoing star formation event. The correction from flux to a newly formed stellar mass relies on the assumption that whichever IMF is adopted is fully sampled. However, while the mass function is continuous, the stellar population is discrete. In starbursts with a low total mass $M < 10^5\,\mathrm{M}_\odot$, the predicted number of very massive stars ($M > 80\,\mathrm{M}_\odot$) formed can be substantially less than one. The formation of these stars is therefore stochastic: they are either present in a population, or not. This stochasticity can lead to a variation in the ionising emission from different populations with the same IMF and the same total mass. Further, it leads to a relation between the size of a starburst and the mass of its most massive star (e.g. Cerviño et al., 2013).

This stochasticity is difficult to deal with in stellar population-synthesis codes, since it requires a probability distribution to be sampled for each collapsing molecular cloud and stellar cluster within a larger stellar population. One galaxy modelling code, Stochastically

12 Code Investigating GALaxy Emission; https://cigale.lam.fr
13 Multi-wavelength Analysis of Galaxy Physical Properties; http://astronomy.swinburne.edu.au/~ecunha/ecunha/MAGPHYS.html

Lighting Up Galaxies (SLUG[14] da Silva et al., 2012), attempts to do this, using Bayesian inference to reconstruct the posterior probability of the stellar-population parameters given an assumed initial cluster-mass function. As da Silva et al. (2014) discusses, the resulting models show a substantial scatter in inferred SFR assuming standard calibrations, particularly at low luminosities and SFRs.

For a single-star population the stochastic effects can be severe. For low-mass star clusters the presence or absence of a single massive star can significantly change the luminous output of that cluster in the ultraviolet. The surprising impact of including binaries in a population while assuming stochastic sampling of the IMF is that binary populations are less susceptible to variation in their ionising and UV output. This was shown using BPASS v1 by Eldridge (2012) and Hermanowicz et al. (2013). While it might be thought that with the huge range of binary parameter space to choose from for each star the eventual outcome will be more variable, the typical binary interactions of mass loss, mass gain, and stellar mergers mean that even initially low-mass stars in a binary are likely to interact and produce more ionising and UV light. This effectively reproduces the output of massive stars, but from lower-mass stars that have a higher probability of existing due to the IMF. While it would be interesting to revisit this with more recent BPASS models, the basic result is unlikely to change.

3.6.3 Metallicity and Abundances

Stellar-evolution tracks and stellar-atmosphere models have now been generated at a wide range of initial metallicities and compositions. These have been used to generate stellar population synthesis models and calibrate SFR indicators as a function of metallicity. However, there remain outstanding issues.

The first is one of calibration. The prescriptions used by stellar-evolution models for both atmosphere opacities and the subsequent winds and mass loss are calibrated on observations of individual stars in the local universe. The Milky Way and its Local Group does include stellar populations with a range of ages and metallicities, but inevitably some regions of the parameter space are less well populated than others. In particular, many of the lowest metallicity stars in the local universe are old, low-mass relics of high redshift star-formation events (e.g. Chiaki et al., 2017). Inevitably, the robustness of extrapolations in evolution and atmosphere models to low metallicity has to be questioned. As adaptive-optics systems become more generally available, and in the epoch of extremely large telescopes, the range of local environments probed by calibration samples will continue to grow.

Similarly, constraints on any IMF variation with metallicity will improve over time. However, in the case of binary population synthesis, the determination of initial binary parameters (period, mass ratio, binary fraction, eccentricity etc.) will likely remain uncertain for some time to come. Work on binary systems to date has demonstrated just how difficult these are to constrain, even given large samples in the local universe. Evidence from

[14] Stochastically Lighting Up Galaxies; www.slugsps.com/home

large stellar radial-velocity measurement surveys already suggests that the multiplicity fraction for field stars may well be higher at low metallicities than at solar abundances (Gao et al., 2017; Badenes et al., 2018). Refinement of the orbital distribution parameters and their metallicity evolution will probably require both detailed observational study and modelling of molecular cloud fragmentation in the low metallicity regime.

Finally, given well-constrained models for low (or high) iron abundances, uncertainties remain on the relative abundance of different elements and their impact on stellar evolution and atmospheres. In particular, elements formed by the α-process are likely enhanced in high redshift (or old, low redshift) populations. These will have some effect on the stellar evolution through modifications to the core burning rates, although iron dominates the opacity in stellar envelopes. Some synthesis codes, such as FSPS (Conroy et al., 2018), are now adopting sets of α-enhanced stellar-evolution models. To date, this has not been attempted in detailed binary spectral synthesis models. A hybrid approach, combining stellar-evolution models with standard abundance patterns with α-enhanced atmosphere models may provide a useful bridging step on the path towards producing a more physical binary-population model grid. The impact of this on SFR indicators, particularly at low metallicity, remains to be assessed.

3.7 Looking to the Future...

In this chapter we have outlined how stellar-evolution models and stellar-atmosphere models are combined to produce spectral-synthesis models. We also highlighted the many uncertainties, assumptions, and caveats involved. Current research strives to resolve and remove these from population synthesis. It is too much for any one group to solve all of these problems, but as progress is made, and comparison between the groups increase, it becomes apparent which areas should be prioritised.

Of the many aspects above, we consider that accounting for interacting binary stars is essential for reliable spectral synthesis and thus for reliable calibration of SFR indicators. The helium stars that are created in binary interactions, as well as the blue stragglers that arise from mass gain, are a vital part of a stellar population which provide ionising photons at late times. Spectral synthesis models based on single-star models alone lack their important contribution. Also important is a more complex understanding of abundances and compositions of the stellar models that go into the spectral synthesis. We know that different groups of elements are produced by different phases of stellar evolution. A true synthesis model should have a composition that changes its abundance ratios, not just iron abundance, with age.

Finally we note that we are living in a fortunate time. Many of the existing spectral synthesis codes are the result of infrastructure initially created at the turn of the twenty-first century and already this has been superseded by faster, more powerful facilities. It is only with today's computational resources, and those we may anticipate into the future, that we can strive to calculate more accurate SFR indicator calibrations, built on detailed synthetic models of the observed universe.

References

Asplund, M., Grevesse, N., Sauval, A. J. et al. 2009. The Chemical Composition of the Sun. *Annual Review of Astronomy and Astrophysics*, **47**(Sept.), 481–522.

Badenes, C., Mazzola, C., Thompson, T. A. et al. 2018. Stellar Multiplicity Meets Stellar Evolution and Metallicity: The APOGEE View. *Astrophysical Journal*, **854**(Feb.), 147.

Belczynski, K., Holz, D. E., Bulik, T. et al. 2016. The First Gravitational-Wave Source from the Isolated Evolution of Two Stars in the 40-100 Solar Mass Range. *Nature*, **534**(June), 512–515.

Belczynski, K., Kalogera, V., Rasio, F. A. et al. 2008. Compact Object Modeling with the StarTrack Population Synthesis Code. *Astrophysical Journal Supplement Series*, **174**(Jan.), 223–260.

Belkus, H., Van Bever, J., Vanbeveren, D. et al. 2003. The Effects of Binaries on the Evolution of UV Spectral Features in Massive Starbursts. *Astronomy and Astrophysics*, **400**(Mar.), 429–436.

Bersten, M. C., Benvenuto, O. G., Folatelli, G. et al. 2014. iPTF13bvn: The First Evidence of a Binary Progenitor for a Type Ib Supernova. *Astronomical Journal*, **148**(Oct.), 68.

Boquien, M., Kennicutt, R., Calzetti, D. et al. 2016. Towards Universal Hybrid Star Formation Rate Estimators. *Astronomy and Astrophysics*, **591**(June), A6.

Bray, J. C., and Eldridge, J. J. 2016. Neutron Star Kicks and Their Relationship to Supernovae Ejecta Mass. *Monthly Notices of the Royal Astronomical Society*, **461**(Oct.), 3747–3759.

Bray, J. C., and Eldridge, J. J. 2018. Neutron Star Kicks – II. Revision and Further Testing of the Conservation of Momentum 'Kick' Model. *Monthly Notices of the Royal Astronomical Society*, **480**(4), 5657–5672.

Bruzual, G., and Charlot, S. 2003. Stellar Population Synthesis at the Resolution of 2003. *Monthly Notices of the Royal Astronomical Society*, **344**(Oct.), 1000–1028.

Burgarella, D., Buat, V., and Iglesias-Páramo, J. 2005. Star Formation and Dust Attenuation Properties in Galaxies from a Statistical Ultraviolet-to-Far-Infrared Analysis. *Monthly Notices of the Royal Astronomical Society*, **360**(July), 1413–1425.

Byler, N., Dalcanton, J. J., Conroy, C. et al. 2017. Nebular Continuum and Line Emission in Stellar Population Synthesis Models. *Astrophysical Journal*, **840**(May), 44.

Caffau, E., Ludwig, H.-G., Steffen, M. et al. P. 2011. Solar Chemical Abundances Determined with a CO5BOLD 3D Model Atmosphere. *Solar Physics*, **268**(Feb.), 255–269.

Cantiello, M., Yoon, S.-C., Langer, N. et al. 2007. Binary Star Progenitors of Long Gamma-ray Bursts. *Astronomy and Astrophysics*, **465**(Apr.), L29–L33.

Cerviño, M., Román-Zúñiga, C., Luridiana, V. et al. 2013. Crucial Aspects of the Initial Mass Function. I. The Statistical Correlation between the Total Mass of An Ensemble of Stars and Its Most Massive Star. *Astronomy and Astrophysics*, **553**(May), A31.

Cerviño, M., Bongiovanni, A., and Hidalgo, S. 2016. Recent SFR Calibrations and the Constant SFR Approximation. *Astronomy and Astrophysics*, **589**(May), A108.

Chabrier, G. 2003. Galactic Stellar and Substellar Initial Mass Function. *Publications of the Astronomical Society of the Pacific*, **115**(July), 763–795.

Chiaki, G., Tominaga, N., and Nozawa, T. 2017. Classification of Extremely Metal-Poor Stars: Absent Region in A(C)-[Fe/H] Plane and the Role of Dust Cooling. *Monthly Notices of the Royal Astronomical Society*, **472**(Nov.), L115–L119.

Condon, J. J. 1992. Radio Emission from Normal Galaxies. *Annual Review of Astronomy and Astrophysics*, **30**, 575–611.

Conroy, C. 2013. Modeling the Panchromatic Spectral Energy Distributions of Galaxies. *Annual Review of Astronomy and Astrophysics*, **51**(Aug.), 393–455.

Conroy, C., Gunn, J. E., and White, M. 2009. The Propagation of Uncertainties in Stellar Population Synthesis Modeling. I. The Relevance of Uncertain Aspects of Stellar Evolution and the Initial Mass Function to the Derived Physical Properties of Galaxies. *Astrophysical Journal*, **699**(July), 486–506.

Conroy, C., Villaume, A., van Dokkum, P. G. et al. 2018. Metal-Rich, Metal-Poor: Updated Stellar Population Models for Old Stellar Systems. *Astrophysical Journal*, **854**(Feb.), 139.

da Cunha, E., Charlot, S., and Elbaz, D. 2008. A Simple Model to Interpret the Ultraviolet, Optical and Infrared Emission from Galaxies. *Monthly Notices of the Royal Astronomical Society*, **388**(Aug.), 1595–1617.

da Silva, R. L., Fumagalli, M., and Krumholz, M. 2012. SLUG – Stochastically Lighting Up Galaxies. I. Methods and Validating Tests. *Astrophysical Journal*, **745**(Feb.), 145.

da Silva, R. L., Fumagalli, M., and Krumholz, M. R. 2014. SLUG - Stochastically Lighting Up Galaxies - II. Quantifying the Effects of Stochasticity on Star Formation Rate Indicators. *Monthly Notices of the Royal Astronomical Society*, **444**(Nov.), 3275–3287.

Davies, B., Kudritzki, R.-P., Plez, B. et al. 2013. The Temperatures of Red Supergiants. *Astrophysical Journal*, **767**(Apr.), 3.

De Marco, O., and Izzard, R. G. 2017. Dawes Review 6: The Impact of Companions on Stellar Evolution. *Publications of the Astronomical Society of Australia*, **34**(Jan.), e001.

de Mink, S. E., Langer, N., Izzard, R. G. et al. 2013. The Rotation Rates of Massive Stars: The Role of Binary Interaction through Tides, Mass Transfer, and Mergers. *Astrophysical Journal*, **764**(Feb.), 166.

Ekström, S., Georgy, C., Eggenberger, P. et al. 2012. Grids of Stellar Models with Rotation. I. Models from 0.8 to 120 $M_{\odot}$ at Solar Metallicity ($Z = 0.014$). *Astronomy and Astrophysics*, **537**(Jan.), A146.

Eldridge, J. J. 2012. Stochasticity, A Variable Stellar Upper Mass Limit, Binaries and Star Formation Rate Indicators. *Monthly Notices of the Royal Astronomical Society*, **422**(May), 794–803.

Eldridge, J. J., and Maund, J. R. 2016. The Disappearance of the Helium-Giant Progenitor of the Type Ib Supernova iPTF13bvn and Constraints on Its Companion. *Monthly Notices of the Royal Astronomical Society*, **461**(Sept.), L117–L121.

Eldridge, J. J., and Stanway, E. R. 2009. Spectral Population Synthesis Including Massive Binaries. *Monthly Notices of the Royal Astronomical Society*, **400**(Dec.), 1019–1028.

Eldridge, J. J., and Stanway, E. R. 2012. The Effect of Stellar Evolution Uncertainties on the Rest-Frame Ultraviolet Stellar Lines of C IV and He II in High-Redshift Lyman-Break Galaxies. *Monthly Notices of the Royal Astronomical Society*, **419**(Jan.), 479–489.

Eldridge, J. J., and Stanway, E. R. 2016. BPASS Predictions for Binary Black Hole Mergers. *Monthly Notices of the Royal Astronomical Society*, **462**(Nov.), 3302–3313.

Eldridge, J. J., Izzard, R. G., and Tout, C. A. 2008. The Effect of Massive Binaries on Stellar Populations and Supernova Progenitors. *Monthly Notices of the Royal Astronomical Society*, **384**(Mar.), 1109–1118.

Eldridge, J. J., Fraser, M., Smartt, S. J. et al. 2013. The Death of Massive Stars - II. Observational Constraints on the Progenitors of Type Ibc Supernovae. *Monthly Notices of the Royal Astronomical Society*, **436**(Nov.), 774–795.

Eldridge, J. J., Stanway, E. R., Xiao, L. et al. 2017. Binary Population and Spectral Synthesis Version 2.1: Construction, Observational Verification, and New Results. *Publications of the Astronomical Society of Australia*, **34**(Nov.), e058.

Ferland, G. J., Korista, K. T., Verner, D. A. et al. 1998. CLOUDY 90: Numerical Simulation of Plasmas and Their Spectra. *Publications of the Astronomical Society of the Pacific*, **110**(July), 761–778.

Fioc, M., and Rocca-Volmerange, B. 1997. PEGASE: A UV to NIR Spectral Evolution Model of Galaxies. Application to the Calibration of Bright Galaxy Counts. *Astronomy and Astrophysics*, **326**(Oct.), 950–962.

Gao, S., Zhao, H., Yang, H. et al. 2017. The Binarity of Galactic Dwarf Stars along with Effective Temperature and Metallicity. *Monthly Notices of the Royal Astronomical Society*, **469**(July), L68–L72.

Georgy, C. 2012. Yellow Supergiants as Supernova Progenitors: An Indication of Strong Mass Loss for Red Supergiants? *Astronomy and Astrophysics*, **538**(Feb.), L8.

Georgy, C., Ekström, S., Meynet, G. et al. 2012. Grids of Stellar Models with Rotation. II. WR Populations and Supernovae/GRB Progenitors at Z = 0.014. *Astronomy and Astrophysics*, **542**(June), A29.

Götberg, Y., de Mink, S. E., and Groh, J. H. 2017. Ionizing Spectra of Stars that Lose Their Envelope through Interaction with a Binary Companion: Role of Metallicity. *Astronomy and Astrophysics*, **608**(Nov.), A11.

Götberg, Y., de Mink, S. E., Groh, J. H. et al. 2018. Spectral Models for Binary Products: Unifying Subdwarfs and Wolf-Rayet Stars as a Sequence of Stripped-Envelope Stars. *Astronomy and Astrophysics*, **615**(Jul), A78.

Greis, S. M. L., Stanway, E. R., Levan, A. J. et al. 2017. Radio Observations Confirm Young Stellar Populations in Local Analogues to $z \sim 5$ Lyman Break Galaxies. *Monthly Notices of the Royal Astronomical Society*, **470**(Sept.), 489–499.

Grevesse, N., and Sauval, A. J. 1998. Standard Solar Composition. *Space Science Reviews*, **85**(May), 161–174.

Grevesse, N., Asplund, M., and Sauval, A. J. 2007. The Solar Chemical Composition. *Space Science Reviews*, **130**(June), 105–114.

Groh, J. H., Oliveira, A. S., and Steiner, J. E. 2008. The qWR Star HD 45166. II. Fundamental Stellar Parameters and Evidence of a Latitude-Dependent Wind. *Astronomy and Astrophysics*, **485**(July), 245–256.

Groh, J. H., Meynet, G., Ekström, S. et al. 2014. The Evolution of Massive Stars and Their Spectra. I. A Non-Rotating 60 $M_\odot$ Star from the Zero-Age Main Sequence to the Pre-Supernova Stage. *Astronomy and Astrophysics*, **564**(Apr.), A30.

Gustafsson, B., Edvardsson, B., Eriksson, K. et al. 2008. A Grid of MARCS Model Atmospheres for Late-Type Stars. I. Methods and General Properties. *Astronomy and Astrophysics*, **486**(Aug.), 951–970.

Gutkin, J., Charlot, S., and Bruzual, G. 2016. Modelling the Nebular Emission from Primeval to Present-Day Star-Forming Galaxies. *Monthly Notices of the Royal Astronomical Society*, **462**(Oct.), 1757–1774.

Hainich, R., Pasemann, D., Todt, H. et al. 2015. Wolf-Rayet Stars in the Small Magellanic Cloud. I. Analysis of the Single WN Stars. *Astronomy and Astrophysics*, **581**(Sept.), A21.

Heger, A., Langer, N., and Woosley, S. E. 2000. Presupernova Evolution of Rotating Massive Stars. I. Numerical Method and Evolution of the Internal Stellar Structure. *Astrophysical Journal*, **528**(Jan.), 368–396.

Heger, A., Woosley, S. E., and Spruit, H. C. 2005. Presupernova Evolution of Differentially Rotating Massive Stars Including Magnetic Fields. *Astrophysical Journal*, **626**(June), 350–363.

Henry, A., Berg, D. A., Scarlata, C. et al. 2018. A Close Relationship between Lyα and Mg II in Green Pea Galaxies. *Astrophysical Journal*, **855**(Mar.), 96.

Hermanowicz, M. T., Kennicutt, R. C., and Eldridge, J. J. 2013. Hα to FUV Ratios in Resolved Star-Forming Region Populations of Nearby Spiral Galaxies. *Monthly Notices of the Royal Astronomical Society*, **432**(July), 3097–3116.

Herwig, F. 2005. Evolution of Asymptotic Giant Branch Stars. *Annual Review of Astronomy and Astrophysics*, **43**(Sept.), 435–479.

Hillier, D. J. 2011. The Atomic Physics Underlying the Spectroscopic Analysis of Massive Stars and Supernovae. *Astrophysics and Space Science*, **336**(Nov.), 87–93.

Hirschi, R., Meynet, G., and Maeder, A. 2004. Stellar Evolution with Rotation. XII. Pre-Supernova Models. *Astronomy and Astrophysics*, **425**(Oct.), 649–670.

Hobbs, G., Lorimer, D. R., Lyne, A. G. et al. 2005. A Statistical Study of 233 Pulsar Proper Motions. *Monthly Notices of the Royal Astronomical Society*, **360**(July), 974–992.

Hunter, I., Brott, I., Lennon, D. J. et al. 2008. The VLT FLAMES Survey of Massive Stars: Rotation and Nitrogen Enrichment as the Key to Understanding Massive Star Evolution. *Astrophysical Journal*, **676**(Mar.), L29.

Hurley, J. R., Tout, C. A., and Pols, O. R. 2002. Evolution of Binary Stars and the Effect of Tides on Binary Populations. *Monthly Notices of the Royal Astronomical Society*, **329**(Feb.), 897–928.

Ivanova, N., Justham, S., Chen, X. et al. 2013. Common Envelope Evolution: Where We Stand and How We Can Move Forward. *Astronomy and Astrophysics Reviews*, **21**(Feb.), 59.

Izzard, R. G., Glebbeek, E., Stancliffe, R. J. et al. 2009. Population Synthesis of Binary Carbon-Enhanced Metal-Poor Stars. *Astronomy and Astrophysics*, **508**(Dec.), 1359–1374.

Kennicutt, R. C., and Evans, N. J. 2012. Star Formation in the Milky Way and Nearby Galaxies. *Annual Review of Astronomy and Astrophysics*, **50**(Sept.), 531–608.

Kim, H.-J., Yoon, S.-C., and Koo, B.-C. 2015. Observational Properties of Type Ib/c Supernova Progenitors in Binary Systems. *Astrophysical Journal*, **809**(Aug.), 131.

Kroupa, P. 2001. On the Variation of the Initial Mass Function. *Monthly Notices of the Royal Astronomical Society*, **322**(Apr.), 231–246.

Kroupa, P., Tout, C. A., and Gilmore, G. 1993. The Distribution of Low-Mass Stars in the Galactic Disc. *Monthly Notices of the Royal Astronomical Society*, **262**(June), 545–587.

Lagattuta, D. J., Mould, J. R., Forbes, D. A. et al. 2017. Evidence of a Bottom-Heavy Initial Mass Function in Massive Early-Type Galaxies from Near-Infrared Metal Lines. *Astrophysical Journal*, **846**(Sept.), 166.

Le Borgne, D., Rocca-Volmerange, B., Prugniel, P. et al. 2004. Evolutionary Synthesis of Galaxies at High Spectral Resolution with the Code PEGASE-HR. Metallicity and Age Tracers. *Astronomy and Astrophysics*, **425**(Oct.), 881–897.

Leitherer, C., Schaerer, D., Goldader, J. D. et al. 1999. Starburst99: Synthesis Models for Galaxies with Active Star Formation. *Astrophysical Journal Supplement Series*, **123**(July), 3–40.

Leitherer, C., Ekström, S., Meynet, G. et al. 2014. The Effects of Stellar Rotation. II. A Comprehensive Set of Starburst99 Models. *Astrophysical Journal Supplement Series*, **212**(May), 14.

Levesque, E. M., Massey, P., Olsen, K. A. G. et al. 2006. The Effective Temperatures and Physical Properties of Magellanic Cloud Red Supergiants: The Effects of Metallicity. *Astrophysical Journal*, **645**(July), 1102–1117.

Li, Z., and Han, Z. 2008. How Binary Interactions Affect Spectral Stellar Population Synthesis. *Astrophysical Journal*, **685**(Sept.), 225–234.

Maeder, A., and Meynet, G. 2001. Stellar Evolution with Rotation. VII. Low Metallicity Models and the Blue to Red Supergiant Ratio in the SMC. *Astronomy and Astrophysics*, **373**(July), 555–571.

Maeder, A., and Meynet, G. 2004. Stellar Evolution with Rotation and Magnetic Fields. II. General Equations for the Transport by Tayler-Spruit Dynamo. *Astronomy and Astrophysics*, **422**(July), 225–237.

Maraston, C. 2005. Evolutionary Population Synthesis: Models, Analysis of the Ingredients and Application to High-Z Galaxies. *Monthly Notices of the Royal Astronomical Society*, **362**(Sept.), 799–825.

McClelland, L. A. S., and Eldridge, J. J. 2016. Helium Stars: Towards an Understanding of Wolf-Rayet Evolution. *Monthly Notices of the Royal Astronomical Society*, **459**(June), 1505–1518.

Meynet, G., and Maeder, A. 1997. Stellar Evolution with Rotation. I. The Computational Method and the Inhibiting Effect of the μ-Gradient. *Astronomy and Astrophysics*, **321**(May), 465–476.

Meynet, G., and Maeder, A. 2000. Stellar Evolution with Rotation. V. Changes in All the Outputs of Massive Star Models. *Astronomy and Astrophysics*, **361**(Sept.), 101–120.

Meynet, G., and Maeder, A. 2005. Stellar Evolution with Rotation. XI. Wolf-Rayet Star Populations at Different Metallicities. *Astronomy and Astrophysics*, **429**(Jan.), 581–598.

Moe, M., and Di Stefano, R. 2017. Mind Your Ps and Qs: The Interrelation between Period (P) and Mass-Ratio (Q) Distributions of Binary Stars. *Astrophysical Journal Supplement Series*, **230**(June), 15.

Nieva, M.-F., and Przybilla, N. 2012. Present-Day Cosmic Abundances: A Comprehensive Study of Nearby Early B-Type Stars and Implications for Stellar and Galactic Evolution and Interstellar Dust Models. *Astronomy and Astrophysics*, **539**(Mar.), A143.

Podsiadlowski, P., Joss, P. C., and Hsu, J. J. L. 1992. Presupernova Evolution in Massive Interacting Binaries. *Astrophysical Journal*, **391**(May), 246–264.

Potter, A. T., Chitre, S. M., and Tout, C. A. 2012. Stellar Evolution of Massive Stars with a Radiative α-Ω Dynamo. *Monthly Notices of the Royal Astronomical Society*, **424**(Aug.), 2358–2370.

Rix, S. A., Pettini, M., Leitherer, C. et al. 2004. Spectral Modeling of Star-Forming Regions in the Ultraviolet: Stellar Metallicity Diagnostics for High-Redshift Galaxies. *Astrophysical Journal*, **615**(Nov.), 98–117.

Rosenfield, P., Marigo, P., Girardi, L. et al. 2014. Evolution of Thermally Pulsing Asymptotic Giant Branch Stars. IV. Constraining Mass loss and Lifetimes of Low Mass, Low Metallicity AGB Stars. *Astrophysical Journal*, **790**(July), 22.

Sana, H., de Mink, S. E., de Koter, A. et al. 2012. Binary Interaction Dominates the Evolution of Massive Stars. *Science*, **337**(July), 444.

Sander, A., Shenar, T., Hainich, R. et al. 2015. On the Consistent Treatment of the Quasi-Hydrostatic Layers in Hot Star Atmospheres. *Astronomy and Astrophysics*, **577**(May), A13.

Schneider, F. R. N., Ramírez-Agudelo, O. H., Tramper, F. et al. 2018. The VLT-FLAMES Tarantula Survey. XXIX. Massive Star Formation in the Local 30 Doradus Starburst. *Astronomy and Astrophysics*, **618**(Oct), A73.

Schröder, K.-P., and Cuntz, M. 2007. A Critical Test of Empirical Mass Loss Formulas Applied to Individual Giants and Supergiants. *Astronomy and Astrophysics*, **465**(Apr.), 593–601.

Simón-Díaz, S., and Stasińska, G. 2008. The Ionizing Radiation from Massive Stars and Its Impact on HII Regions: Results from Modern Model Atmospheres. *Monthly Notices of the Royal Astronomical Society*, **389**(Sept.), 1009–1021.

Stanway, E. R., and Eldridge, J. J. 2018. Re-evaluating Old Stellar Populations. *Monthly Notices of the Royal Astronomical Society*, **479**(Sept.), 75–93.

Stanway, E. R., Eldridge, J. J., and Becker, G. D. 2016. Stellar Population Effects on the Inferred Photon Density at Reionization. *Monthly Notices of the Royal Astronomical Society*, **456**(Feb.), 485–499.

Steidel, C. C., Strom, A. L., Pettini, M. et al. 2016. Reconciling the Stellar and Nebular Spectra of High-Redshift Galaxies. *Astrophysical Journal*, **826**(Aug.), 159.

Tabatabaei, F. S., Schinnerer, E., Krause, M. et al. 2017. The Radio Spectral Energy Distribution and Star-Formation Rate Calibration in Galaxies. *Astrophysical Journal*, **836**(Feb.), 185.

Tauris, T. M., Kramer, M., Freire, P. C. C. et al. 2017. Formation of Double Neutron Star Systems. *Astrophysical Journal*, **846**(Sept.), 170.

Topping, M. W., and Shull, J. M. 2015. The Efficiency of Stellar Reionization: Effects of Rotation, Metallicity, and Initial Mass Function. *Astrophysical Journal*, **800**(Feb.), 97.

Tremblay, P.-E., Gianninas, A., Kilic, M. et al. 2015a. 3D Model Atmospheres for Extremely Low-Mass White Dwarfs. *Astrophysical Journal*, **809**(Aug.), 148.

Tremblay, P.-E., Ludwig, H.-G., Freytag, B. et al. 2015b. Calibration of the Mixing-Length Theory for Convective White Dwarf Envelopes. *Astrophysical Journal*, **799**(Feb.), 142.

Tremblay, P.-E., Ludwig, H.-G., Steffen, M. et al. 2011. Solution to the Problem of the Surface Gravity Distribution of Cool DA White Dwarfs from Improved 3D Model Atmospheres. *Astronomy and Astrophysics*, **531**(July), L19.

Van Bever, J., and Vanbeveren, D. 2003. The Effects of Binaries on the Evolution of Wolf-Rayet Type Spectral Features in Starbursts. *Astronomy and Astrophysics*, **400**(Mar.), 63–79.

Van Dyk, S. D., Zheng, W., Brink, T. G. et al. 2018. SN 2017ein and the Possible First Identification of a Type Ic Supernova Progenitor. *Astrophysical Journal*, **860**(June), 90.

Vanbeveren, D., Van Bever, J., and Belkus, H. 2007. The Wolf-Rayet Population Predicted by Massive Single Star and Massive Binary Evolution. *Astrophysical Journal*, **662**(June), L107–L110.

Vanbeveren, D., Mennekens, N., Shara, M. M. et al. 2018. Spin Rates and Spin Evolution of O Components in WR+O Binaries. *Astronomy and Astrophysics*, **615**(July), A65.

Vaughan, S. P., Davies, R. L., Zieleniewski, S. et al. 2018. The Stellar Population and Initial Mass Function of NGC 1399 with MUSE. *Monthly Notices of the Royal Astronomical Society*, **479**(Sept.), 2443–2456.

Vazdekis, A., Koleva, M., Ricciardelli, E. et al. 2016. UV-Extended E-MILES Stellar Population Models: Young Components in Massive Early-Type Galaxies. *Monthly Notices of the Royal Astronomical Society*, **463**(Dec.), 3409–3436.

Vink, J. S., de Koter, A., and Lamers, H. J. G. L. M. 2001. Mass-Loss Predictions for O and B Stars as a Function of Metallicity. *Astronomy and Astrophysics*, **369**(Apr.), 574–588.

Wofford, A., Charlot, S., Bruzual, G. et al. 2016. A Comprehensive Comparative Test of Seven Widely Used Spectral Synthesis Models against Multi-Band Photometry of Young Massive-Star Clusters. *Monthly Notices of the Royal Astronomical Society*, **457**(Apr.), 4296–4322.

Xiao, L., Stanway, E. R., and Eldridge, J. J. 2018. Emission-Line Diagnostics of Nearby H II Regions Including Interacting Binary Populations. *Monthly Notices of the Royal Astronomical Society*, **477**(June), 904–934.

Xiao, L., Galbany, L., Eldridge, J. J. et al. 2019. Core-Collapse Supernovae Ages and Metallicities from Emission-Line Diagnostics of Nearby Stellar Populations. *Monthly Notices of the Royal Astronomical Society*, **482**(1), 384–401.

Yoon, S.-C. 2017. Towards a Better Understanding of the Evolution of Wolf-Rayet Stars and Type Ib/Ic Supernova Progenitors. *Monthly Notices of the Royal Astronomical Society*, **470**(Oct.), 3970–3980.

Yoon, S.-C., Gräfener, G., Vink, J. S. et al. 2012. On the Nature and Detectability of Type Ib/c Supernova Progenitors. *Astronomy and Astrophysics*, **544**(Aug.), L11.

Zapartas, E., de Mink, S. E., Izzard, R. G. et al. 2017. Delay-Time Distribution of Core-Collapse Supernovae with Late Events Resulting from Binary Interaction. *Astronomy and Astrophysics*, **601**(May), A29.

Zhang, F., Li, L., and Han, Z. 2005. Evolutionary Population Synthesis for Binary Stellar Population at High Spectral Resolution: Integrated Spectral Energy Distributions and Absorption-Feature Indices. *Monthly Notices of the Royal Astronomical Society*, **364**(Dec.), 503–514.

Zhang, F., Li, L., Cheng, L. et al. 2015. Radiation Fields of Intermediate-Age Stellar Populations with Binaries as Ionizing Sources of H II Regions. *Monthly Notices of the Royal Astronomical Society*, **447**(Feb.), L21–L25.

4

Dust Extinction, Attenuation, and Emission

KARL D. GORDON

4.1 Introduction

Dust impacts observations of stars and gas in galaxies by absorbing and scattering photons and generally re-emitting the absorbed photons in the infrared. Correctly accounting for the effects of dust allows for more accurate studies of a galaxy's stars and gas while also enabling the study of the dust grains themselves.

The impact of dust on measurements of individual stars in a galaxy can be straightforwardly modeled as extinguishing the stellar light. Dust extinction towards a star is defined as the combined effect of absorption of photons and scattering of photons out of the line-of-sight towards the star.

For integrated measurements of regions of galaxies, or whole galaxies that contain multiple stars intermixed with dust, the effects of dust are termed attenuation and are harder to model. Integrated measurements include two effects that are not present for individual stars, namely: stars extinguished with different amounts of dust *and* scattering of photons into the measurement aperture. The infrared dust emission powered by the absorbed photons provides a vital measurement of the amount of energy absorbed by dust. This infrared measurement is not possible for individual stars and provides an important constraint in modeling the effects of dust on integrated measurements.

The aim of this chapter is to provide the details of dust extinction, attenuation, and emission and recommendations for how to model the effects of dust on observations.

4.2 Extinction

The best known measurement of the impact of foreground dust on the observations of a background source is extinction. Extinction is also commonly referred to as reddening since extinction usually makes objects with foreground dust have colors redder than their intrinsic colors. Extinction formally refers to the combination of two processes: absorption of photons and scattering of photons out of the line-of-sight. Measuring extinction is straightforward for individual stars. Measurements towards two effectively identical stars are made where one star has significant foreground dust and the other has little or no foreground dust. The ratio of the reddened (extinguished) star's flux to the unreddened (comparison) star's flux measures the dust extinction. Finding such effectively identical stars is straightforward, they just need to have the same spectral types and abundances. This ensures that the two stars have the same surface physics making their spectra identical except for the effects of dust extinction. The geometry of the system with the foreground dust distributed along the line-of-sight to the star and the diameter of the star being much, much smaller than the distance to the star means that the non-zero diameter of the star can be neglected and the amount of photons scattered into the line-of-sight to the star is negligible.

Dust extinction is usually measured relative to the V band and presented in magnitude units. Explicitly, the color excess

$$E(\lambda - V) = m(\lambda - V)_r - m(\lambda - V)_c \tag{4.1}$$

where $m(\lambda - V)$ is the difference in magnitudes between the flux at λ and V band, r refers to the reddened star, and c refers to the comparison star. Often, the extinction curve is normalized to $E(B - V)$ to allow curves taken through different dust columns to be compared. The extinction is measured relative to a reference wavelength (e.g. V band) as this removes the different distances to the two stars. In order to convert dust-extinction curves from color-excess measurements to a more useful direct measurement at a single wavelength, the total-to-selective extinction in the V band, $R(V) = A(V)/E(B - V)$, is needed. Specifically the conversion is

$$\frac{A(\lambda)}{A(V)} = \frac{E(\lambda - V)}{E(B - V)} \frac{1}{R(V)} + 1 \tag{4.2}$$

where $A(\lambda)$ is the extinction at λ. Measuring $R(V)$ is done by measuring $E(\lambda - V)$ at the longest wavelength possible and extrapolating this measurement to infinite wavelength using a model of the extinction shape. In practice, the longest wavelength is often the K band and the extrapolation is on the order of 10% of the $E(V - K)/E(B - V)$ value (Whittet et al., 1976; Fitzpatrick and Massa, 2009). The effect of dust extinction on a star's flux is then

$$F_{\text{red}}(\lambda) = e^{-\tau(\lambda)} F_{\text{intrinsic}} \tag{4.3}$$

where

$$\tau(\lambda) = 0.921 A(\lambda). \tag{4.4}$$

Extinction curves have been measured towards stars in our Galaxy and other galaxies in the Local Group. Measurements outside of the Local Group are challenging as accurate extinction measurements require individual stars to be resolved from their neighbors. In general, extinction-curve measurements have focused on the ultraviolet (UV) wavelength range as this is where the most diagnostic power exists in understanding dust grains. The ultraviolet is where the 2175 Å extinction bump is located, one of the most prominent dust-extinction features. The focus on the UV range by dust researchers is fortuitous for star-formation research as the UV provides one of the direct measures of massive star formation.

Many extinction curve measurements have been made in the Milky Way with the largest studies made by Valencic et al. (2004) and Fitzpatrick and Massa (2007). In total, extinction along around 450 sightlines has been measured in the Milky Way, providing a wealth of information. The majority of the shape variation seen between the Milky Way curves can be described by a family of curves dependent on only a single parameter, usually chosen to be $R(V)$. This result was first seen by Cardelli et al. (1989) and has been confirmed by many later works (Valencic et al., 2004; Fitzpatrick and Massa, 2007; Gordon et al., 2009). The $R(V)$-dependent extinction-curve relationship does not encompass the full shape variation seen along different lines-of-sight (Mathis and Cardelli, 1992; Valencic et al., 2004; Fitzpatrick and Massa, 2007), instead it should be seen as describing the *average* behavior as a function of $R(V)$.

Measurements of UV extinction curves in the Large and Small Magellanic Clouds (LMC and SMC) show both curves that look like those seen in the Milky Way as well as curves that are quite different. Measurements of UV extinction curves beyond the Magellanic Clouds is currently in progress with the first results in M31 showing curves similar to those seen in the Milky Way and LMC (Clayton et al., 2015). That LMC and SMC extinction curves have shapes not generally seen in the Milky Way has been known for many years (e.g. Prevot et al., 1984; Clayton and Martin, 1985). The quantitative disagreement of many of the LMC and SMC curves with the $R(V)$ dependent relationship was shown by Gordon et al. (2003). Recently, Gordon et al. (2016) has proposed a mixture model dependent on two variables to describe the average behavior of extinction curves in the Milky Way and Magellanic Clouds. This new model mixes the $R(V)$-dependent relationship derived from Milky Way measurements with the SMC Bar (Gordon et al., 2003) average extinction curve where the mixture coefficient is $f_{\mathcal{A}}$. The behavior of this mixture model is shown in Fig. 4.1. There are other formulations of the extinction-curve shapes that use more than two parameters (e.g. Fitzpatrick and Massa, 1990; Pei, 1992) where these parametrizations are most useful in describing individual observed extinction curves.

The detailed shape of UV extinction curves in the Local Group has been well characterized through extensive spectral observations with IUE[1] and *HST*[2]/STIS.[3] This is not the case for the optical, near-infrared, and mid-infrared wavelength ranges where the majority of extinction curves are based on photometric measurements. One rare case of spectroscopic based optical extinction curves is that by Cardelli and Clayton (1988) which focused on the Orion nebula. The mid-infrared extinction has been spectroscopically measured for fairly high dust columns showing the 3.4 μm hydrocarbon and longer-wavelength silicate absorption features (Pendleton and Allamandola, 2002; Chiar and Tielens, 2006).

[1] *International Ultraviolet Explorer*; http://archive.stsci.edu/iue/
[2] *Hubble Space Telescope*; www.stsci.edu/hst
[3] Space Telescope Imaging Spectrograph; www.stsci.edu/hst/stis

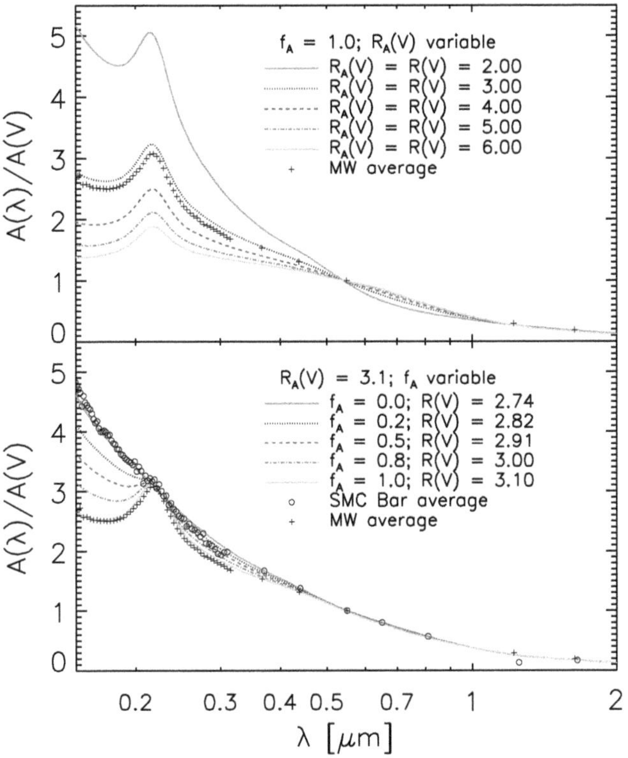

Figure 4.1 Figure 3 from Gordon et al. (2016) shows the $R(V)$ and $f_{\mathcal{A}}$ dependent extinction curve model along with observed average extinction curves for the Milky Way (Gordon et al., 2009) and SMC Bar (Gordon et al., 2003). $R(V)_{\mathcal{A}}$ is the $R(V)$ value of the A component. ©AAS. Reproduced with permission. Color version available online.

Unfortunately, it is currently not possible to measure the ultraviolet extinction for such high dust column sightlines. This limits our understanding of the detailed shape of dust extinction to quite broad wavelength bins at some wavelengths or requires combining measurement from different environments. There is ongoing work to derive and characterize optical to mid-infrared extinction curves spectroscopically for common sightlines using ground based, *HST*/STIS (Fitzpatrick et al., in prep.), and *Spitzer*[4]/IRS[5] (Gordon et al., in prep) observations. This gives hope that in the near future our understanding of dust extinction be at the spectroscopic level from the UV to the mid-IR.

[4] *Spitzer Space Telescope*; www.spitzer.caltech.edu
[5] Infrared Spectrograph; https://irsa.ipac.caltech.edu/data/SPITZER/docs/irs/

4.3 Attenuation

Attenuation refers to the total effect of dust absorption and scattering on the emitted Spectral Energy Distribution (SED) of a system. An attenuation curve describes the effect of adding dust to a system and can be quantified as the ratio of the dust-affected spectrum divided by the dust-free spectrum. By this definition, extinction curves are attenuation curves. But the converse is not true, not all attenuation curves are extinction curves. An extinction measurement is restricted to the specific geometry of dust in the foreground of a star and includes dust absorption and scattering out of the line-of-sight. The main advantage of extinction curves is that their normalized shapes are invariant to the amount of dust along the line-of-sight. This is not the case for any other dust and star geometry. The impact of dust in more complex geometries results in the attenuation-curve shape varying with the amount of dust in the system (e.g. Witt et al., 1992; Witt and Gordon, 2000). This can be due to having multiple stars attenuated by different amounts of dust or scattering into the line-of-sight or both. Another way of stating this is that extinction curves are directly related the dust-grain properties alone (size, shape, composition, alignment) while attenuation curves include the dust and star geometry as well.

Measuring attenuation curves for complex systems is non-trivial. While for individual stars it is possible to observe two identical stars, one with dust and one dust-free, this is not the case for most complex systems. For example, a region of a galaxy has a unique star-formation and dynamical history leading to an unique star and dust geometry. It is unlikely that an identical region in the same galaxy or another galaxy can be found that does *not* have dust. Thus, measuring an attenuation curve for an individual region or galaxy using the pair method used to measure extinction curves is not possible. It is possible to measure average attenuation curves for samples of galaxies. The starburst attenuation curve (Calzetti et al., 1994; Calzetti, 1997) is a good example of such an average attenuation curve. This attenuation curve describes the average behavior for UV bright starburst galaxies and is appropriate for modeling or correcting for the effects of dust attenuation of similar galaxies *on average*. It is not theoretically correct to use this curve to correct an individual galaxy as that galaxy may not have the average properties of the original sample.

Theoretically predicting attenuation curves for complex systems is non-trivial. Radiative transfer through dust involves solving a six dimensional integro-differential equation where each location in the model is coupled through dust scattering and emission with every other location (Steinacker et al., 2013; see also Chapter 9). This complexity results in attenuation curves with complex and non-linear behaviors that depend on the amount of dust and geometry of the stars and dust. In principle, the attenuation curve of a mixed system of stars and dust is unique to that system.

The complex behavior of attenuation curves in a very simple system is illustrated in Fig. 4.2. This system consists of a slab of uniform density dust with two identical stars embedded at different depths in the dust slab. The resulting attenuation curves for this

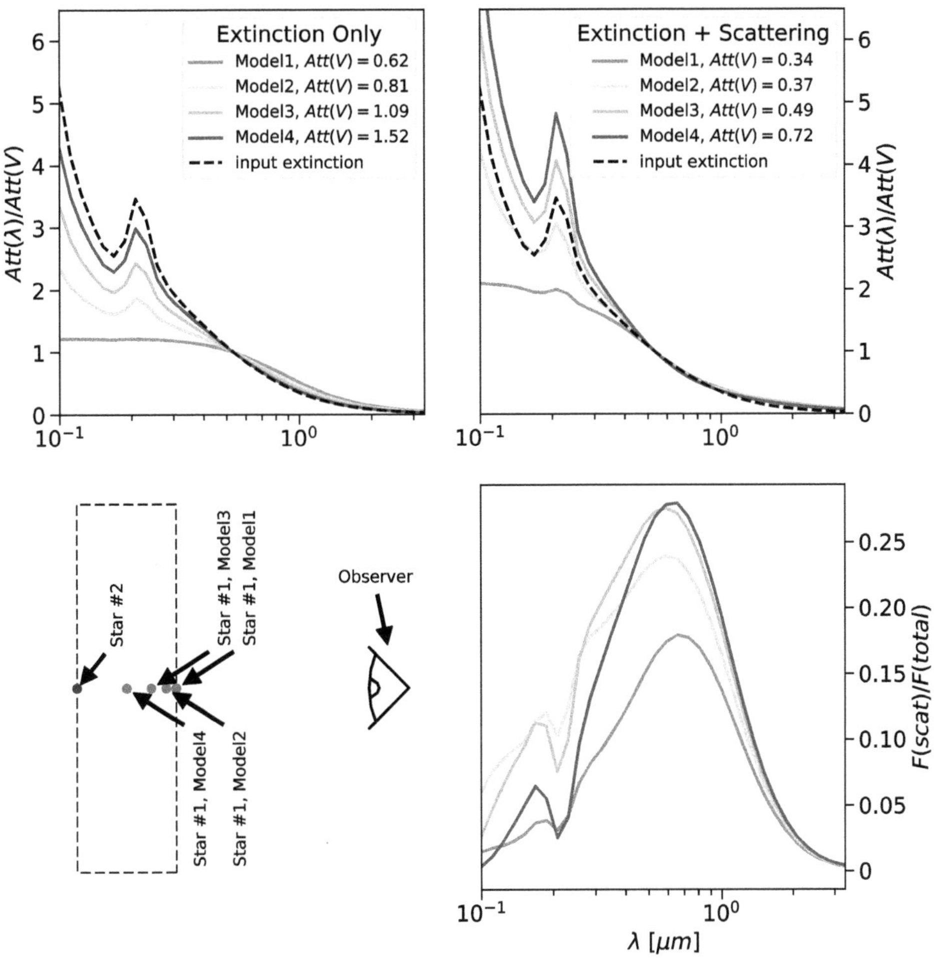

Figure 4.2 Example attenuation curves from a simple two star mixture model are shown. A visualization of this system is given in the lower left. The model has a uniform density slab of dust and two stars both with the same intrinsic luminosity and spectral shape. The total amount of dust is given by $A(V) = 2$ that is measured along the line-of-sight from the observer to the 2nd star. The amount of dust in front of the 1st star varies with values of 0, 0.2, 0.5, and 1, while the 2nd star is behind the full dust column. The dust has the properties of the Milky Way average grains from Weingartner and Draine (2001) with the input extinction curve shown. The "Extinction Only" plots gives the solution for extinction alone (absorption and scattering out of the observer line-of-sight. The "Extinction+Scattering" plot gives the full solution including scattering into the observer line-of-sight. The lower-right plots gives the scattered light fraction (into the observer line-of-sight) for the different models. Color version available online.

system were computed using the DIRTY[6] model (Gordon et al., 2001; Misselt et al., 2001; Law et al., 2018) with a variant of the TRUST[7] Slab benchmark geometry (Gordon et al., 2017).

The first complex geometry effect is illustrated in the "Extinction Only" plot in Fig. 4.2. Model1 shows the simplest case where the 1st star has no dust in front of it. In this case, the attenuation curve is almost flat for the UV and most of the optical as the 1st star completely dominates the integrated flux and $Att(V) = 0.62$ mag as almost 1/2 of the flux is missing at V band. The attenuation decreases starting around 1 μm as the flux from the 2nd star rises as the optical depth of the system becomes increasingly transparent at these wavelengths. The three models show the effect of increasing the amount of dust in front of the 1st star by embedding it deeper in the slab. These attenuation curves become steeper, approaching the input extinction curve. All the attenuation curves are shallower (grayer) than the input extinction curve.

The second effect is illustrated in the "Extinction+Scattering" plot in Fig. 4.2. The attenuation curves in this plot include the first effect as well as the impact of the flux scattered by the dust into the observer's line-of-sight. The fraction of the total flux due to light scattered into the measurement beam is shown in the lower-right plot. The overall amount of dust attenuation is reduced compared to the "Extinction Only" case as the scattered light adds to the system flux counteracting the direct extinction. The attenuation curve for Model1 is much less flat for the UV and optical and has a clear slope with higher attenuation at shorter wavelength. The attenuation curves for the other three models show a similar behavior with all being steeper than their equivalent "Extinction Only" curves. Note that both Model3 and Model4 show the cases where the normalized attenuation curve is steeper than the input extinction curve due to multiple scatterings amplifying the absorption of photons. The scattered flux can be a substantial component of the total flux and peaks at over 25% for this simple example.

The behavior of the scattered light in a complex system is non-linear with the amount of dust as the peak dust-scattered flux is seen at optical depths of $\sim$1. From zero to $\sim$1 optical depth, the scattered-light flux scales linearly due to the addition of more scattering sites. Above $\sim$1 optical depth, the scattered-light flux drops with increasing dust column due to multiple scatterings providing more opportunities for the photons to be absorbed instead of scattered. Thus, the impact of scattered photons in a complex system peaks at wavelengths where the optical depth is $\sim$1 and is lower at longer and shorter wavelengths.

Full radiative-transfer calculations can and have been performed using star/dust geometries approximating regions of galaxies and entire galaxies. Witt and Gordon (2000) calculated attenuation curves for the galactic environments first introduced by Witt et al. (1992) for dust distributions that are locally clumpy. One panel from fig. 9 of Witt and Gordon (2000) is shown in Fig. 4.3 showing the comparison between some of the radiative-transfer attenuation curves and the observationally derived Calzetti attenuation curve

[6] DustI Radiative Transfer, Yeah!; https://dirty-dustrt.readthedocs.io/en/latest/
[7] Transport of Radiation through a DUSTy Medium; https://ipag.osug.fr/RT13//RTTRUST/

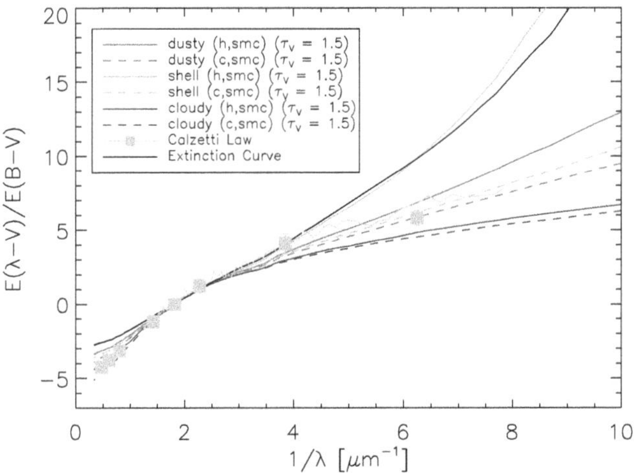

Figure 4.3 Figure 9 from Witt and Gordon (2000) shows the radiative transfer calculated attenuation curves for the three galactic environments (dusty, shell, cloudy) and local dust distributions (h = homogeneous, c = clumpy). All the curves were calculated with SMC-type dust and the input extinction curve is shown. The Calzetti law is plotted for direct comparison to the radiative-transfer calculations. © AAS. Reproduced with permission. Color version available online.

(Calzetti, 1997). This figure illustrates the dependence on the assumed global star/dust geometry and local dust geometry (homogeneous versus clumpy). In addition, this plot illustrates that the Calzetti curve is fit well by a clumpy shell model with "SMC" type dust grains and a radial V-band optical depth of ~1.5. Attenuation curves for similar models with lower V-band optical depths are steeper than the Calzetti curve and models with higher V-band optical depths are shallower than the Calzetti curve (Witt and Gordon, 2000). This indicates that using the Calzetti curve for individual galaxies should work if the galaxies have intrinsic V-band optical depths ~1.5 and are similar to those used to derive the Calzetti curve. Using the Calzetti curve for galaxies with lower or higher intrinsic V band optical depths will introduce systematic biases.

In general, attenuation curves computed from radiative-transfer models become grayer as the amount of dust in a system is increased (Chevallard et al., 2013). This has recently been empirically shown by Salmon et al. (2016). The top panel in Fig. 4.4 illustrates the graying of attenuation curves from the Witt and Gordon (2000) clumpy SMC shell model with increasing amounts of dust. The bottom panel in Fig. 4.4 compares measurements of the slope of the attenuation curve empirically measured for galaxies with redshifts of 1.5 to 3 and the Witt and Gordon (2000) models. The empirically measured curves were fit using the Noll et al. (2009) model where the overall slope is parameterized by δ, where lower values are for steeper curves. The Calzetti (1997) curve has $\delta = 0$. The overlap is quite good, especially considering that the Noll et al. (2009) model does not work well for the very steep attenuation curves for those models with very low amounts of dust (Witt and Gordon, 2000).

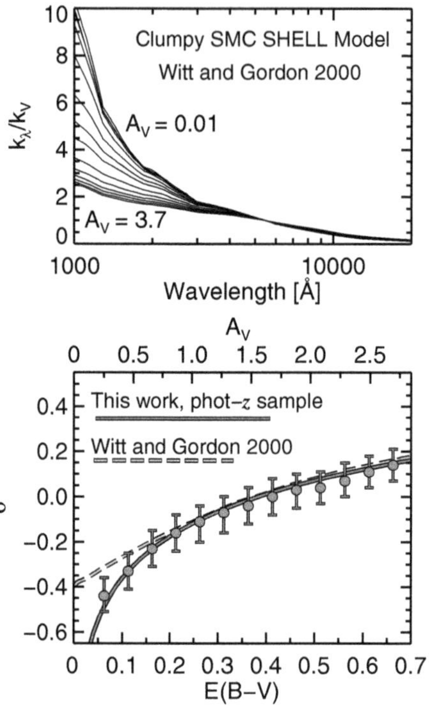

Figure 4.4 Figure 13 from Salmon et al. (2016) compares the empirical measurements of CANDELS[8] $z \sim 1.5$–3 galaxies with theoretical predictions from Witt and Gordon (2000). The UV slope of the dust attenuation (δ) closely follows the theoretical predictions. The disagreement at low $E(B - V)$ values is due to the theoretical curves not being well described by the empirical attenuation model of Noll et al. (2009). Note that $k_\lambda = A(\lambda)$. © AAS. Reproduced with permission. Color version available online.

4.4 Emission

The energy that dust grains absorb is generally thermalized and re-emitted in the IR. Observations of this emission provide a straightforward measurement of the amount of energy absorbed by dust. Combining observations in the UV/optical with observations of the total IR dust emission provides a way to correct the UV/optical observations for dust attenuation (Gordon et al., 2000). Additionally, combining observations in the UV or Hα with total IR or 24 μm provides a composite tracer of star formation that directly accounts for the unattenuated and attenuated star formation, respectively (Calzetti et al., 2007; Kennicutt et al., 2009).

Going beyond just using the dust emission as a probe of the total energy absorbed by dust requires modeling the emission spectrum for dust-grains of different sizes.

[8] Cosmic Assembly Near-infrared Deep Extragalactic Legacy Survey

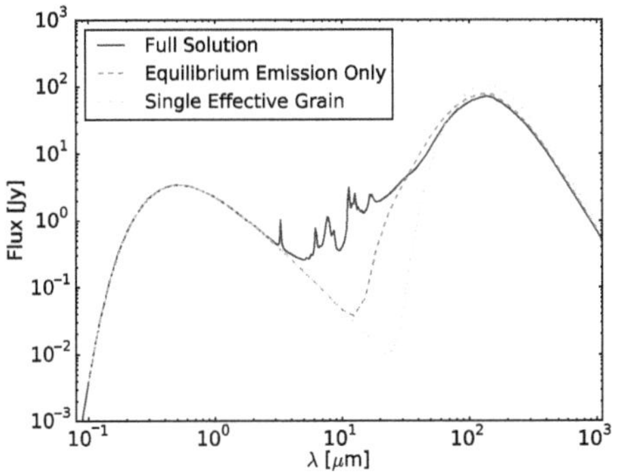

Figure 4.5 Output SEDs from the DIRTY radiative transfer model for a simple system with a hot star illuminating an external slab of dust. The full solution includes non-equilibrium emission by small grains. The equilibrium emission only case gives the approximation assuming all the grains are in equilibrium. The single effective grain SED gives the case where the dust emission is done from a single effective grain and not the full grain size distribution. Credit: Gordon et al. (2017), reproduced with permission © ESO. Color version available online.

Interstellar-dust grains have a range of sizes ranging from nanometers to microns. In general, emission from dust grains is modeled as blackbodies modified by dust-emissivity laws. The total emission is then the sum of the emission from the different grain sizes and compositions. Large dust grains (e.g. 0.1 μm radius) are generally in thermal equilibrium with the radiation field and so their thermal emission is characterized by a modified blackbody with a single temperature. Small dust grains are generally not in equilibrium with the radiation field as the time between absorbing photons can be large enough to allow the grain time to cool (Guhathakurta and Draine, 1989). Thus, the emission from these small grains is characterized by the average emission from modified blackbodies with a range of temperatures. Calculating the emission from grains in equilibrium is significantly easier than from those in non-equilibrium and assuming that all grains are in equilibrium is a common approximation. Such an approximation can lead to significant errors especially in the mid-IR as is illustrated in Fig. 4.5.

Self consistently modeling the dust emission from stellar populations for a large range of stellar populations is computationally challenging. Recently, Law et al. (2018) have done just such calculations to create the DirtyGrid[9] based on the star and dust geometries from Witt and Gordon (2000). What makes the DirtyGrid important is that it is the first time that galactic SED templates have been generated with the coupling between the dust absorption and emission has been dictated solely by the radiative-transfer physics. Thus, not only is energy conserved, but the shape and strength of the radiation field from the stellar population directly effects the shape and strength of the dust IR emission. Figure 4.6 illustrates some of the possible variations based on the stellar and dust parameters. Of particular

[9] https://zenodo.org/record/1214271#.Xmqe9S2B2g8

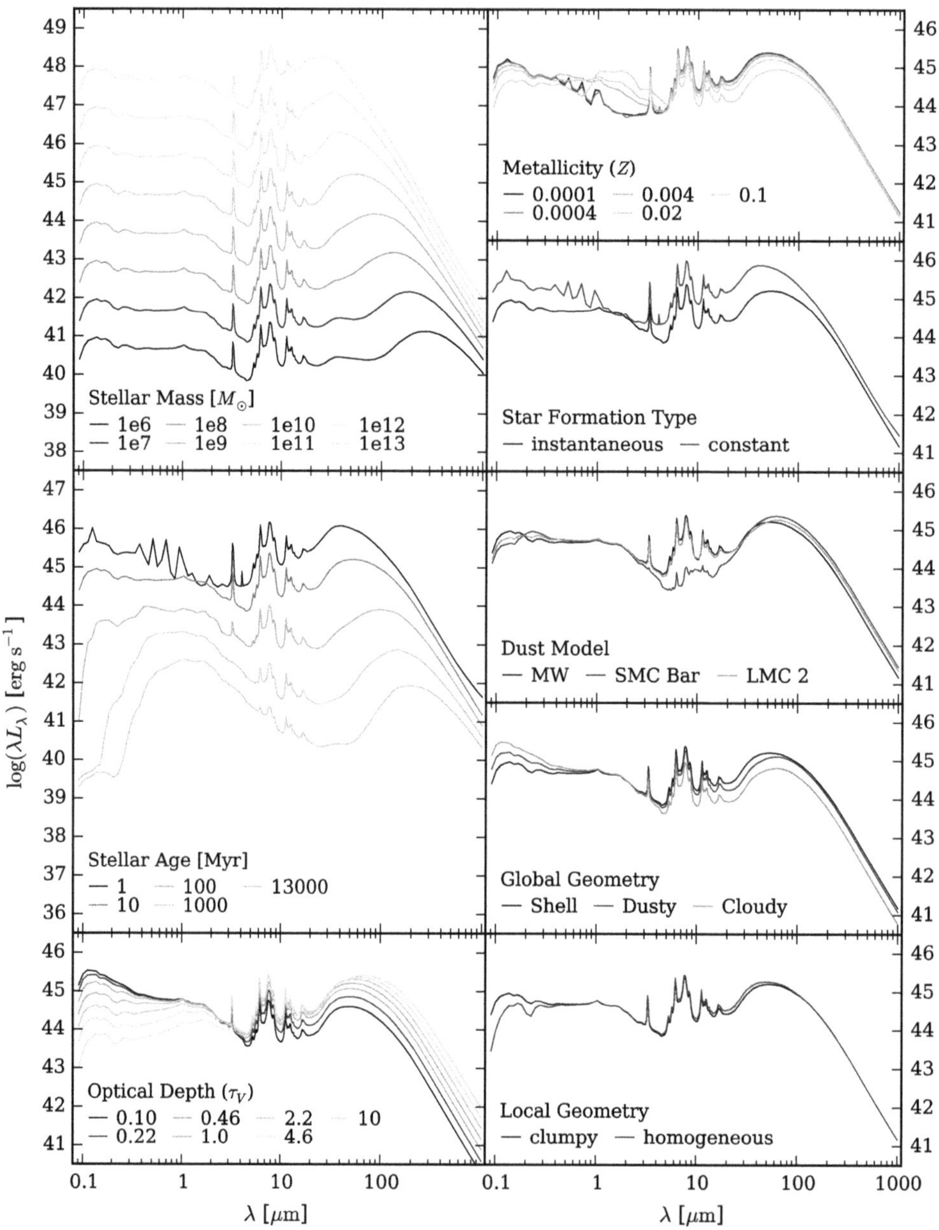

Figure 4.6 Figure 9 from Law et al. (2018) illustrates the variation in the full UV to far-IR SEDs in the DirtyGrid. For each plot, one parameter is varied and the rest fixed to age = 10 Myr, mass = 10^{10} solar masses, metallicity = solar, star-formation type = instantaneous, $\tau(V) = 1$, dust-type = MW type, geometry = SHELL, radius = 10 kpc, and local geometry = clumpy. © AAS. Reproduced with permission. Color version available online.

note are the dust-emission spectrum changes in shape as well as intensity as the different parameters are varied. For example, as seen in the upper left panel, the peak of the IR dust emission shifts to longer wavelengths as the stellar age increases. DirtyGrid models are available in the open-source `dirtygrid`[10] python package.

4.5 Recommendations – Resolved Stellar Populations

For resolved stellar populations, it is relatively straightforward to account for the effects of dust as the dust extinction towards each star can be modeled separately. The best solution is to fit the UV to NIR SED of each star simultaneously for the stellar and dust-extinction properties. This can be done star by star (Maíz-Apellániz, 2004; Da Rio and Robberto, 2012; Gordon et al., 2016) or for ensembles of stars (Dolphin, 2002). In the best case, the observations cover a broad wavelength range allowing for the dust properties to be determined separately from the stellar properties (Gordon et al., 2016). Otherwise, assumptions about the stellar and/or dust properties need to be made. An often used dust assumption is that the dust-extinction curve shape is set by the average found in the Milky Way or nearby galaxies. For galaxies like our own, the Milky Way average curve is a good assumption. For galaxies with low metallicities or high star-formation rates, the SMC Bar extinction may be a better. In this case, the type of extinction curve assumed should be varied to understand the impact on the derived properties.

Dust-extinction models are available in the open-source "`dust_extinction`"[11] python package.

4.6 Recommendations – Integrated Observations

For integrated measurements of regions of or entire galaxies, the needed modeling is necessarily more complex given the mixing of the stars and dust. The recommendation of how to account for the effects of dust depends both on the observations and the complexity of modeling appropriate in the context of the study.

Ultraviolet to Far-IR

Ideally, observations from the UV to the far-IR are available. The full ultraviolet (UV) to far-IR SED encodes significant information about the stellar populations, amount and kind of dust, and the stellar and dust geometry. Combining such observations with dust radiative-transfer models allows for the full information inherent in the observations to be extracted. In addition to the SED, there is significant insight to be gained from analyzing spatial variations in the SED across a galaxy. Using the full wavelength and spatial information in particular can provide additional constraints on the star and dust geometry. Disk galaxies are often targeted for such modeling (e.g. Bianchi and Xilouris, 2011; Viaene et al., 2017).

[10] https://github.com/karllark/dirtygrid
[11] https://github.com/karllark/dust_extinction

Radiative-transfer models encompassing all types of galaxies generally assume simpler geometries (Witt et al., 1992; Witt and Gordon, 2000; Seon and Draine, 2016; Law et al., 2018). Moving to the next level of approximation, fitting can be done using models that assume an attenuation curve and only use the far-IR information as a constraint on the total energy absorbed by dust (e.g. Gordon et al., 2000; Noll et al., 2009).

Ultraviolet to Near-IR

With only UV to NIR data (or some subset), constraining models becomes more difficult. The degeneracies between the effects of varying the stellar populations and the dust attenuation are difficult to disentangle without independent knowledge of the amount of energy absorbed by the dust. This is illustrated by the lack of good correlation between the UV slope of the SED, β, and ratio of the FIR to UV emission as has been discussed in the literature extensively (Meurer et al., 1999; Witt and Gordon, 2000; Kong et al., 2004). Ideally, the full range of possible attenuation-curve shapes allowed should be used to ensure that the lack of the full constraints is quantified in the stellar properties. Often, a particular attenuation-curve shape is assumed and scaled for different amounts of dust. This can be a dangerous assumption as radiative-transfer models have shown that the shape of the attenuation curve is correlated with the amount of dust where, in general, the larger the amount of dust, the grayer the attenuation curve (Chevallard et al., 2013). One solution is to assume the attenuation-curve shape can vary through some parametric model (Charlot and Fall, 2000; Noll et al., 2009). This is clearly better than assuming an average shape, but it does not capture the correlation between the amount of dust and shape. Another solution is to use the results of radiative-transfer model grids (e.g. Law et al., 2018). For example, the clumpy SHELL geometry may be a good approximation for star-formation regions especially given that such a geometry with $\tau(V) \sim 1.5$ reproduces the Calzetti average attenuation curve (Witt and Gordon, 2000).

Dust-attenuation models are available in the open-source `dust_attenuation` python package.[12]

Infrared

When only IR data are available, the information that can be determined is mainly focused on the dust properties. The most common model used to fit IR SEDs of galaxies or regions of galaxies is a simple blackbody modified by a dust emissivity assumed to be proportional to $\lambda^{-\beta}$ (Hildebrand, 1983). More complex models, including a blackbody with broken emissivity law and multiple blackbodies, can be used as well (e.g. Gordon et al., 2014). The most complex models for IR fitting are those that use a full grain model to describe the dust properties (size, shape, and optical properties) and heating from a range of radiation fields (Draine et al., 2007; Chastenet et al., 2017). In general, such IR fitting focuses on dust masses with some studies going into more detail on the dust-grain properties and radiation-field heating (e.g. Draine et al., 2014; Gordon et al., 2014).

[12] https://github.com/karllark/dust_attenuation

References

Bianchi, S., and Xilouris, E. M. 2011. The Extent of Dust in NGC 891 from Herschel/SPIRE Images. *Astronomy and Astrophysics*, **531**(July), L11.

Calzetti, D. 1997. Reddening and Star Formation in Starburst Galaxies. *Astronomical Journal*, **113**(Jan.), 162–184.

Calzetti, D., Kinney, A. L., and Storchi-Bergmann, T. 1994. Dust Extinction of the Stellar Continua in Starburst Galaxies: The Ultraviolet and Optical Extinction Law. *Astronomical Journal*, **429**(July), 582–601.

Calzetti, D., Kennicutt, R. C., Engelbracht, C. W. et al. 2007. The Calibration of Mid-Infrared Star Formation Rate Indicators. *Astronomical Journal*, **666**(Sept.), 870–895.

Cardelli, J. A., and Clayton, G. C. 1988. An Environmental Impact Study of Orion Nebula Dust. *Astronomical Journal*, **95**(Feb.), 516–525.

Cardelli, J. A., Clayton, G. C., and Mathis, J. S. 1989. The Relationship between Infrared, Optical, and Ultraviolet Extinction. *Astrophysical Journal*, **345**(Oct.), 245–256.

Charlot, S., and Fall, S. M. 2000. A Simple Model for the Absorption of Starlight by Dust in Galaxies. *Astrophysical Journal*, **539**(Aug.), 718–731.

Chastenet, J., Bot, C., Gordon, K. D. et al. 2017. Modeling Dust Emission in the Magellanic Clouds with Spitzer and Herschel. *Astronomy and Astrophysics*, **601**(May), A55.

Chevallard, J., Charlot, S., Wandelt, B. et al. 2013. Insights into the Content and Spatial Distribution of Dust from the Integrated Spectral Properties of Galaxies. *Monthly Notices of the Royal Astronomical Society*, **432**(July), 2061–2091.

Chiar, J. E., and Tielens, A. G. G. M. 2006. Pixie Dust: The Silicate Features in the Diffuse Interstellar Medium. *Astrophysical Journal*, **637**(Feb.), 774–785.

Clayton, G. C., and Martin, P. G. 1985. Interstellar Dust in the Large Magellanic Cloud. *Astrophysical Journal*, **288**(Jan.), 558–568.

Clayton, G. C., Gordon, K. D., Bianchi, L. C. et al. 2015. New Ultraviolet Extinction Curves for Interstellar Dust in M31. *Astrophysical Journal*, **815**(Dec.), 14.

Da Rio, N., and Robberto, M. 2012. TA-DA: A Tool for Astrophysical Data Analysis. *Astronomical Journal*, **144**(Dec.), 176.

Dolphin, A. E. 2002. Numerical Methods of Star Formation History Measurement and Applications to Seven Dwarf Spheroidals. *Monthly Notices of the Royal Astronomical Society*, **332**(May), 91–108.

Draine, B. T., Aniano, G., Krause, O. et al. 2014. Andromeda's Dust. *Astrophysical Journal*, **780**(Jan.), 172.

Draine, B. T., Dale, D. A., Bendo, G. et al. 2007. Dust Masses, PAH Abundances, and Starlight Intensities in the SINGS Galaxy Sample. *Astrophysical Journal*, **663**(July), 866–894.

Fitzpatrick, E. L., and Massa, D. 1990. An Analysis of the Shapes of Ultraviolet Extinction Curves. III – An Atlas of Ultraviolet Extinction Curves. *Astrophysical Journal Supplement Series*, **72**(Jan.), 163–189.

Fitzpatrick, E. L., and Massa, D. 2007. An Analysis of the Shapes of Interstellar Extinction Curves. V. The IR-through-UV Curve Morphology. *Astrophysical Journal*, **663**(July), 320–341.

Fitzpatrick, E. L., and Massa, D. 2009. An Analysis of the Shapes of Interstellar Extinction Curves. VI. The Near-IR Extinction Law. *Astrophysical Journal*, **699**(July), 1209–1222.

Gordon, K. D., Cartledge, S., and Clayton, G. C. 2009. FUSE Measurements of Far-Ultraviolet Extinction. III. The Dependence on R(V) and Discrete Feature Limits from 75 Galactic Sightlines. *Astrophysical Journal*, **705**(Nov.), 1320–1335.

Gordon, K. D., Baes, M., Bianchi, S. et al. 2017. TRUST. I. A 3D Externally Illuminated Slab Benchmark for Dust Radiative Transfer. *Astronomy and Astrophysics*, **603**(July), A114.

Gordon, K. D., Clayton, G. C., Misselt, K. A. et al. 2003. A Quantitative Comparison of the Small Magellanic Cloud, Large Magellanic Cloud, and Milky Way Ultraviolet to Near-Infrared Extinction Curves. *Astrophysical Journal*, **594**(Sept.), 279–293.

Gordon, K. D., Clayton, G. C., Witt, A. N. et al. 2000. The Flux Ratio Method for Determining the Dust Attenuation of Starburst Galaxies. *Astrophysical Journal*, **533**(Apr.), 236–244.

Gordon, K. D., Fouesneau, M., Arab, H. et al. 2016. The Panchromatic Hubble Andromeda Treasury. XV. The BEAST: Bayesian Extinction and Stellar Tool. *Astrophysical Journal*, **826**(Aug.), 104.

Gordon, K. D., Misselt, K. A., Witt, A. N. et al. 2001. The DIRTY Model. I. Monte Carlo Radiative Transfer through Dust. *Astrophysical Journal*, **551**(Apr.), 269–276.

Gordon, K. D., Roman-Duval, J., Bot, C. et al. 2014. Dust and Gas in the Magellanic Clouds from the HERITAGE Herschel Key Project. I. Dust Properties and Insights into the Origin of the Submillimeter Excess Emission. *Astrophysical Journal*, **797**(Dec.), 85.

Guhathakurta, P., and Draine, B. T. 1989. Temperature Fluctuations in Interstellar Grains. I – Computational Method and Sublimation of Small Grains. *Astrophysical Journal*, **345**(Oct.), 230–244.

Hildebrand, R. H. 1983. The Determination of Cloud Masses and Dust Characteristics from Submillimetre Thermal Emission. *Quarterly Journal of the Royal Astronomical Society*, **24**(Sept.), 267.

Kennicutt, Jr., R. C., Hao, C.-N., Calzetti, D. et al. 2009. Dust-Corrected Star Formation Rates of Galaxies. I. Combinations of Hα and Infrared Tracers. *Astrophysical Journal*, **703**(Oct.), 1672–1695.

Kong, X., Charlot, S., Brinchmann, J. et al. 2004. Star Formation History and Dust Content of Galaxies Drawn from Ultraviolet Surveys. *Monthly Notices of the Royal Astronomical Society*, **349**(Apr.), 769–778.

Law, K.-H., Gordon, K. D., and Misselt, K. A. 2018. DirtyGrid I: 3D Dust Radiative Transfer Modeling of Spectral Energy Distributions of Dusty Stellar Populations. *Astrophysical Journal Supplement Series*, **236**(June), 32.

Maíz-Apellániz, J. 2004. CHORIZOS: A χ^2 Code for Parameterized Modeling and Characterization of Photometry and Spectrophotometry. *Publications of the Astronomical Society of the Pacific*, **116**(Sept.), 859–875.

Mathis, J. S., and Cardelli, J. A. 1992. Deviations of Interstellar Extinctions from the Mean R-Dependent Extinction Law. *Astrophysical Journal*, **398**(Oct.), 610–620.

Meurer, G. R., Heckman, T. M., and Calzetti, D. 1999. Dust Absorption and the Ultraviolet Luminosity Density at z ~ 3 as Calibrated by Local Starburst Galaxies. *Astrophysical Journal*, **521**(Aug.), 64–80.

Misselt, K. A., Gordon, K. D., Clayton, G. C. et al. 2001. The DIRTY Model. II. Self-Consistent Treatment of Dust Heating and Emission in a Three-dimensional Radiative Transfer Code. *Astrophysical Journal*, **551**(Apr.), 277–293.

Noll, S., Burgarella, D., Giovannoli, E. et al. 2009. Analysis of Galaxy Spectral Energy Distributions from far-UV to Far-IR with CIGALE: Studying a SINGS Test Sample. *Astronomy and Astrophysics*, **507**(Dec.), 1793–1813.

Pei, Y. C. 1992. Interstellar Dust from the Milky Way to the Magellanic Clouds. *Astrophysical Journal*, **395**(Aug.), 130–139.

Pendleton, Y. J., and Allamandola, L. J. 2002. The Organic Refractory Material in the Diffuse Interstellar Medium: Mid-Infrared Spectroscopic Constraints. *Astrophysical Journal Supplement Series*, **138**(Jan.), 75–98.

Prevot, M. L., Lequeux, J., Prevot, L. et al. 1984. The Typical Interstellar Extinction in the Small Magellanic Cloud. *Astronomy and Astrophysics*, **132**(Mar.), 389–392.

Salmon, B., Papovich, C., Long, J. et al. 2016. Breaking the Curve with CANDELS: A Bayesian Approach to Reveal the Non-Universality of the Dust-Attenuation Law at High Redshift. *Astrophysical Journal*, **827**(Aug.), 20.

Seon, K.-I., and Draine, B. T. 2016. Radiative Transfer Model of Dust Attenuation Curves in Clumpy, Galactic Environments. *Astrophysical Journal*, **833**(Dec.), 201.

Steinacker, J., Baes, M., and Gordon, K. D. 2013. Three-Dimensional Dust Radiative Transfer*. *Annual Review of Astronomy and Astrophysics*, **51**(Aug.), 63–104.

Valencic, L. A., Clayton, G. C., and Gordon, K. D. 2004. Ultraviolet Extinction Properties in the Milky Way. *Astrophysical Journal*, **616**(Dec.), 912–924.

Viaene, S., Baes, M., Tamm, A. et al. 2017. The Herschel Exploitation of Local Galaxy Andromeda (HELGA). VII. A SKIRT Radiative Transfer Model and Insights on Dust Heating. *Astronomy and Astrophysics*, **599**(Mar.), A64.

Weingartner, J. C., and Draine, B. T. 2001. Dust Grain-Size Distributions and Extinction in the Milky Way, Large Magellanic Cloud, and Small Magellanic Cloud. *Astrophysical Journal*, **548**(Feb.), 296–309.

Whittet, D. C. B., van Breda, I. G., and Glass, I. S. 1976. Infrared Photometry, Extinction Curves and R Values for Stars in the Southern Milky Way. *Monthly Notices of the Royal Astronomical Society*, **177**(Dec.), 625–643.

Witt, A. N., and Gordon, K. D. 2000. Multiple Scattering in Clumpy Media. II. Galactic Environments. *Astrophysical Journal*, **528**(Jan.), 799–816.

Witt, A. N., Thronson, Jr., H. A., and Capuano, Jr., J. M. 1992. Dust and the Transfer of Stellar Radiation within Galaxies. *Astrophysical Journal*, **393**(July), 611–630.

Part II

SFR Measurements

5
Star-Formation Rates from Resolved Stellar Populations

JOHN. S. GALLAGHER III, ANDREW COLE,
AND ELENA SABBI

5.1 Introduction

The study of resolved stellar populations grew out of Galactic astronomy, where stellar investigations are almost entirely based on observations of individual stars (e.g. Struve and Zebergs, 1962). In many ways, this is the best method for exploring properties of stellar populations. However, the method requires measurements to be obtained for individual stars, and this rapidly becomes experimentally challenging as the distance to extragalactic systems increases. In this chapter we provide a brief overview of the history and some of the techniques used to derive star-formation rates (SFRs) and the associated star-formation histories (SFHs) of galaxies through observations of their resolved stars.

The depths of resolved stellar samples in galaxies are primarily limited by the levels of stellar fluxes and effects of crowding. Both of these constraints become more severe with increasing distance, and currently most resolved stellar-population studies are constrained to galaxies within a distance of about 20 Mpc. Fortunately, the short-lived massive stars, whose numbers trace SFRs, are luminous and thus among the most readily observed, especially when they are not obscured by interstellar dust. Ideally, high-quality observations allow the number of stars above a fiducial luminosity in a set of spectroscopic bandpasses to be counted and corrected for incomplete sampling. The distribution of these stars is compared to expectations of stellar-population models to derive estimates for the observed mass in the form of stars detected in the data. Further modeling provides an interpretation in terms of stellar masses within age bins.

In this approach stellar-evolution tracks are utilized to define the masses and ages of stars that occupy effective temperature-luminosity cells in the theoretical Hertzsprung-Russell (H-R) diagram. Thus by combining the tracks of stars with varying metallcity, ages, and masses, model predictions can be made for the density of stars in each ($\Delta \log L$, $\Delta \log T_{\rm eff}$) cell where L is the luminosity and $T_{\rm eff}$ is the stellar effective temperature. The theoretical stellar information comes in the form of stellar-mass M, $T_{\rm eff}$, and L for evolutionary phase i, time intervals $\tau_i d\tau$, and the numbers of stars in each generation usually are set by assuming a time-independent form for the stellar initial mass function (IMF, Bastian et al., 2010). The theoretical predictions can be converted to observed magnitudes and the distributions of stellar numbers in magnitude bins m_λ at wavelength λ matched to the data. This process then yields the numbers, and thus masses, of stars born over a span of times.

For example, consider a corrected photometric sample for a nearby galaxy that contains a reliable estimate for the number of main sequence stars with $M \geq 15\,M_\odot$. Stars with an initial mass of $15\,M_\odot$ have main sequence lifetimes of $\tau_{MS} \approx 12\,Myr$, thus we are counting all of the stars that formed in the last $\tau_{SF} = 12\,Myr$. This can then be used to give $M_{star}(\geq 15)$, the mass in stars with $M \geq 15\,M_\odot$. $M_{star}(\geq 15)$ can be corrected to a total stellar mass M_{star} formed during the past 12 Myr using the assumed form of the IMF. The SFR then is simply M_{star}/τ_{SF}.

While the basic technique for deriving a SFR (or a SFH that is the SFR as a function of time) is straightforward, reliable measurements require attention to many details. We present an overview of several of the main issues in the remainder of this chapter. More information can be found in books describing studies of stellar populations, such as those by Greggio and Renzini (2011) or Salaris and Cassisi (2006). In Section 5.2 we give a brief historical account of the development of techniques for quantitatively assessing properties of stellar populations. Section 5.3 deals with measurements of current SFRs, which we take to be measurements over time spans of less than typical galactic dynamical time scales, or $\tau_* \leq 100\,Myr$. Methods for deriving SFHs covering longer time spans are presented in Section 5.4 and a discussion of future prospects is given in Section 5.5.

5.2 Brief Historical Overview

Techniques for determining the distributions of stars as functions of mass, age, metallicity, and location were initally developed to meet the needs of Galactic astronomy. In many respects studies of resolved stellar populations in galaxies are extensions of the methods used in the Milky Way (see van den Bergh, 1975). For example, basic ages can come from the locations of main sequence turnoffs (MSTOs), and the critical issue of the form of the IMF is mainly based on Galactic results (e.g. Chomiuk and Povich, 2011), as are frequencies of binary and multiple stars (Duquennoy et al., 1991; Chini et al., 2012; Duchêne and Kraus, 2013). Thus, a key step in extragalactic resolved stellar-population research is to assume, and then test for, the universality of intrinsic stellar-population properties as compared to those in the Milky Way. The main disadvantages in assessing properties of extragalactic stellar populations stem from the impact of much greater distances and from having to deal with different obscuration effects by interstellar dust. There also is an advantage that in most cases all of the stars can be considered to be located at the same distance, which removes lingering uncertainties associated with distances that bedeviled many traditional Galactic stellar-population studies.

Although the Magellanic Clouds were the first galaxies to be resolved into stars (see Herschel, 1847), nearby galaxies only were resolved once the combination of larger-aperture reflectors and sensitive photographic plates became available in the early twentieth century (Hubble, 1929). Thus, it soon was accepted that galaxies were stellar systems, which contained stars similar to those in the Milky Way. However, the analysis of their stellar populations had to await the combination of better observations and the development of stellar-evolution models.

A major breakthrough came from W. Baade's resolution of red giant stars in M31 and its companions with the 100-inch Hooker telescope (Baade, 1944a,b). This pair of papers is noteworthy for their discussion of key issues in resolving stars against the background light of a galaxy. The roles of the delivered angular resolution, background intensity levels, choice of wavelength bandpass, and effects of crowding are clearly discussed. Baade's papers thus continue to offer useful introductions to technical issues associated with the detection of individual stars in crowded fields (see also Tonry and Schneider, 1988; Dolphin, 2000). Baade's results propelled the concept that stellar populations covering a range of ages exist in galaxies, a conclusion that became clearer as numerical stellar-evolution models began to mature in the 1950s (see Hodge, 1989; Osterbrock, 1995).

The quantitative analysis of extragalactic stellar populations rested on a combination of major improvements in observational techniques and theoretical models that produced a revolution in the late twentieth century. Modern large ground-based reflectors, such as the Canada–France–Hawaii Telescope, recovered the ability to routinely produce sub-arcsecond images that once had been the domain of large refractors (e.g. Freedman, 1988) and thus exceeded Baade's best efforts. The repaired *Hubble Space Telescope* (*HST*)[1] offered an order of magnitude gain in resolution and depth over ground-based telescopes, as well as access across the spectrum from the vacuum ultraviolet to near-infrared. The development of digital photometric techniques to enable efficient and accurate photometry of large numbers of stars in crowded images also was crucial (Stetson, 1987; Butcher, 1977; Dolphin, 2000). Thus, by the 1990s, observations could readily yield Hess diagrams,[2] primarily for a variety of nearby dwarf galaxies (e.g. Dohm-Palmer et al., 1997; Gallagher et al., 1998; Grebel, 1997; Aloisi et al., 1999; Gallart et al., 1999).

Equally important was the development of theoretical frameworks for the evolution of stellar populations in galaxies (Tinsley, 1968, 1980; Maeder and Conti, 1994; Massey, 2003; Gallart et al., 2005). These demonstrated how stellar ages and metallicities combine to produce regions in color-magnitude space where stellar colors and magnitudes from stars of different types can overlap. Thus, the nature of the age-metallicity relationship (AMR) emerged as a significant factor for modeling the long-term evolution of stellar populations (see discussions in Aaronson and Mould, 1982; Cole et al., 2005). This point was recognized early on, and was encapsulated, for example, in the "stellar-population box" with SFR, age and metallicity axes presented by Hodge (1971) in his pioneering review of evolutionary patterns in dwarf galaxies.

The last step in deriving the SFR(t) for an observed stellar population is to match the observed data to stellar-population models. This requires the statistical techniques of stellar-population synthesis to find best fits to the data. Pioneering work on this problem by

[1] *Hubble Space Telescope*; www.stsci.edu/hst

[2] A Hess diagram plots the density of stars on a color-magnitude plane and thus explicitly presents the densities of stars in terms of temperatures (colors) and luminosities.

Tosi et al. (1991) and Aparicio et al. (1996) set forth the basic steps for comparing model and observed stellar populations. These and other initial studies focused on dwarf galaxies, where, due to their low stellar densities, the best data could be obtained. Using clever observational techniques, stellar-population synthesis was also applied in a wholesale manner by Harris and Zaritsky to the main stellar bodies of the Large and Small Magellanic Clouds yielding both the SFRs and SFHs (Harris and Zaritsky, 2004, 2009).

Over the next decade, a variety of improvements including the exploration of maximum likelihood estimators and the use of Bayesian inference[3] to appropriately weight the information contained in relatively rare but important classes of stars (Tolstoy and Saha, 1996) were introduced by various authors. These gradually led to increasingly powerful techniques as summarized in the classic paper by Dolphin (2002) and form the basis for many of the stellar population-synthesis codes currently used to analyze resolved stellar populations (e.g. Dohm-Palmer et al., 1997; Aparicio and Hidalgo, 2009; Tolstoy et al., 2009; Cignoni and Tosi, 2010; Weisz et al., 2011; Grocholski et al., 2012; Lewis et al., 2015; Sacchi et al., 2016).

5.3 Star-Formation Rates

The instantaneous SFR of a galaxy is a theoretical concept. Star formation tends to be extensive in space and time, thus, even if a single "star forming event" could be defined, it is likely to last for a longer time than the lifetimes of its most massive stars (Elmegreen, 2000). If we wish to determine how the stellar content of a galaxy evolved, then we should average the SFR over a sufficient time to allow for fluctuations due to the stochastic nature of star formation, but for less than a dynamical timescale $\tau_{\rm dyn}$ for a galaxy to undergo global evolution (Gallagher et al., 1984). The choice of the star-formation timescale $\tau_{\rm SF}$ depends on the lifetime of the tracer of the young stellar mass and the level of star forming activity (Kennicutt and Evans, 2012). For example, in a the central region of a galaxy, the situation could arise where $\tau_{\rm SF} \geq \tau_{\rm dyn}$ due to the short dynamical timescales at small galactocentric radii. Although this possibility is taken into account by models, it still complicates the interpretation of SFRs in systems that are experiencing rapid evolution, as in some types of starbursts (Lee et al., 2009; McQuinn et al., 2010b).

In the remainder of this chapter we will discuss measurements of stellar populations with varying characteristic ages as a means to determine SFRs over time. Table 5.1 gives an approximate overview of the stellar-population parameter space that goes into these models. Naturally, the details of stellar evolution are more complex than the simple matchings listed in Table 5.1, but this illustrates the basic ideas behind the use of stellar-population synthesis to quantitatively explore the stellar-histories of galaxies.

[3] In Bayesian inference, the probability of a model being true given the observed data is inferred from the product of the probability of the data given the model and the prior probability of the model, allowing maximum likelihood solutions to accommodate expectations from physical paradigms such as a preference for a constant or exponentially decreasing SFR.

Table 5.1. *Basic stellar star-formation rate/history tracers*

Maximum Timescale	$M_*/M_\odot$	Main Sequence	Evolved	Application
7Myr	≥25	O-stars	Hypergiants, WR-stars LBVs, ccSNe	SFR
30 Myr	8–25	B-stars,	RSGs, Blue Loop ccSNe, Cepheids	SFR
∼2 Gyr	2–8	B,A-stars	AGB stars Miras, LPVs	SFH
∼8 Gyr	∼ 1–2	F & G stars	RGB, Red Clump	SFH
∼ 13 Gyr	∼ 0.7–1	G stars; Oldest MSTO	RGB, HB RR Lyrae	SFH

Note: The first line in the evolvedstar column gives the primary stellar types and the second line gives examples of variable stars.
Acronyms for stellar types WR = Wolf–Rayet; LBV = luminous blue variable star; ccSNE = core collapse supernovae; RSG = red supergiants; AGB = asymptotic giant branch stars; LPVs = long period variables; RGB = red giant branch; HB = horizontal branch.
References: General texts: Prialnik (2009); Kippenhahn et al. (2012). OB stars: Salaris and Cassisi (2006). A-G stars: Iben (2013a,b).

5.3.1 Photometric Studies

Photometry currently is preferred for SFR determinations based on resolved stars as it allows the properties of large samples of stars to be measured efficiently. Ideally the SFR should be calculated from the population statistics for a complete sample of massive stars with main-sequence lifetimes $\tau_{MS} \leq \tau_{SF}$. Unfortunately, such samples usually cannot be obtained, and the observed numbers of massive stars normally are lower limits to their true numbers. One issue stems from the photometric completeness of the sample. For example, impact of crowding can be assessed via inclusion of artificial stars with known fluxes whose recovery determines the efficiency in measuring the real stars. However, these tests do not include the inevitable loss of stars in dense clusters or in binary/multiple star systems. Thus, crowding reduces photometric completeness (e.g. Dohm-Palmer et al., 1997).

Varying levels of obscuration by interstellar dust present another set of issues that lead to some stars being missed. Although the effects of interstellar dust can be minimized for massive stars, for example by fitting the observed spectral energy distribution for filters covering a suitable range in wavelength with models that include dust opacity, the results are not unique (Gordon et al., 2016). At present SFRs derived from resolved stars therefore are most reliable for galaxies with modest stellar densities and low extinction levels, such as dwarfs or the outer parts of low inclination spiral disks.

Determining the masses of observed stars that populate various color-magnitude planes also can be problematic. Even when far-ultraviolet data are available, stars with $T_{eff} \geq 30{,}000$ K are only observable in the Rayleigh-Jeans tails of their spectral energy distributions (Massey, 1993). As a result their colors are not sensitive to T_{eff} or to their mass and evolutionary state. This problem is compounded by the high and varying levels of interstellar extinction in the ultraviolet. Photometric studies therefore are not particularly sensitive to main sequence stars with initial masses M greater than about 15–20 $M_{\odot}$. Fortunately the presence of such stars normally can be detected through their cooler supergiant descendants, and by the presence of Wolf–Rayet stars that have unique spectra, as well as through indirect methods, such as SFRs based on hydrogen emission-line luminosities (see Kennicutt and Evans, 2012).

The best stellar candidates for measuring SFRs in galaxies are stars where mass and the observed luminosity are closely correlated, and where T_{eff} varies monotonically with mass. These conditions assure that lifetimes can be estimated with reasonable accuracy, and thus even a quick examination of the resolved populations yields information on recent star-formation products (e.g. Sabbi et al., 2009). This situation most clearly applies to stars with the same metallicity on the hydrogen-burning main sequence. The most luminous main-sequence stars, however, have $M_V \geq -5$ to -6 over a large range of initial masses, and even so are difficult to trace at distances beyond about 10 Mpc. However, in situations where the stellar metallicity is low, massive stars with core helium burning in the "blue loop" evolutionary phase also follow a simple mass-luminosity relationship (Tang et al., 2014). In these cases the numbers of stars per unit luminosity bin can be used to derive a luminosity function and estimate of the ages and masses in young stars. The stellar mass depends to some degree on the stellar models and on the choice of IMF, and the SFR, which follows from the stellar mass divided by the timescale from the massive stellar population, then also depends on these choices.

5.3.2 Fitting the Data

In a "simple" stellar population, such as an idealized star cluster, stellar models in the form of isochrones of the appropriate metallicity can simply be shifted to the appropriate distance and reddening of the system to be analyzed. The magnitude of the MSTO and subgiant branch then yield the age of the population. Only one color is required, although it is advisable to use as long of a color baseline as instrumental sensitivity and model reliability permit to provide the largest leverage on temperature and reddening effects. Depending on the photometric depth, number of evolved stars, and the type and quality of ancillary data (e.g. multiple filters over a long wavelength baseline, spectroscopic metallicity measures), it may or may not be desirable to include evolved stars in the SFH determination.

Assuming that the calculated isochrones are correct, issues like the form of the IMF and the effects of binary stars are minimized. In this idealized case, the relative errors on derived quantities such as the timing, duration, and amplitude of star-formation episodes depend only on the number of well-measured stars which have measurably evolved away from the zero-age main sequence, so that the models have some leverage on the age determination.

By contrast, in most real galaxies old and young populations are layered one over the other in varying proportions, and there is rarely the luxury of finding discrete bursts of star formation well-separated by fallow periods (c.f. the Carina dwarf spheroidal, Smecker-Hane et al., 1994). Therefore, in virtually all observed galaxies the derivation of the SFH becomes the probabilistic, inverse problem of inferring the best possible linear sum of (imperfect) isochrones plus (non-Gaussian) noise to synthesize the observed Hess diagram.

The derivation of SFRs and metallicities as a function of time is an optimization problem. Unfortunately, this problem is generally ill-defined for various reasons. For example, the choice of what range of stellar ages to include, and how to group them into age bins is not obvious in many cases. There is also the issue of how to treat stars with colors and magnitudes that are not predicted by the models, which can strongly influence the numerical value of any figure of merit for deciding what is a "good" fit (c.f. Dolphin, 2002).

When considering how to bin the models into ranges of age and metallicity during optimization, there is a bias-variance tradeoff: The finer the divisions and the more models considered, the more closely the "best" fit is to approximate the true SFH, but the larger the range of possibilities and hence the larger the error bars. For these reasons and more, there is not yet consensus in the literature on the best way to optimize the synthetic Color-Magnitude Diagrams (CMDs) to infer star-formation histories.

However, all methods proceed in much the same way. Photometric packages provide lists of stars containing their positions and magnitudes in a given set of filters, i.e. lists of (x, y, m_f) for each star and filter $f = 1, \ldots, k$. These data can be plotted on a Hess diagram to give the numbers of stars observed in cells $i = 1, \ldots, c$ that are chosen to cover the observed region. Thus, the observations give N_{oi} stars for each cell i located at (m_f, col_{f1f2}), where the color is $col_{f1f2} = m_{f1} - m_{f2}$ and f1 and f2 are distinct filters.

The basic inputs consist of a form for the stellar IMF and model stellar isochrones which are assumed to be astrophysically correct. The isochrones need to be converted from the theoretical ($L/L_\odot$,T_{eff}) to the observational (m_f, col_{f1f2}) plane for the comparison; the transformation relies on sound knowledge of the instrumental response and on realistic stellar atmospheres or empirical relations. Distance and wavelength-dependent obscuration by interstellar dust may be assumed or solved for as part of the fit process (e.g. Gordon et al., 2016). Finally the effects of chemical abundances need to be considered. Fortunately, in most cases chemical abundances are relatively constant during the time interval when SFRs are measured, but must be included when deriving longer term SFHs.

A partial model is created from each isochrone, with a predicted number of stars in each cell of the Hess diagram. Synthetic CMDs are generated by summing the partial models with scalings applied such that the best model Hess diagram optimally matches the observations, with N_{Mi} stars in each cell. The critical step requiring the most care is in the treatment of the errors on the photometric data. A stellar measurement gives a color, magnitude, and likely error in each quantity, but these errors are in general not normally distributed. Furthermore, because of crowding and completeness effects, the expectation value of the measured color and magnitude are not necessarily centered on the true value of those quantities in the absence of noise. For example, a star that is nominally below the

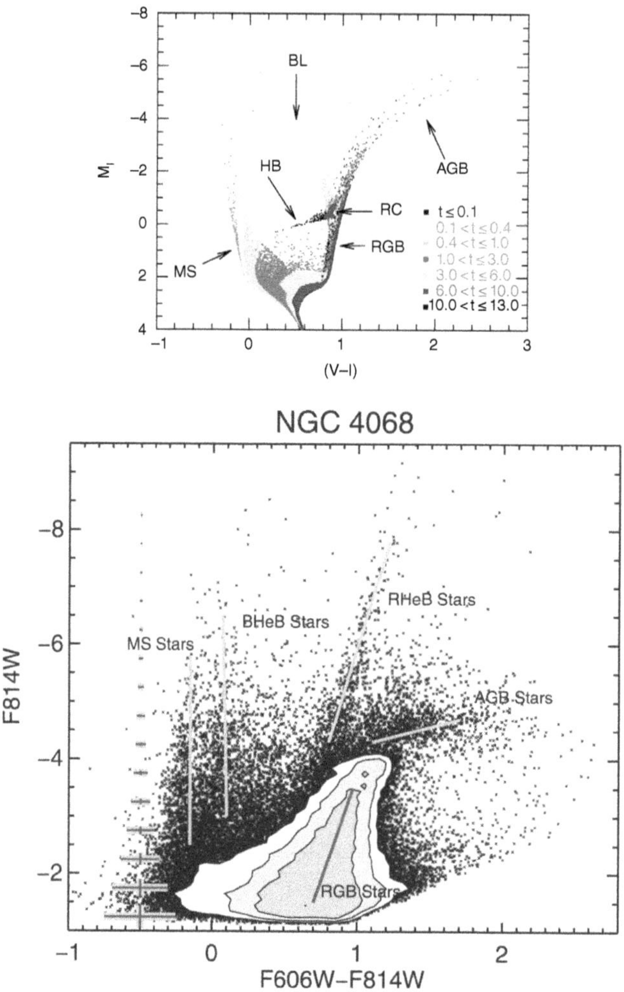

Figure 5.1 *Top:* An example of the age layering of composite stellar population derived from isochrones by Aparicio and Gallart (2004). The main sequence (MS) sets the blue color boundary of the Hess diagram. Stars along the MS follow an age-mass-luminosity correlation. In a low metallicity system, such as that modeled here, high mass stars in their helium core-burning post-main sequence blue loop (BL) evolutionary phase spend most of their lives with either blue or red colors. The red giant branch (RGB) contains stars with ages older than 2 Gyr that feed into the red clump (RC) and horizontal branch (HB) core helium-burning phases. The asymptotic giant branch contains stars with masses of less than $\sim 8\, M_\odot$. SFR measurements focus on the analysis of luminous, short-lived stars while SFH studies require an analysis of the long-lived, lower-mass stars.
Bottom: Example of an observed Hess diagram from McQuinn et al. (2015a) showing both the red (RHeB) and blue (BHeB) branches of massive core helium-burning blue-loop stars. This also displays how errors, reddening, and crowding/multiple stars broaden the observed distributions of stars.

detection limit could appear in the CMD due to being crowded with a similar-brightness star, or due to random error that scatters up the star in the CMD. Because the surface density of stars increases steeply with decreasing luminosity, many more stars will be scattered up into the detected region of the CMD than will be scattered down out of the detected CMD.

It is, therefore, of vital importance to accurately simulate the probability of recovery and the distribution function of recovered colors and magnitudes for stars. These will be strong functions of color, magnitude, and surface density of stars. The only way to achieve this is to calculate many artificial star tests, in which synthetic stars with the same point spread function as the data are injected into the imaging frames and measured in exactly the same way as the data, across a very fine grid of color and magnitude in order to accurately capture the noise parameters of the system. The net effect is to shift and broaden the stellar sequences derived from the isochrones, and to impose a color-dependent shallowing of the luminosity function at the faint end. If there is to be any hope of matching the models to the data, these factors must be appropriately taken into account (the deeper the data and the lower the crowding, the less the conclusions will depend on these issues, e.g. Gallart et al., 1996; Weisz et al., 2011).

The SFH is inferred from the synthetic Hess diagram that best matches the data. The matching is determined by maximizing some goodness of fit parameter, such as a χ^2, Cash (1979) statistic, or other estimate of the likelihood that the data could have been obtained from the models. The optimization may be done by any of the usual methods, including, for example, downhill simplex, Markov Chain Monte Carlo, or simulated annealing. Each has various advantages and disadvantages, trading off between speed, reliability, and accurate characterization of the uncertainties in the solution (see e.g. Dolphin, 2002; Skillman and Gallart, 2002; Cole et al., 2007; Aparicio and Hidalgo, 2009).

It can be difficult to link subtle features of the SFH solutions (e.g. bursts or gaps in SFR(t)) to specific structures in the Hess diagram. This is because the solutions tend to be dominated by regions of color-magnitude space where the number of stars is greatest (i.e. the lower main sequence). In addition, the isochrones for long-lived low-mass stars lie close together and therefore are nearly degenerate with age. The number of main sequence stars in a given system below the oldest MSTO is approximately given by $N_*(M,dM) \approx$ Ψ M$^{-2.3}$ dM τ(M, dM), where Ψ is the average SFR over τ(M, dM), the stellar lifetime of the star of mass M. Thus the number of stars increase as M decreases due to the IMF and then is additionally boosted by the longer stellar lifetimes and larger τ(M, dM) for lower masses.

The distribution of stars by number on the Hess diagram therefore is dominated by lower-mass main-sequence stars and by the adopted incompleteness and crowding corrections, which in turn control a simple χ^2 statistic. This can result in models that, while fitting in a statistical sense, do not include key SFR indicators, for example in the form of short-lived supergiant stars. A variety of methods thus now are in use that are designed to make effective use of all of the information that is contained in a series of Hess diagrams observed in a variety of filters (e.g. Dolphin, 2002; Tolstoy et al., 2009; Cole et al., 2014; Lewis et al., 2015; Sacchi et al., 2016). These may involve solving for different age ranges using different regions of the CMD; over-weighting, de-weighting, or excluding features such as the red giant branch or AGB stars; or using custom grids of bin shapes and sizes in the Hess diagram.

5.3.3 Examples of SFR Measurements from Resolved Stars

In most cases the coverage of galaxies is limited by instrumental fields of view and is incomplete. Therefore, most of the measurements of integrated SFRs derived from resolved stellar photometry with *HST* are for dwarf galaxies. These systems have small angular sizes that allow for efficient observations with *HST*. Dwarfs also have low stellar densities and reduced levels of interstellar dust obscuration that further simplifies the anaylsis of the photometry in terms of SFRs and SFHs.

The Magellanic Clouds, however, are sufficiently nearby that ground-based observations can resolve stars in much of the galaxy and reach below the oldest MSTO. The analysis of the total stellar populations of the main bodies of the Magellanic Clouds by Harris and Zaritsky (2004) and Harris and Zaritsky (2009) stand as a major milestone in ground-based studies of extragalactic resolved stellar populations. Their deep Hess diagrams allowed a systematic selection and identification of stellar-population components. Using stellar-population synthesis models to fit their multi-color Hess diagrams, they found SFRs averaged over approximately 30 Myr of 0.4 and $0.15\,\mathrm{M_\odot\,yr^{-1}}$ for the Large and Small Magellanic Cloud, respectively. More recently, deeper multi-wavelength surveys are extending this approach (Cioni et al., 2011).

At distances beyond the Magellanic Clouds high angular resolution becomes increasingly important, especially in the denser regions that frequently support much of galactic star formation. Thus, *HST* has been the instrument of choice for SFR measurements in dwarfs out to the distance of the I Zw 18 extremely metal-poor system at a distance of ~20 Mpc (Annibali et al., 2013). An important series of studies carried out by K. McQuinn and collaborators examined different approaches to the determination of SFRs. This included careful comparisons between SFRs derived from Hess diagrams based on optical *HST* photometry with those derived from ultraviolet and mid-infrared luminosities (McQuinn et al., 2015a). The results generally show good agreement, with the SFRs based on resolved stars giving SFRs that are somewhat higher than those derived from integrated ultraviolet photometry.

An added benefit of deep *HST* observations lies in the ability to measure both SFRs and the total stellar mass from the Hess diagrams. A mean SFR can then be defined as $\overline{SFR} = M_{\rm tot}/\tau_{\rm cosmic}$. The results from *HST* show that in the majority of star-forming dwarf galaxies, the ratio $\overline{SFR}/SFR \approx 0.5$–2; starbursts with substantially enhanced SFRs thus are uncommon (Lee et al., 2009; McQuinn et al., 2010a, 2015b). This result thus supports the early results from Hunter and Gallagher (1985) that low-mass star-forming dwarfs have approximately constant SFRs over their lifetimes, albeit with fluctuations of a factor of a few. This behavior in turn requires that the SFR is controlled via a feedback mechanism, as it is unlikely that external gas resupply alone would operate in such a consistent manner across galaxies covering a range in mass and existing in a variety of local environments (e.g. Gallagher et al., 1984).

Resolved-star studies of the denser disks of spiral galaxies requires special efforts to deal with the dual problems introduced by crowding and interstellar obscuration. The study of about half of the main body of the M31 spiral by PHAT is one such heroic effort (Dalcanton et al., 2012). This project obtained multi-band photometry of 117 million stars leading to estimates of both an SFR and SFH. Lewis et al. (2015) derive a SFR for the region observed by PHAT, while the SFH indicates that significant fluctuations in the SFR occurred, that likely are associated with past merger events.

Investigations of the young stellar content in large regions of galaxy disks can be extended by making observations in the ultraviolet. Young stars with low dust obscuration stand out in this part of the spectrum and the level of the background produced by old stellar populations is greatly reduced. However, dust extinction levels are higher in the ultraviolet, and so dust obscuration corrections are critical. However, ongoing studies, especially using ultraviolet images from *HST* produced by the Legacy Galaxy Ultraviolet Survey (Calzetti et al., 2015; Sabbi et al., 2018, LEGUS;), hold a promise for providing more cases where SFRs based on resolved stellar populations can be compared with those based on total luminosities (see Fig. 5.2).

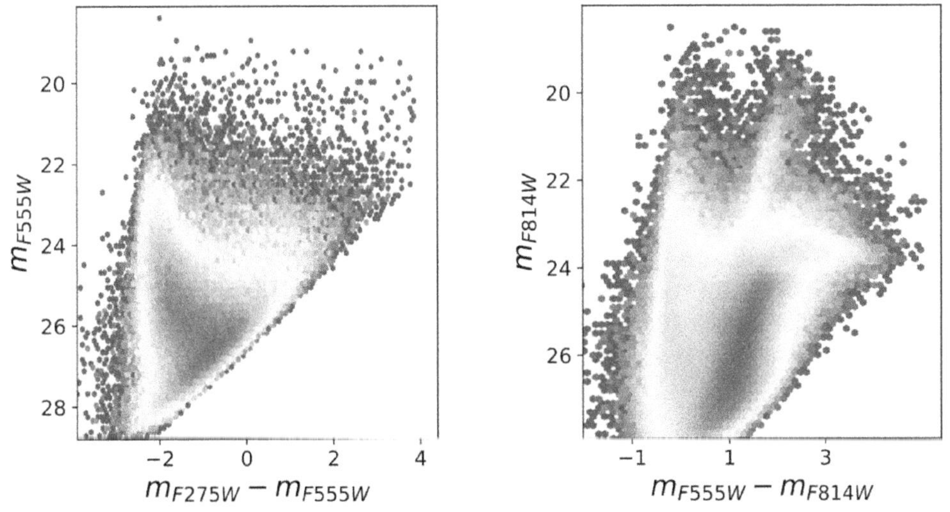

Figure 5.2 Hess diagrams from HST/LEGUS stellar photometry of the Magellanic galaxy NGC 4449 (Sabbi et al., 2018). The plot on the *left* shows the distribution of stars based on a mid-ultraviolet to visual color. The peak of the distribution shows as a plume of stars that become bluer with higher luminosity, and thus emphasizes massive stars that are on or near the main sequence. On the *right* is a more traditional optical Hess diagram. The main sequence is less pronounced and only a modest correlation exists between main sequence color and magnitude. Ultraviolet photometry is important for measuring properties of young, massive stars. The recent SFR for NGC'4449 based on LEGUS data is $\approx 0.4 \pm 0.2\, M_{\odot}\, yr^{-1}$ (Sacchi et al., 2018). Color version available online.

5.3.4 Variable Stars

Photometric mapping can be designed to measure properties of variable stars. Variable stars with relatively high-amplitude flux variations offer a cost effective tool for determining the properties of stellar populations. The key issue is to relate patterns of variability to stellar lifetimes and masses, which allows determinations of SFRs or SFHs (see Table 5.1). This approach is not widely used as the necessary time domain data often are incomplete or not available, but in several instances variable stars have proven to be powerful tracers of the histories of stellar populations. This situation, however, is changing as surveys increasingly target stellar variability, an area that will be further advanced by the operation of the *Large Synoptic Survey Telescope (LSST)*[4].

Core Collapse Supernova

The most extreme star forming events occur in dense regions where most individual stars are hidden. However, supernovae often can be detected, either in the optical-near infrared or as radio sources. Since core collapse supernovae (ccSNe) arise from massive stars, supernova rates, when corrected for the presence of type Ia supernova that come from white dwarfs and the potential for failed supernovae (Horiuchi et al., 2014), provides an estimate of the SFR (Xiao and Eldridge, 2015). In addition, one also can use, with proper cautions (see Smith et al., 1998; Maoz and Badenes, 2010), supernova remnants (SNRs) as proxies for the supernova rate. As ccSNe and SNRs can be detected in the radio, they allow SFRs to be measured in dense environments where high levels of dust opacity obscure optical and even near infrared measurements (see also Chapter 11).

This method has been applied to a few extreme starburst galaxies by using radio interferometry to detect supernovae and supernova remnants and estimate SFRs. The initial application was to M82 (Kronberg and Wilkinson, 1975), but since then very long baseline techniques allowed deeper measurements in M82 (e.g. Fenech et al., 2008; Kimani et al., 2016) as well as several other nearby starburst systems, including Arp 220 (Lonsdale et al., 2006; Parra et al., 2007; Batejat et al., 2011; Varenius et al., 2019), Arp 299 (Bondi et al., 2012), NGC 253 (Antonucci and Ulvestad, 1988; Rampadarath et al., 2014), and NGC 2146 (Tarchi et al., 2000). The resulting SFRs are reasonably consistent with estimates based on integrated luminosities using standard models (e.g. Lenc and Tingay, 2006; Lacki et al., 2011; Yoast-Hull et al., 2014; Eichmann and Becker Tjus, 2016).

High Mass X-ray Binaries

High mass X-ray binary stars (HMXBs) are products of massive star evolution where one component evolved to a compact object, a neutron star, or stellar-mass black hole (see Lewin et al., 1997). Gas accretion for the compact star's companion produces high luminosity X-rays with a wide range of patterns of variability. Because HMXBs are relatively

[4] *Large Synoptic Survey Telescope*; www.lsst.org

rare and the backgrounds in X-rays usually are relatively low, HMXBs are a potentially useful tracer of recent levels of star-forming activity (see also Chapter 11).

Grimm et al. (2003) demonstrated that SFRs measured by other means and the populations of HMXBs are correlated, thereby establishing the potential for X-ray studies to reveal rates of formation of very massive stars. However, Mineo et al. (2012) later found that the relationship between SFR measurements and properties of HMXBs displayed extra dispersion that suggested astrophysical factors in addition to SFRs are involved. The metallicity of the stellar populations is one such factor, since metal content affects mass loss and stellar radii (Brorby et al., 2016). Correcting for this effect leads to an improved and more promising SFR correlation that merits further development (Antoniou and Zezas, 2016).

Cepheid Variables and Patterns of Star Formation

Cepheid variables originate when stars following the main-sequence phase rapidly traverse the Hertzsprung gap where they become unstable to high amplitude pulsations (e.g. Chiosi et al., 1993, and references therein). Cepheids with pulsation periods longer than a few days originate from B-spectral class main sequence stars. The fundamental pulsation period of a star scales approximately as $\overline{\rho_*}^{-1/2}$ and thus depends on the stellar mass and, therefore, also stellar age. Observationally this information is encoded in terms of a Cepheid period-age-color relationship (Bono et al., 2005; Senchyna et al., 2015; Jacyszyn-Dobrzeniecka et al., 2016).

Maps of the locations and numbers of Cepheid variables as a function of pulsation period and color therefore offers a superb tracer of star-formation histories. Payne-Gaposchkin (1974) first applied this approach to map the locations of star-forming activity with position across the Magellanic Clouds (Grebel and Brandner, 1998). Later, Becker et al. (1977) demonstrated how the distribution of Cepheids with pulsation period could be utlilized to measure the variation of SFRs over time. Recent studies using Cepheids as tracers of galactic SFRs still focus on the Magellanic Clouds (e.g. Subramanian and Subramaniam, 2015; Ripepi et al., 2017) but, with the advent of deeper wide-area survey, this approach can be extended to a variety of nearby galaxies.

5.4 Lifetime Star-Formation Histories

In principle, deriving the SFH from the main-sequence luminosity function can be extended to the maximum age allowed by the absolute depth of a given dataset. This lookback time-depth correlation arises because the MSTO and subgiant branch get fainter with age (Sandage, 1956). In practice, this approach is complicated by the reduced detection efficiency at fainter magnitudes, which is exacerbated by the strongly increasing surface density of stars of lower mass due to the IMF.

The decreased brightness and increased crowding at faint magnitudes make the limiting useful depth of photometry an extremely strong function of instrumental resolution and target surface brightness. It is this circumstance that allowed so many gains to be made in understanding after the launch of *HST*, with its very high angular resolution and sharp point spread function. Progress in resolving cool stellar populations also has come from ground-based adaptive optics observations in the near-infrared of small fields of view. However, at the present, this approach has been applied to only a few galaxies with *HST* wide field photometry (e.g. Olsen et al., 2006).

Purely based on the quality of data, after two decades of *HST* our understanding of the star-formation history of M31 and some of its satellites (d $\sim$ 0.8–1 Mpc) (Brown et al., 2003; Dalcanton et al., 2012; Lewis et al., 2015) would be at the same stage as our understanding of the Magellanic Clouds (d $\approx$ 50 kpc) was, from ground-based, natural-seeing data (culminating in Harris and Zaritsky, 2004, 2009). In fact, our state of knowledge of M31 circa 2018 is more advanced than our knowledge of the LMC circa 1998 because of improvements to both stellar evolutionary tracks and analytical approaches over the intervening decades.

The vast majority of stars lie within the main sequence, which is also the best-modeled phase of stellar evolution. The mass-luminosity relationship that results from the physics of hydrostatic fusion of hydrogen into helium given the stellar equation of state allows the direct conversion of main-sequence luminosity functions into SFRs as a function of time.

The required ingredients are knowledge of the IMF and chemical composition of the stars (the former is usually assumed to be universal, while the latter must be be determined for each population to be studied, and is often both multivalued at a given age, and a function of age). With deep enough, high-quality data these factors may be determined simultaneously with the age distribution of stars to produce a complete star-formation history with a high degree of reliability (Cignoni et al., 2013; Weisz et al., 2013). The greater the complexity of the population in terms of the range of ages and metallicities represented, line-of-sight depth, and differential reddening, the more ancillary information is required to achieve the same relative precision in age dating and absolute SFR.

Further complications which have been addressed with varying degrees of success are the presence of bright pre-main sequence stars, the impact of binary stars (as extreme examples of crowding effects that alter the measured stellar brightness and color), rotation in massive stars, and the end-products of binary-star evolution. In practice, these systematic effects are almost impossible to tightly constrain using the same datasets which are used to derive the SFH (e.g. the discussions in Cignoni et al., 2016, 2018).

These systematic effects, together with uncertainties in the physical ingredients of stellar evolutionary tracks and color-temperature relations, conspire to limit the precision and accuracy of the inferred SFH in the nearest and simplest galaxies. In more distant systems, the systematic uncertainties due to stellar models can be brought to a minimum through achieving the greatest possible photometric depth and lowest possible crowding noise (see also Johnson et al., 2013); the other effects are best treated by applying a range of models and assumptions to the data in order to explore the full range of possible SFHs admissible to any dataset (e.g. Monelli et al., 2010b; Skillman et al., 2014).

5.4.1 Evolved Stars

If we imagine a simplified pathway of stellar evolution in which stars instantaneously vanish at the end of their main-sequence lifetime, then no information about the SFH could be derived for ages older than the age corresponding to the faintest reliably measured MSTO. In real systems, evolved stars make up a very small percentage of the total stellar mass, but disproportionately contribute to the integrated light of a system. For stars older than $\approx$1 Gyr, the post-main-sequence stages are more luminous (frequently by a large factor) than their main-sequence progenitors. This gives researchers something of a handle on the SFH at older ages than that of the faintest observed MSTO.

Red Giants

Stars with $0.5 \lesssim \mathsf{M} \lesssim 2$ develop fully degenerate helium cores after the main sequence, burning hydrogen in a shell outside the core via the CNO cycle. The shell-burning structure causes them to develop a deep convective envelope and expand in radius. As nearly fully convective stars, they evolve upward along the Hayashi track for their metallicity until the degenerate helium core becomes hot enough to ignite helium-burning through the 3α process.

This terminates red-giant branch evolution for all low-mass stars at nearly the same luminosity, which creates the CMD feature known as the RGB tip (TRGB). The TRGB, at $\mathrm{M}_I \approx -4$, is an extremely powerful distance indicator for populations older than $\approx$1 Gyr. However, the bulk properties of the upper RGB (e.g. color, TRGB magnitude) are not strong functions of stellar mass and the RGB, therefore, only provides weak constraints on SFR(t). The color of the RGB has a very strong dependence on metallicity and only a weak dependence on age.

As the convective envelope deepens during the ascent of a star up the RGB, it may encounter composition discontinuities left behind by hydrogen burning processes, temporarily reversing the upward evolution in the CMD. This produces a local feature in the RGB luminosity function known as the RGB-bump. The luminosity of the RGB-bump is a function of stellar mass and metallicity, and is often used as a test of the validity of stellar-evolution models.

The radius and therefore $\mathrm{T}_{\mathrm{eff}}$ of an RGB star is not calculable from first principles, and so the color of an RGB star as a function of metallicity (and to a lesser extent, age) must be empirically calibrated. This remains a point of disagreement between various stellar-evolution models and therefore a potential source of systematic error in attempts to derive galaxy SFHs from CMD data including the RGB. The RGB is therefore best used in conjunction with spectroscopic metallicity data to constrain the age-metallicity relation, and to support SFH inferences based on deep main sequence data (Monelli et al., 2010a).

CMDs that do not reach the TRGB can only provide integrated stellar-mass information for the lifetime SFH of a system. This immediately follows from convergence of evolved stars from the lower main sequence on similar red giant branches whose properties thus are not particularly sensitive to stellar ages. For all but the youngest stars, integrated light methods (colors, spectra) will provide better results.

Red Clump and Horizontal Branch

For stars in which helium-burning commences under non-degenerate conditions ($M \gtrsim 2$), the stars fall along a well-defined mass-luminosity relation, significantly brighter than the main sequence but much less populous, for their core helium-burning lifetime (the blue loops, q.v.). In low-mass stars, this mass-luminosity relation is terminated at $M_V \approx +0.5$ at the faint end by the physics of the electron degenerate helium core at the TRGB. Low-mass stars over a factor of $\approx$4 in total mass burn helium at nearly a common luminosity; the initial size of the helium core is nearly constant for these stars, although they differ in the mass of overlying hydrogen-rich material, and therefore in the percentage of their luminosity produced in a hydrogen-burning shell (Sweigart and Gross, 1976; Seidel et al., 1987; Girardi, 2016).

The funneling of low-mass stars with a wide range of masses into a small region of the CMD produces a feature known as the red clump (Cannon, 1970). As the most populous post-main sequence phase of stellar evolution, star counts in the red clump are a good indicator of the total star formation, integrated over the lifetime of the galaxy from $\approx$1–13 Gyr.

The red clump is not a precise standard candle, but its properties vary in a subtle but regular way with metallicity and stellar mass. The thicker the overlying envelope and the higher the metallicity of the star, the lower the T_{eff}, and the closer the star remains to the RGB during helium burning. At fixed stellar mass, lower-metallicity stars have higher luminosity; at fixed metallicity, lower-mass stars have lower luminosity. These properties mean that for sufficiently high quality data, the red clump is an extremely useful addition to SFH determinations. In conjunction with the RGB color (metallicity) and main sequence luminosity function ($\Psi(t, Z)$), the red clump can play a critical role in ensuring that self-consistent metallicity, reddening, and age distributions are inferred.

At the low-mass and low-metallicity end, the envelope of material above the helium-burning core is insufficient to support a deep convective envelope, and the star trends to higher T_{eff}, leading to much bluer colors than the RGB. Among the oldest and most metal-poor stars, factors other than initial stellar mass control the envelope mass and hence the color of the star during core helium-burning. This causes stars of the same age and initial composition to spread out in the CMD, smearing the red clump into a horizontal branch (most prominently observed in globular clusters, classical dwarf spheroidal galaxies, and galaxy halos). Because the horizontal branch color and spread of colors is controlled by things such as mass loss during the red giant phase, it is in general difficult to quantitatively derive SFH information from the HB properties. The HB is extremely useful though, as a clear signpost pointing to the existence of stellar populations older than $\approx$10 Gyr, which are otherwise quite difficult to detect if younger populations are also present.

CMDs that do not reach the level of the red clump/horizontal branch have very little age-resolution for ages older than $\approx$1 Gyr and are strongly subject to age-metallicity-reddening degeneracies. Rough star-formation histories can be determined, especially if independent metallicity information is available.

Asymptotic Giant Branch (AGB)

After the exhaustion of helium in the core of a star, it will tend to evolve back to cooler T_{eff} as the core contracts and shell burning outside the core ignites. In low-mass stars, a double-shell structure outside the degenerate carbon-oxygen core is established, with alternating phases of helium- and hydrogen-burning in shells. Such double-shell stars asymptotically approach the RGB as the convective envelope expands, although they lack a helium core, so the evolution is not terminated at the TRGB. These stars are more luminous and extend to much lower T_{eff} than first-ascent red giants, and are hence easily detectable by infrared observations.

The complicated structure and low T_{eff} of AGB stars makes them difficult to model, so quantitative SFH information is rather uncertain and model dependent. However, the number of luminous AGB stars compared to RGB stars is a decreasing function of age (for ages older than ≈ 0.5 Gyr), so a strong AGB population is prima facie evidence for populations aged ≈ 1 Gyr. The deep convective envelopes of AGB stars can dredge CNO-processed material to the surface of the star, changing the abundances. This process is responsible for production of carbon-rich stars, which have easily identifiable characteristics in near-infrared CMDs. The efficiency of C star production is a function of both age and metallicity, so the ratio of C- to M-type AGB stars can be taken as a rough SFH indicator (e.g. Harbeck et al., 2004; Cioni et al., 2008).

5.4.2 Examples and Key Results

The use of deep CMDs, analyzed by producing synthetic Hess diagrams from stellar models and optimized using statistical methods, gives the most complete and reliable determinations of star-formation history over the lifetime of a galaxy (e.g. Aparicio et al., 1996; Dolphin, 2002; Tolstoy et al., 2009). This field has burgeoned in the last 20 years, driven by *HST* imaging and improving stellar isochrones.

At the time this chapter was prepared, a large fraction of galaxies within the Local Group have imaging reaching to the oldest MSTO; many galaxies out to the distance of the M81 group have imaging reaching to the HB/red clump. In the former case, inferences on the lifetime star-formation history of the galaxies are limited by the angular size of the galaxies compared to the *HST* field of view. In the case of more distant galaxies, further progress is limited by the steeply increasing time cost of deeper photometry and the crowding limits set by the pixel scale of *HST* and the galaxy surface brightnesses.

As *HST* approaches the end of its mission and preparations begin for the *James Webb Space Telescope* (*JWST*),[5] it is useful to consider an overview of the key insights into galaxy formation and evolution delivered by deep CMD analyses and point to the areas in which future progress may be expected. The strengths of Hess diagram-fitting measures of

[5] *James Webb Space Telescope* www.stsci.edu/jwst/

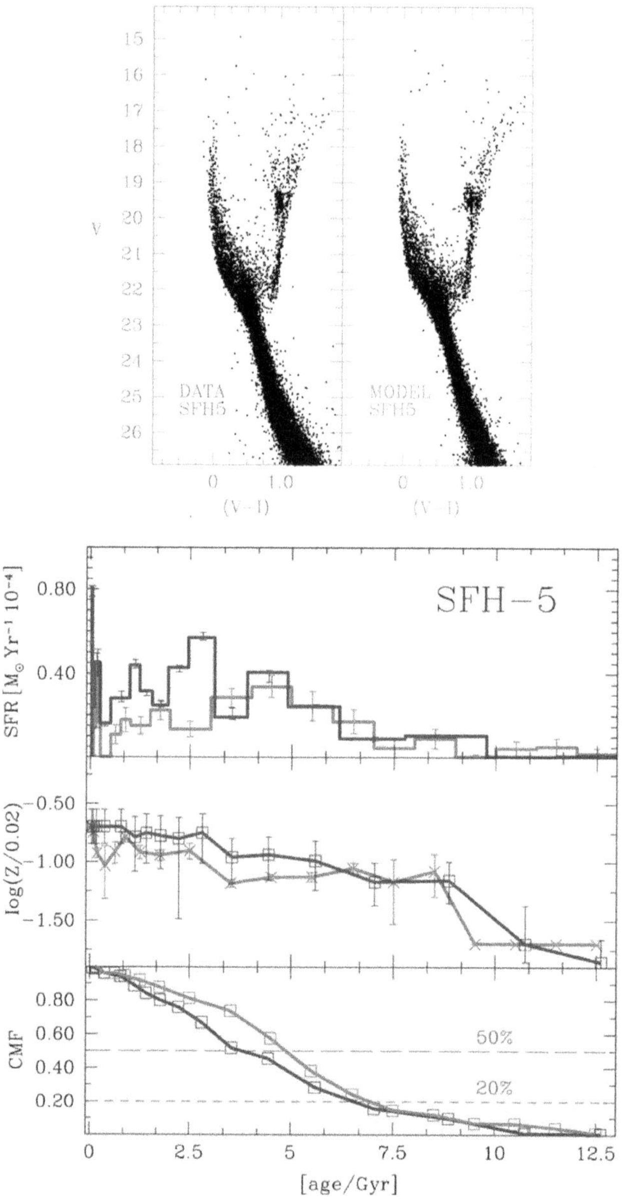

Figure 5.3 These diagrams illustrate results from SFHs derived by fitting models to Hess diagrams obtained with *HST*.
Top: Observed Hess diagram for an *HST* pointing in the Small Magellanic Cloud, and predicted CMD from a SFH model to its immediate right.
Bottom: Output from the SFH model for field 5.
From Cignoni et al. (2013). © AAS. Reproduced with permission.

SFH are clear. Direct evidence for the timing and duration of past bursts of star formation is relatively easily gathered, along with information concerning the time evolution of metallicity in the system. Galaxies with very unusual star-formation histories, subject to large previous bursts or delays in the onset of star formation can be identified.

This creates a link between galaxy-evolution theory and direct measures of the link between galaxy assembly, environment, star formation, and chemical enrichment. However, the *location* of the stars at the time of their formation cannot be easily measured, as it is not obvious whether a given population formed in situ or was accreted post-formation.

The absolute timing of bursts is model-dependent; at older ages in particular this may thwart efforts to, for example link the initial bursts or early quenching of star formation to environmental factors like photoheating of gas after cosmological reionization. Depth and angular resolution are essential, as is thorough areal coverage of the target systems in order to account for population gradients. When these factors are appropriately considered, Hess diagram-fitting allows the exploration of connections between the lifetime average star-formation rate, galaxy morphology and gas content, and the timing and duration of major star-formation events over a Hubble time.

Case Study: The Carina Dwarf Spheroidal

The Carina dSph (d $\approx$100 kpc) provides an excellent example of the improving measures of SFH gained by increased CMD depth. At the time of this galaxy's discovery in 1977, dwarf spheroidals were considered to be closely analogous to the globular clusters of the Milky Way halo, i.e. purely old and metal-poor systems. As very low mass systems, they have correspondingly few evolved stars from which to easily characterize their SFH. Cannon et al. (1981) serendipitously discovered carbon stars near the TRGB in Carina, which was the first suggestion that its populations differed from the Milky Way halo. Mould and Aaronson (1983) made the first CCD-based CMD of Carina and concluded based on the relatively bright MSTO that Carina was dominated by intermediate-age stars ($\approx$7.5 Gyr). Smecker-Hane et al. (1994) painstakingly obtained imaging down to $\approx$ 1 mag below the horizontal branch and were able to clearly identify red clump and horizontal-branch features corresponding to both ancient ($\gtrsim$ 10 Gyr) and intermediate-age ($\approx$ 6 Gyr) populations. The first photometry deep enough to reach the old MSTO (Hurley-Keller et al., 1998) revealed the existence and relative proportions of three distinct bursts of star formation and showed that the majority of stars formed in the intermediate-age burst $\approx$7 Gyr ago. This picture has been further refined by the combination of very deep *HST* imaging over a small field, deep wide-field imaging from the ground, quantitative CMD-fitting, and spectroscopically determined metallicities (Koch et al., 2008; de Boer et al., 2014); this has revealed the timing, duration, and amplitude of the main bursts of star formation, the age-metallicity relation, and their radial variations.

Star-Formation Histories: Summary

A number of high level, general conclusions can be drawn based on the sum total of studies in the literature. Regardless of the specific stellar population or instrument, these can be presented as principles to guide future studies of galaxy star-formation histories.

1. Conclusions about the long-term star-formation history are dominated by star counts at the faint end of any given dataset. It is absolutely critical to have a thorough understanding of the effects of crowding on the photometric errors and completeness of the data, modeled through extensive artificial star tests.
2. Systematic effects dominate the uncertainties in the long-term SFH for CMDs shallower than the oldest MSTO (Weisz et al., 2011). Random errors are a strong function of the total number of stars sampled brighter than the oldest MSTO. Both may be reduced by ancillary information such as multiple colors or independent metallicity measurements.
3. Given sufficiently accurate evolutionary tracks and color-temperature transformations, it is possible to glean information about the SFH at ages older than the oldest cleanly resolved MSTO on the basis of evolved stars. However, it must be recognized that the age resolution will be severely degraded compared to main-sequence-based measures, and the start and end times of inferred episodes of star formation will be dominated by systematic errors including age-metallicity-reddening degeneracy and uncertainties in the stellar modeling. Furthermore, in low-density systems where few evolved stars are present, such inferences will be strongly limited by shot noise. The best way to add reliability to evolved-star inference is to incorporate metallicity information from other means such as spectroscopy of large samples of stars.
4. In the words of Hodge (1989), "it is difficult to distinguish an [unevolved, main sequence] low-mass old star from a low-mass young star." However, it is still of great use to obtain resolved-star photometry below the oldest MSTO, because in this fashion the slope of the initial mass function at low masses can be directly measured. The entire enterprise of deriving relative SFRs as a function of age hinges on knowledge of the IMF (Sandage, 1956). While it is observationally expensive to obtain good data for the faintest stars in a system ($M_V \gtrsim +6$), doing so has the additional benefit of better completeness and better characterization of crowding effects at the oldest MSTO.

5.5 Future Work

In the next several years improvements are likely to be made in at least five critical technical areas: angular resolution, spectroscopic surveys, deep surveys for variable sources, stellar-model reliability, and statistical sophistication. These advances will support the further extension of resolved stellar-population studies into several key areas, including birth rates of massive binaries as progenitors of compact binaries and neutron stars that can merge, understanding the nature of the IMF in a wider range of astrophysical environments, and the properties of stellar populations in low-density regions of the nearby universe.

The forefront of angular resolution most likely will come from the combination of optical interferometers and ground-based telescopes equipped with multi-conjugate adaptive optics. Existing ground-based telescopes already have apertures/baselines exceeding those of space telescopes that are foreseeable in the next decade, and 20–30m aperture telescopes are in development. The reach of large aperture ground-based telescopes will be further enhanced through the continued advancement of integral field spectroscopy.

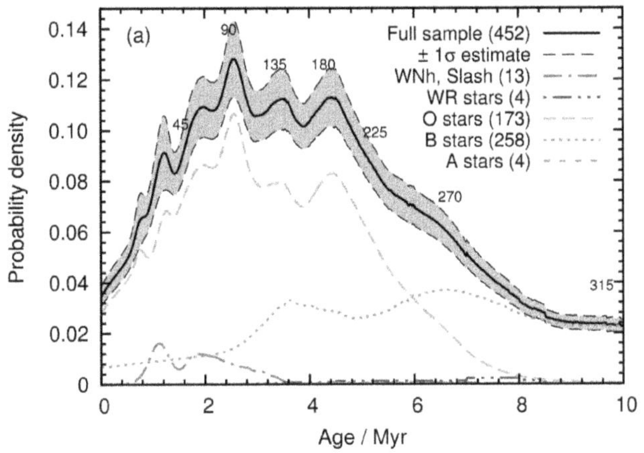

Figure 5.4 SFR versus time for the wider 30 Doradus region in the Large Magellanic Cloud based on a spectroscopic survey of massive stars. This investigation offers a preview of future studies where the young stellar populations can be assessed through spectroscopic observations that provide information on the numbers and nature of the most massive stars (see also Evans et al., 2011). Credit: Schneider et al. (2018), reproduced with permission © ESO. Color version available online.

This combination can allow spectroscopic studies of substantial populations of stars in galaxies beyond the Local Group.

Pioneering studies of this type are ongoing in the Magellanic Clouds. For example, a spectroscopic SFR determination based on ESO VLT observations was obtained as part of an investigation of massive stars in the 30 Doradus region (Evans et al., 2011; Schneider et al., 2018, see Fig. 5.4). Another program to observe and analyze massive stars in the Large Magellanic Cloud (LMC) N206 complex illustrates how high-quality spectra can be combined with detailed stellar-atmosphere models to measure feedback from massive stars (Ramachandran et al., 2018). In addition, spectroscopic surveys are exploring the massive stellar content in Local Group galaxies with increasing depth (e.g. Massey et al., 2016). These projects demonstrate the essential role of spectra in defining the properties of high-mass stellar populations, as in both instances a few very massive stars, which would be difficult to identify using photometric techniques, dominate photoionization rates and mechanical-energy inputs. Spectra in the ultraviolet to near-infrared offer the only reliable way to ascertain the properties of the most-massive stars as the peaks of their energy distributions are inevitably hidden by absorption in the Lyman continuum in addition to probing the frequency and nature of binary stars.

The likely increase in ground-based infrared capabilities to match the best performance of adaptive-optics systems may promote a shift in the choice of stars for studies of SFRs, Recent SFRs, for example, could be keyed to numbers and luminosities of red supergiants (e.g. Alexander et al., 2009), while intermediate time spans can be explored through measurements of AGB star populations (Olszewski et al., 1996; Cioni et al., 2008;

Rezaeikh et al., 2014). Given the narrow range in T_{eff} of cool evolved stars, a combination of spectroscopy, photometry, updated stellar models, and variability studies will be needed to ascertain initial masses that hold the key for determinations of SFRs and SFHs (Davies et al., 2013; Tabernero et al., 2018). In addition, surveys such as that planned for the *LSST* offer the potential to further utilize the properties of long-period variable stars and eclipsing binaries to probe stellar populations (e.g. Battinelli and Demers, 2011).

The *JWST* will offer a leap in capabilities in terms of spectral coverage and sensitivity in the near to mid-infrared, thus opening a new frontier for studies of resolved cool stars. We can, therefore, anticipate better utilization of AGB stars and red supergiants as evolutionary tracers that can be accessed even in relatively dust-obscured regions. The *JWST* also will offer unique opportunities for area spectroscopy that can improve our ability to characterize the properties of luminous cool stellar populations. In nearer galaxies, including the Magellanic Clouds, the *JWST* will offer greatly increased access to stars on the lower main sequence and to protostars (e.g. Sabbi et al., 2007; Whitney et al., 2008; Carlson et al., 2011; De Marchi et al., 2017) that will allow SFRs to be determined based on the types of stars that contain most of the stellar mass in systems beyond our Galactic family.

Measurements of SFRs for external galaxies also rely on the quality of the astrophysical calibration of stellar observables, and especially their spectra (Lançon et al., 2016). The results from *Gaia*[6] as well as ongoing massive spectroscopic surveys of Galactic stars are building a new foundation for stellar-population research. These and other projects focused on the Milky Way and its satellites are yielding essential insights into the properties of multiple stars, chemical abundance patterns, and correlations between stellar and dynamical properties of stellar populations.

Several initiatives are in progress to measure the properties of larger samples of resolved massive stars. These projects will observe OB stars in the far ultraviolet and especially target sub-solar metallicities. For example, *Hubble Space Telescope's* Ultraviolet Legacy Library of Young Stars as Essential Standards (ULLYSES) project will provide far ultraviolet spectra for large samples of nearby OB stars. This will serve as a basis for deriving better physical parameters and evolutionary models for this critical class of stars and in turn support better techniques for determining SFRs. We also anticipate the *James Webb Space Telescope* will add to the store of information on resolved populations of deeply embedded massive stars, thereby enhancing our ability to connect unobscured stellar populations to SFRs.

As a result the empirical framework for the analysis of observations in galaxies should soon be much more robust than what we currently have. This, in turn, will allow better estimations of errors that are a critical factor in producing stellar population models that yield SFRs and SFHs, as well as key properties, such as the form of the upper stellar IMF in galaxies throughout the Local Supercluster. We will be in a much stronger position to also combine information from resolved and unresolved components of galaxies to produce uni-

[6] https://sci.esa.int/web/gaia

fied evolutionary models. The combination of deeper surveys and better multi-wavelength observational capabilities also can lead to improved use of supernovae as tracers of SFRs in starbursts and other high SFR systems that are not well represented in samples of nearby galaxies. These advances will allow us to make quantitative tests of evolutionary models of extragalactic stellar systems reaching over a wider range of space and time.

References

Aaronson, M., and Mould, J. 1982. The Extended Giant Branches of Intermediate Age Globular Clusters in the Magellanic Clouds. II. *Astrophysical Journal Supplement Series*, **48**(Feb.), 161–184.

Alexander, M. J., Kobulnicky, H. A., Clemens, D. P. et al. 2009. The Discovery of a Massive Cluster of Red Supergiants with GLIMPSE. *Astronomical Journal*, **137**(June), 4824–4833.

Aloisi, A., Tosi, M., and Greggio, L. 1999. The Star Formation History of I ZW 18. *Astronomical Journal*, **118**(July), 302–322.

Annibali, F., Cignoni, M., Tosi, M. et al. 2013. The Star Formation History of the Very Metal-Poor Blue Compact Dwarf I Zw 18 from HST/ACS Data. *Astronomical Journal*, **146**(Dec.), 144.

Antoniou, V., and Zezas, A. 2016. Star Formation History and X-ray Binary Populations: The Case of the Large Magellanic Cloud. *Monthly Notices of the Royal Astronomical Society*, **459**(June), 528–553.

Antonucci, R. R. J., and Ulvestad, J. S. 1988. A Large Family of Compact Radio Sources in the Starburst Nucleus of NGC 253. *Astrophysical Journal*, **330**(July), L97–L100.

Aparicio, A., and Gallart, C. 2004. IAC-STAR: A Code for Synthetic Color-Magnitude Diagram Computation. *Astronomical Journal*, **128**(Sept.), 1465–1477.

Aparicio, A., and Hidalgo, S. L. 2009. IAC-Pop: Finding the Star Formation History of Resolved Galaxies. *Astronomical Journal*, **138**(Aug.), 558–567.

Aparicio, A., Gallart, C., Chiosi, C. et al. 1996. Model Color-Magnitude Diagrams for Hubble Space Telescope Observations of Local Group Dwarf Galaxies. *Astrophysical Journal*, **469**(Oct.), L97.

Baade, W. 1944a. NGC 147 and NGC 185, Two New Members of the Local Group of Galaxies. *Astrophysical Journal*, **100**(Sept.), 147.

Baade, W. 1944b. The Resolution of Messier 32, NGC 205, and the Central Region of the Andromeda Nebula. *Astrophysical Journal*, **100**(Sept.), 137.

Bastian, N., Covey, K. R., and Meyer, M. R. 2010. A Universal Stellar Initial Mass Function? A Critical Look at Variations. *Annual Review of Astronomy and Astrophysics*, **48**(Sept.), 339–389.

Batejat, F., Conway, J. E., Hurley, R. et al. 2011. Resolution of the Compact Radio Continuum Sources in Arp220. *Astrophysical Journal*, **740**(Oct.), 95.

Battinelli, P., and Demers, S. 2011. Long Period Variable Survey in NGC 6822. *Astronomy and Astrophysics*, **525**(Jan.), A69.

Becker, S. A., Iben, Jr., I., and Tuggle, R. S. 1977. On the Frequency-Period Distribution of Cepheid Variables in Galaxies in the Local Group. *Astrophysical Journal*, **218**(Dec.), 633–653.

Bondi, M., Pérez-Torres, M. A., Herrero-Illana, R. et al. 2012. The Nuclear Starburst in Arp 299-A: From the 5.0 GHz VLBI Radio Light-Curves to Its Core-Collapse Supernova Rate. *Astronomy and Astrophysics*, **539**(Mar.), A134.

Bono, G., Marconi, M., Cassisi, S. et al. 2005. Classical Cepheid Pulsation Models. X. The Period-Age Relation. *Astrophysical Journal*, **621**(Mar.), 966–977.

Brorby, M., Kaaret, P., Prestwich, A. et al. 2016. Enhanced X-ray Emission from Lyman Break Analogues and a Possible L_X-SFR-Metallicity Plane. *Monthly Notices of the Royal Astronomical Society*, **457**(Apr.), 4081–4088.

Brown, T. M., Ferguson, H. C., Smith, E. et al. 2003. Evidence of a Significant Intermediate-Age Population in the M31 Halo from Main-Sequence Photometry. ApJ, **592**(July), L17–L20.

Butcher, H. 1977. A Main-Sequence Luminosity Function for the Large Magellanic Cloud. *Astrophysical Journal*, **216**(Sept.), 372–380.

Calzetti, D., Lee, J. C., Sabbi, E. et al. 2015. Legacy Extragalactic UV Survey (LEGUS) with the Hubble Space Telescope. I. Survey Description. *Astronomical Journal*, **149**(Feb.), 51.

Cannon, R. D. 1970. Red Giants in Old Open Clusters. *Monthly Notices of the Royal Astronomical Society*, **150**, 111–135.

Cannon, R. D., Niss, B., and Norgaard-Nielsen, H. U. 1981. The Discovery of Two Carbon Stars in the Carina Dwarf Galaxy. *Monthly Notices of the Royal Astronomical Society*, **196**(July), 1P.

Carlson, L. R., Sewiło, M., Meixner, M. et al. 2011. A Panchromatic View of NGC 602: Time-Resolved Star Formation with the Hubble and Spitzer Space Telescopes. *Astrophysical Journal*, **730**(Apr.), 78.

Cash, W. 1979. Parameter Estimation in Astronomy through Application of the Likelihood Ratio. *Astrophysical Journal*, **228**(Mar.), 939–947.

Chini, R., Hoffmeister, V. H., Nasseri, A. et al. 2012. A Spectroscopic Survey on the Multiplicity of High-Mass Stars. *Monthly Notices of the Royal Astronomical Society*, **424**(Aug.), 1925–1929.

Chiosi, C., Wood, P. R., and Capitanio, N. 1993. Theoretical Models of Cepheid Variables and their BVI(c) Colors and Magnitudes. *Astrophysical Journal Supplement Series*, **86**(June), 541–598.

Chomiuk, L., and Povich, M. S. 2011. Toward a Unification of Star Formation Rate Determinations in the Milky Way and Other Galaxies. *Astronomical Journal*, **142**(Dec.), 197.

Cignoni, M., and Tosi, M. 2010. Star Formation Histories of Dwarf Galaxies from the Colour-Magnitude Diagrams of Their Resolved Stellar Populations. *Advances in Astronomy*, **2010**, 158568.

Cignoni, M., Cole, A. A., Tosi, M. et al. 2013. Mean Age Gradient and Asymmetry in the Star Formation History of the Small Magellanic Cloud. *Astronomical Journal*, **775**(Oct.), 83.

Cignoni, M., Sabbi, E., van der Marel, R. P. et al. 2016. Hubble Tarantula Treasury Project V. The Star Cluster Hodge 301: The Old Face of 30 Doradus. *Astrophysical Journal*, **833**(Dec.), 154.

Cignoni, M., Sacchi, E., Aloisi, A. et al. 2018. Star Formation Histories of the LEGUS Dwarf Galaxies. I. Recent History of NGC 1705, NGC 4449, and Holmberg II. *Astrophysical Journal*, **856**(Mar.), 62.

Cioni, M.-R. L., Clementini, G., Girardi, L. et al. 2011. The VMC Survey. I. Strategy and First Data. *Astronomy and Astrophysics*, **527**(Mar.), A116.

Cioni, M.-R. L., Irwin, M., Ferguson, A. M. N. et al. 2008. AGB Stars as Tracers of Metallicity and Mean Age across M 33. *Astronomy and Astrophysics*, **487**(Aug.), 131–146.

Cole, A. A., Tolstoy, E., Gallagher, III, J. S. et al. 2005. Spectroscopy of Red Giants in the Large Magellanic Cloud Bar: Abundances, Kinematics, and the Age-Metallicity Relation. *Astronomical Journal*, **129**(Mar.), 1465–1482.

Cole, A. A., Skillman, E. D., Tolstoy, E. et al. 2007. Leo A: A Late-Blooming Survivor of the Epoch of Reionization in the Local Group. *Astrophysical Journal*, **659**(Apr.), L17–L20.

Cole, A. A., Weisz, D. R., Dolphin, A. E. et al. 2014. Delayed Star Formation in Isolated Dwarf galaxies: Hubble Space Telescope Star Formation History of the Aquarius Dwarf Irregular. *Astrophysical Journal*, **795**(Nov.), 54.

Dalcanton, J. J., Williams, B. F., Lang, D. et al. 2012. The Panchromatic Hubble Andromeda Treasury. *Astrophysical Journal Supplement Series*, **200**(June), 18.

Davies, B., Kudritzki, R.-P., Plez, B. et al. 2013. The Temperatures of Red Supergiants. *Astrophysical Journal*, **767**(Apr.), 3.

de Boer, T. J. L., Tolstoy, E., Lemasle, B. et al. 2014. The Episodic Star Formation History of the Carina Dwarf Spheroidal Galaxy. *Astronomy and Astrophysics*, **572**(Dec.), A10.

De Marchi, G., Panagia, N., and Beccari, G. 2017. Photometric Determination of the Mass Accretion Rates of Pre-Main-Sequence Stars. V. Recent Star Formation in the 30 Dor Nebula. *Astrophysical Journal*, **846**(Sept.), 110.

Dohm-Palmer, R. C., Skillman, E. D., Saha, A. et al. 1997. The Dwarf Irregular Galaxy Sextans A. II. Recent Star Formation History. *Astronomical Journal*, **114**(Dec.), 2527.

Dolphin, A. E. 2000. WFPC2 Stellar Photometry with HSTPHOT. *Publications of the Astronomical Society of the Pacific*, **112**(Oct.), 1383–1396.

Dolphin, A. E. 2002. Numerical Methods of Star Formation History Measurement and Applications to Seven Dwarf Spheroidals. *Monthly Notices of the Royal Astronomical Society*, **332**(May), 91–108.

Duchêne, G., and Kraus, A. 2013. Stellar Multiplicity. *Annual Review of Astronomy and Astrophysics*, **51**(Aug.), 269–310.

Duquennoy, A., Mayor, M., and Halbwachs, J.-L. 1991. Multiplicity among Solar Type Stars in the Solar Neighbourhood. I – CORAVEL Radial Velocity Observations of 291 Stars. *Astronomy and Astrophysics Supplement Series*, **88**(May), 281–324.

Eichmann, B., and Becker Tjus, J. 2016. The Radio-Gamma Correlation in Starburst Galaxies. *Astrophysical Journal*, **821**(Apr.), 87.

Elmegreen, B. G. 2000. Star Formation in a Crossing Time. *Astrophysical Journal*, **530**(Feb.), 277–281.

Evans, C. J., Taylor, W. D., Hénault-Brunet, V. et al. 2011. The VLT-FLAMES Tarantula Survey. I. Introduction and Observational Overview. *Astronomy and Astrophysics*, **530**(June), A108.

Fenech, D. M., Muxlow, T. W. B., Beswick, R. J. et al. 2008. Deep MERLIN 5GHz Radio Imaging of Supernova Remnants in the M82 Starburst. *Monthly Notices of the Royal Astronomical Society*, **391**(Dec.), 1384–1402.

Freedman, W. L. 1988. Stellar Content of Nearby Galaxies. I – BVRI CCD Photometry for IC 1613. *Astronomical Journal*, **96**(Oct.), 1248–1306.

Gallagher, J. S., Tolstoy, E., Dohm-Palmer, R. C. et al. 1998. A Wide Field Planetary Camera 2 Study of the Resolved Stellar Population of the Pegasus Dwarf Irregular Galaxy (DDO 216). *Astronomical Journal*, **115**(May), 1869–1887.

Gallagher, III, J. S., Hunter, D. A., and Tutukov, A. V. 1984. Star Formation Histories of Irregular Galaxies. *Astrophysical Journal*, **284**(Sept.), 544–556.

Gallart, C., Aparicio, A., and Vílchez, J. M. 1996. The Local Group Dwarf Irregular Galaxy NGC 6822. I. The Stellar Content. *Astronomical Journal*, **112**(Nov.), 1928.

Gallart, C., Zoccali, M., and Aparicio, A. 2005. The Adequacy of Stellar Evolution Models for the Interpretation of the Color-Magnitude Diagrams of Resolved Stellar Populations. *Annual Review of Astronomy and Astrophysics*, **43**(Sept.), 387–434.

Gallart, C., Freedman, W. L., Aparicio, A. et al. 1999. The Star Formation History of the Local Group Dwarf Galaxy Leo I. *Astronomical Journal*, **118**(Nov.), 2245–2261.

Girardi, L. 2016. Red Clump Stars. *Annual Review of Astronomy and Astrophysics*, **54**(Sept.), 95–133.

Gordon, K. D., Fouesneau, M., Arab, H. et al. 2016. The Panchromatic Hubble Andromeda Treasury. XV. The BEAST: Bayesian Extinction and Stellar Tool. *Astrophysical Journal*, **826**(Aug.), 104.

Grebel, E. K. 1997. Star Formation Histories of Local Group Dwarf Galaxies. (Ludwig Biermann Award Lecture 1996). Pages 29–60 of: Schielicke, R. E. (ed.), *Reviews in Modern Astronomy*. Reviews in Modern Astronomy, vol. 10.

Grebel, E. K., and Brandner, W. 1998 (Dec). The Recent Star Formation History of the Large Magellanic Cloud. Pages 151–154 of: *Magellanic Clouds and Other Dwarf Galaxies*.

Greggio, L., and Renzini, A. 2011. *Stellar Populations: A User Guide from Low to High Redshift*.

Grimm, H.-J., Gilfanov, M., and Sunyaev, R. 2003. High-Mass X-ray Binaries as a Star Formation Rate Indicator in Distant Galaxies. *Monthly Notices of the Royal Astronomical Society*, **339**(Mar.), 793–809.

Grocholski, A. J., van der Marel, R. P., Aloisi, A. et al. 2012. HST/ACS Photometry of Old Stars in NGC 1569: The Star Formation History of a Nearby Starburst. *Astronomical Journal*, **143**(May), 117.

Harbeck, D., Gallagher, III, J. S., and Grebel, E. K. 2004. WIYN Survey for Carbon Stars in the M31 and Cetus Dwarf Spheroidal Galaxies: Evolutionary Implications. *Astronomical Journal*, **127**(May), 2711–2722.

Harris, J., and Zaritsky, D. 2004. The Star Formation History of the Small Magellanic Cloud. *Astronomical Journal*, **127**(Mar.), 1531–1544.

Harris, J., and Zaritsky, D. 2009. The Star Formation History of the Large Magellanic Cloud. *Astronomical Journal*, **138**(Nov.), 1243–1260.

Herschel, Sir, J. F. W. 1847. *Results of astronomical observations made during the years 1834, 5, 6, 7, 8, at the Cape of Good Hope; being the completion of a telescopic survey of the whole surface of the visible heavens, commenced in 1825.*

Hodge, P. 1989. Populations in Local Group galaxies. *Annual Review of Astronomy and Astrophysics*, **27**, 139–159.

Hodge, P. W. 1971. Dwarf Galaxies. *Annual Review of Astronomy and Astrophysics*, **9**, 35.

Horiuchi, S., Nakamura, K., Takiwaki, T. et al. 2014. The Red Supergiant and Supernova Rate Problems: Implications for Core-Collapse Supernova Physics. *Monthly Notices of the Royal Astronomical Society*, **445**(Nov.), L99–L103.

Hubble, E. P. 1929. A Spiral Nebula as a Stellar System, Messier 31. *Astrophysical Journal*, **69**(Mar.).

Hunter, D. A., and Gallagher, III, J. S. 1985. Star-Forming Properties and Histories of Dwarf Irregular Galaxies – Down but Not Out. *Astrophysical Journal Supplement Series*, **58**(Aug.), 533–560.

Hurley-Keller, D., Mateo, M., and Nemec, J. 1998. The Star Formation History of the Carina Dwarf Galaxy. *Astronomical Journal*, **115**(May), 1840–1855.

Iben, Jr., I. 2013a. *Stellar Evolution Physics, Vol. 1: Physical Processes in Stellar Interiors*.

Iben, Jr., I. 2013b. *Stellar Evolution Physics, Vol. 2: Advanced Evolution of Single Stars*.

Jacyszyn-Dobrzeniecka, A. M., Skowron, D. M., Mróz, P. et al. 2016. OGLE-ing the Magellanic System: Three-Dimensional Structure of the Clouds and the Bridge Using Classical Cepheids. *Acta Astronomica*, **66**(June), 149–196.

Johnson, B. D., Weisz, D. R., Dalcanton, J. J. et al. 2013. Measuring Galaxy Star Formation Rates from Integrated Photometry: Insights from Color-Magnitude Diagrams of Resolved Stars. *Astrophysical Journal*, **772**(July), 8.

Kennicutt, R. C., and Evans, N. J. 2012. Star Formation in the Milky Way and Nearby Galaxies. *Annual Review of Astronomy and Astrophysics*, **50**(Sept.), 531–608.

Kimani, N., Sendlinger, K., Brunthaler, A. et al. 2016. Radio Evolution of Supernova SN 2008iz in M 82. *Astronomy and Astrophysics*, **593**(Aug.), A18.

Kippenhahn, R., Weigert, A., and Weiss, A. 2012. *Stellar Structure and Evolution*.

Koch, A., Grebel, E. K., Gilmore, G. F. et al. 2008. Complexity on Small Scales. III. Iron and α Element Abundances in the Carina Dwarf Spheroidal Galaxy. *Astronomical Journal*, **135**(Apr.), 1580–1597.

Kronberg, P. P., and Wilkinson, P. N. 1975. High-Resolution, Multifrequency Radio Observations of M82. *Astrophysical Journal*, **200**(Sept.), 430–435.

Lacki, B. C., Thompson, T. A., Quataert, E. et al. 2011. On the GeV and TeV Detections of the Starburst Galaxies M82 and NGC 253. *Astrophysical Journal*, **734**(June), 107.

Lançon, A., Prugniel, P., Powalka, M. et al. 2016 (Dec.). Desirable Evolutions of Stellar Spectroscopic Services in the Light of Current Studies of Remote Stellar Populations. Pages 329–332 of: Reylé, C., Richard, J., Cambrésy, L. et al. (eds), *SF2A-2016: Proceedings of the Annual meeting of the French Society of Astronomy and Astrophysics*.

Lee, J. C., Kennicutt, Jr., R. C., Funes, S. J. J. G. et al. 2009. Dwarf Galaxy Starburst Statistics in the Local Volume. *Astrophysical Journal*, **692**(Feb.), 1305–1320.

Lenc, E., and Tingay, S. J. 2006. The Subparsec-Scale Radio Properties of Southern Starburst Galaxies. I. Supernova Remnants, the Supernova Rate, and the Ionized Medium in the NGC 253 Starburst. *Astronomical Journal*, **132**(Sept.), 1333–1345.

Lewin, W. H. G., van Paradijs, J., and van den Heuvel, E. P. J. 1997. *X-ray Binaries*.

Lewis, A. R., Dolphin, A. E., Dalcanton, J. J. et al. 2015. The Panchromatic Hubble Andromeda Treasury. XI. The Spatially Resolved Recent Star Formation History of M31. *Astrophysical Journal*, **805**(June), 183.

Lonsdale, C. J., Diamond, P. J., Thrall, H. et al. 2006. VLBI Images of 49 Radio Supernovae in Arp 220. *Astrophysical Journal*, **647**(Aug.), 185–193.

Maeder, A., and Conti, P. S. 1994. Massive Star Populations in Nearby Galaxies. *Annual Review of Astronomy and Astrophysics*, **32**, 227–275.

Maoz, D., and Badenes, C. 2010. The Supernova Rate and Delay Time Distribution in the Magellanic Clouds. *Monthly Notices of the Royal Astronomical Society*, **407**(Sept.), 1314–1327.

Massey, P. 1993 (Jan.). The Massive Star Content of the Galaxy and Magellanic Clouds: Methods and Madness. Page 168 of: Cassinelli, J. P., and Churchwell, E. B. (eds), *Massive Stars: Their Lives in the Interstellar Medium*. Astronomical Society of the Pacific Conference Series, vol. 35.

Massey, P. 2003. Massive Stars in the Local Group: Implications for Stellar Evolution and Star Formation. *Annual Review of Astronomy and Astrophysics*, **41**, 15–56.

Massey, P., Neugent, K. F., and Smart, B. M. 2016. A Spectroscopic Survey of Massive Stars in M31 and M33. *Astronomical Journal*, **152**(Sept.), 62.

McQuinn, K. B. W., Skillman, E. D., Cannon, J. M. et al. 2010a. The Nature of Starbursts. I. The Star Formation Histories of Eighteen Nearby Starburst Dwarf Galaxies. *Astrophysical Journal*, **721**(Sept.), 297–317.

McQuinn, K. B. W., Skillman, E. D., Cannon, J. M. et al. 2010b. The Nature of Starbursts. II. The Duration of Starbursts in Dwarf Galaxies. *Astrophysical Journal*, **724**(Nov.), 49–58.

McQuinn, K. B. W., Skillman, E. D., Dolphin, A. E. et al. 2015a. Calibrating UV Star Formation Rates for Dwarf Galaxies from STARBIRDS. *Astrophysical Journal*, **808**(Aug.), 109.

McQuinn, K. B. W., Cannon, J. M., Dolphin, A. E. et al. 2015b. Characterizing the Star Formation of the Low-Mass Shield Galaxies from Hubble Space Telescope Imaging. *Astrophysical Journal*, **802**(Mar.), 66.

Mineo, S., Gilfanov, M., and Sunyaev, R. 2012. X-ray Emission from Star-Forming Galaxies – I. High-Mass X-ray Binaries. *Monthly Notices of the Royal Astronomical Society*, **419**(Jan.), 2095–2115.

Monelli, M., Cassisi, S., Bernard, E. J. et al. 2010a. The ACS LCID Project. IV. Detection of the Red Giant Branch Bump in Isolated Galaxies of the Local Group. *Astrophysical Journal*, **718**(Aug.), 707–716.

Monelli, M., Gallart, C., Hidalgo, S. L. et al. 2010b. The ACS LCID Project. VI. The Star Formation History of the Tucana dSph and the Relative Ages of the Isolated dSph Galaxies. *Astrophysical Journal*, **722**(Oct.), 1864–1878.

Mould, J., and Aaronson, M. 1983. The Carina Dwarf Spheroidal: An Intermediate Age Galaxy. *Astrophysical Journal*, **273**(Oct.), 530–538.

Olsen, K. A. G., Blum, R. D., Stephens, A. W. et al. 2006. The Star Formation Histories of the Bulge and Disk of M31 from Resolved Stars in the Near-Infrared. *Astronomical Journal*, **132**(July), 271–289.

Olszewski, E. W., Suntzeff, N. B., and Mateo, M. 1996. Old and Intermediate-Age Stellar Populations in the Magellanic Clouds. *Annual Review of Astronomy and Astrophysics*, **34**, 511–550.

Osterbrock, D. E. 1995. Walter Baade's Discovery of the Two Stellar Populations. Page 21 of: van der Kruit, P. C., and Gilmore, G. (eds), *Stellar Populations*. IAU Symposium, vol. 164.

Parra, R., Conway, J. E., Diamond, P. J. et al. 2007. The Radio Spectra of the Compact Sources in Arp 220: A Mixed Population of Supernovae and Supernova Remnants. *Astrophysical Journal*, **659**(Apr.), 314–330.

Payne-Gaposchkin, C. H. 1974. Period, Color, and Luminosity for Cepheid Variables. *Smithsonian Contributions to Astrophysics*, **17**.

Prialnik, D. 2009. *An Introduction to the Theory of Stellar Structure and Evolution*.

Ramachandran, V., Hainich, R., Hamann, W. R. et al. 2018. Stellar Population of the Superbubble N 206 in the LMC. I. Analysis of the Of-Type Stars. *Astronomy and Astrophysics*, **609**(Jan.), A7.

Rampadarath, H., Morgan, J. S., Lenc, E. et al. 2014. Multi-Epoch Very Long Base-Line Interferometric Observations of the Nuclear Starburst Region of NGC 253: Improved Modeling of the Supernova and Star Formation Rates. *Astronomical Journal*, **147**(Jan.), 5.

Rezaeikh, S., Javadi, A., Khosroshahi, H. et al. 2014. The Star Formation History of the Magellanic Clouds Derived from Long-Period Variable Star Counts. *Monthly Notices of the Royal Astronomical Society*, **445**(Dec.), 2214–2222.

Ripepi, V., Cioni, M.-R. L., Moretti, M. I. et al. 2017. The VMC Survey – XXV. The 3D Structure of the Small Magellanic Cloud from Classical Cepheids. *Monthly Notices of the Royal Astronomical Society*, **472**(Nov.), 808–827.

Sabbi, E., Sirianni, M., Nota, A. et al. 2007. Past and Present Star Formation in the SMC: NGC 346 and Its Neighborhood. *Astronomical Journal*, **133**(Jan.), 44–57.

Sabbi, E., Gallagher, J. S., Tosi, M. et al. 2009. Star Formation History of the Small Magellanic Cloud: Six Hubble Space Telescope/Advanced Camera for Survey Fields. *Astrophysical Journal*, **703**(Sept.), 721–735.

Sabbi, E., Calzetti, D., Ubeda, L. et al. 2018. The Resolved Stellar Populations in the LEGUS Galaxies1. *Astrophysical Journal Supplement Series*, **235**(Mar.), 23.

Sacchi, E., Annibali, F., Cignoni, M. et al. 2016. Stellar Populations and Star Formation History of the Metal-Poor Dwarf Galaxy DDO 68. *Astrophysical Journal*, **830**(Oct.), 3.

Sacchi, E., Cignoni, M., Aloisi, A. et al. 2018. Star Formation Histories of the LEGUS Dwarf Galaxies. II. Spatially Resolved Star Formation History of the Magellanic Irregular NGC 4449. *Astrophysical Journal*, **857**(Apr.), 63.

Salaris, M., and Cassisi, S. 2006. *Evolution of Stars and Stellar Populations*.

Sandage, A. 1956. The Systematics of Color-Magnitude Diagrams and Stellar Evolution. *Publications of the Astronomical Society of the Pacific*, **68**(Dec.), 498.

Schneider, F. R. N., Sana, H., Evans, C. J. et al. 2018. An Excess of Massive Stars in the Local 30 Doradus Starburst. *Science*, **359**(6371), 69–71.

Seidel, E., Demarque, P., and Weinberg, D. 1987. The Evolution of Red Clump Stars – Theoretical Sequences. *Astrophysical Journal Supplement Series*, **63**(Apr.), 917–945.

Senchyna, P., Johnson, L. C., Dalcanton, J. J. et al. 2015. Panchromatic Hubble Andromeda Treasury. XIV. The Period-Age Relationship of Cepheid Variables in M31 Star Clusters. *Astrophysical Journal*, **813**(Nov.), 31.

Skillman, E. D., and Gallart, C. 2002. First Results of the Coimbra Experiment. Page 535 of: Lejeune, T., and Fernandes, J. (eds), *Observed HR Diagrams and Stellar Evolution*. Astronomical Society of the Pacific Conference Series, vol. 274.

Skillman, E. D., Hidalgo, S. L., Weisz, D. R. et al. 2014. The ACS LCID Project. X. The Star Formation History of IC 1613: Revisiting the Over-Cooling Problem. *Astrophysical Journal*, **786**(May), 44.

Smecker-Hane, T. A., Stetson, P. B., Hesser, J. E. et al. 1994. The Stellar Populations of the Carina Dwarf Spheroidal Galaxy. 1: A New Color-Magnitude Diagram for the Giant and Horizontal Branches. *Astronomical Journal*, **108**(Aug.), 507–513.

Smith, H. E., Lonsdale, C. J., and Lonsdale, C. J. 1998. The Starburst-AGN Connection. II. The Nature of Luminous Infrared Galaxies as Revealed by VLBI, VLA, Infrared, and Optical Observations. *Astrophysical Journal*, **492**(Jan.), 137–172.

Stetson, P. B. 1987. DAOPHOT – A Computer Program for Crowded-Field Stellar Photometry. *Publications of the Astronomical Society of the Pacific*, **99**(Mar.), 191–222.

Struve, O., and Zebergs, V. 1962. *Astronomy of the 20th century*.

Subramanian, S., and Subramaniam, A. 2015. Disk of the Small Magellanic Cloud as Traced by Cepheids. *Astronomy and Astrophysics*, **573**(Jan.), A135.

Sweigart, A. V., and Gross, P. G. 1976. Horizontal-Branch Evolution with Semiconvection II. Theoretical Sequences. *Astrophysical Journal Supplement Series*, **32**(Oct.), 367.

Tabernero, H. M., Dorda, R., Negueruela, I. et al. 2018. An LTE Effective Temperature Scale for Red Supergiants in the Magellanic Clouds. *Monthly Notices of the Royal Astronomical Society*, **476**(May), 3106–3123.

Tang, J., Bressan, A., Rosenfield, P. et al. 2014. New PARSEC Evolutionary Tracks of Massive Stars at Low Metallicity: Testing Canonical Stellar Evolution in Nearby

Star-Forming Dwarf Galaxies. *Monthly Notices of the Royal Astronomical Society*, **445**(Dec.), 4287–4305.

Tarchi, A., Neininger, N., Greve, A., Klein, U. et al. 2000. Radio Supernovae, Supernova Remnants and H II Regions in NGC 2146 Observed with MERLIN and the VLA. *Astronomy and Astrophysics*, **358**(June), 95–103.

Tinsley, B. M. 1968. Evolution of the Stars and Gas in Galaxies. *Astrophysical Journal*, **151**(Feb.), 547.

Tinsley, B. M. 1980. Evolution of the Stars and Gas in Galaxies. *Fundamental Cosmic Physics*, **5**, 287–388.

Tolstoy, E., and Saha, A. 1996. The Interpretation of Color-Magnitude Diagrams through Numerical Simulation and Bayesian Inference. *Astrophysical Journal*, **462**(May), 672.

Tolstoy, E., Hill, V., and Tosi, M. 2009. Star-Formation Histories, Abundances, and Kinematics of Dwarf Galaxies in the Local Group. *Annual Review of Astronomy and Astrophysics*, **47**(Sept.), 371–425.

Tonry, J., and Schneider, D. P. 1988. A New Technique for Measuring Extragalactic Distances. *Astronomical Journal*, **96**(Sept.), 807–815.

Tosi, M., Greggio, L., Marconi, G. et al. 1991. Star Formation in Dwarf Irregular Galaxies – Sextans B. *Astronomical Journal*, **102**(Sept.), 951–974.

van den Bergh, S. 1975. Stellar Populations in Galaxies. *Annual Review of Astronomy and Astrophysics*, **13**, 217–255.

Varenius, E., Conway, J. E., Batejat, F. et al. 2019. The Population of SNe/SNRs in the Starburst Galaxy Arp 220. A Self-Consistent Analysis of 20 Years of VLBI Monitoring. *Astronomy and Astrophysics*, **623**(Mar), A173.

Weisz, D. R., Dalcanton, J. J., Williams, B. F. et al. 2011. The ACS Nearby Galaxy Survey Treasury. VIII. The Global Star Formation Histories of 60 Dwarf Galaxies in the Local Volume. *Astrophysical Journal*, **739**(Sept.), 5.

Weisz, D. R., Dolphin, A. E., Skillman, E. D. et al. 2013. Comparing the Ancient Star Formation Histories of the Magellanic Clouds. *Monthly Notices of the Royal Astronomical Society*, **431**(May), 364–371.

Whitney, B. A., Sewilo, M., Indebetouw, R. et al. 2008. Spitzer Sage Survey of the Large Magellanic Cloud. III. Star Formation and ~1000 New Candidate Young Stellar Objects. *Astronomical Journal*, **136**(July), 18–43.

Xiao, L., and Eldridge, J. J. 2015. Core-Collapse Supernova Rate Synthesis within 11 Mpc. *Monthly Notices of the Royal Astronomical Society*, **452**(Sept.), 2597–2605.

Yoast-Hull, T. M., Gallagher, III, J. S., Zweibel, E. G. et al. 2014. Active Galactic Nuclei, Neutrinos, and Interacting Cosmic Rays in NGC 253 and NGC 1068. *Astrophysical Journal*, **780**(Jan.), 137.

6

Star-Formation Measurements in Nearby Galaxies

DANIELA CALZETTI

6.1 Conditions for a Reliable Star-Formation Rate Indicator

When measuring the star-formation rates (SFRs) of galaxies, we mostly rely on the central limit theorem[1] to ensure the robustness of our measurements. The SFR of a system is the mass of newly formed stars per unit time, evaluated over some period of time during which the SFR can be assumed to be constant. For practical reasons, the "mass of newly formed stars" is typically extrapolated from the sum of the mass of massive stars, and the "period of time" is the lifetime of those massive stars ($\approx$10–100 Myr). These choices are dictated by purely observational constraints: massive stars are luminous and can be detected at relatively large distances, and (because of their short lifetimes) they trace the most recent events of star formation. Massive stars are born in crowded environments (star clusters, associations, etc.) and cannot, in general, be resolved individually, especially in external galaxies beyond the Magellanic Clouds. This observational limitation mandates the use of the cumulative light from the massive stars for the purpose of measuring SFRs. The SFR estimators are then calibrated by converting this light to a total number of stars, across the entire mass spectrum, by using an educated guess on the distribution of stars at birth, or stellar Initial Mass Function[2] (IMF).

The three key conditions for calibrating a reliable SFR indicator from the light of unresolved massive stars, therefore, are: (1) the star formation can be considered constant over the relevant timescale ($\sim$10–100 Myr, depending on the SFR indicator); (2) the effects of dust attenuation are averaged across multiple geometrical realizations for the dust-stars distribution (a "central limit" condition for dust attenuation); and (3) enough stars are formed that the stellar IMF is fully sampled across all masses, including the rare massive stars. For the purpose of this chapter, a massive star is defined to have $M > 20\ M_\odot$ (spectral type O9 or earlier), above which mass significant amounts of ionizing photons are produced.

Massive ($>10^9\ M_\odot$), star-forming galaxies satisfy all three conditions. In the local universe, they have a typically SFR $>0.1\ M_\odot\ yr^{-1}$ (Cook et al., 2014), reaching up to

[1] The concept of "central limit" in reference to SFRs of galaxies was introduced by R. C. Kennicutt several years ago.

[2] We adopt a Kroupa (2001) IMF in the mass range 0.1–100 $M_\odot$ for our calculations. This IMF is characterized by a double power law $dN/dM \propto M^\alpha$, with $\alpha = -1.3$ in the mass range 0.1–0.5 $M_\odot$ and $\alpha = -2.3$ at higher stellar masses. We refer to Chapter 2 for a detailed presentation of the IMF.

several tens to a few hundred $M_\odot$ yr^{-1}. Thus, over the timescale of a star-formation event, >10 Myr, these galaxies produce over 10^6 $M_\odot$ in new stars, and over 3000 O stars. Under these conditions, the aggregate of all the star-forming sites within a galaxy is well above any random effects due to incomplete sampling of the stellar IMF (e.g. Cerviño et al., 2002), and typically above systematic effects due to IMF variations (Pflamm-Altenburg et al., 2009). In other words, measures of the SFRs of these galaxies can rely on the full sampling of the stellar IMF.

The characteristic mass of <100 Myr old star clusters in local spiral galaxies is $\sim 10^5$ $M_\odot$ (Larsen, 2009), and young cluster populations are tens to hundreds of systems strong (Adamo and Bastian, 2018). Young star clusters are usually widely distributed spatially, e.g. throughout the spiral arms or other dynamical features, or across the galaxy, in hierarchically clustered structures with sizes of several hundreds parsec (Grasha et al., 2017a,b). In this case, dynamical timescales are long enough that star formation can be considered constant or almost-constant over a length of time that is comparable to the lifetime of massive stars. And because of their large number, clusters and associations will present a range of levels of mixing with the dust and will populate the dust disk with a range of optical depths. Thus, SFR estimators that assume constant star formation and take into account effects of dust attenuation that are averaged across the galaxy (see Chapter 4 and Chapter 7) will usually yield reasonably accurate measures for massive, star-forming galaxies.

6.2 Star-Formation Rates of Star-Forming Regions

Star-forming regions within galaxies, when analyzed in samples containing large numbers, can also average out variations in physical properties, enabling the calibration of "local" SFR indicators. Here a "star-forming region" is defined as a local peak of emission within a galaxy with size between several tens and a few hundred pc. These sizes correspond to large HII regions up to unresolved star-forming complexes. The local peak of emission needs to be observed at wavelengths that are usually associated with recent star formation: UV continuum, optical-infrared hydrogen recombination lines, mid/far-infrared continuum emission (> 3–4 μm), and/or radio (e.g. free-free emission). Massive galaxies are metal-rich: galaxies with MS > 10^9 $M_\odot$ have oxygen abundances $12 + \mathrm{Log(O/H)} > 8.4$[3], i.e. comparable or higher than the oxygen abundance of the Large Magellanic Cloud (Garnett, 1999, LMC; Tremonti et al., 2004). They contain significant amounts of dust that are concentrated in the most active regions of star formation (Lada et al., 2010, 2012). Thus, a significant fraction of the UV-optical light from HII regions and star-forming complexes in massive galaxies is processed by dust in the infrared (IR, $\lambda > 3$–4 μm).

[3] In our convention, the Sun has oxygen abundance $12 + \mathrm{Log(O/H)} = 8.69$ (Asplund et al., 2009).

6.2.1 Single-Band SFR Indicators

Calibrations of IR SFR indicators have been performed for star-forming regions with luminosities $L(\mathrm{H}\alpha) > 10^{38}$ erg s^{-1} (Alonso-Herrero et al., 2006; Pérez-González et al., 2006; Calzetti et al., 2007; Relaño et al., 2007), which correspond to stellar masses of several 1000's $\mathrm{M}_{\odot}$. This is above the limit for stochastic sampling of the IMF (Cerviño et al., 2002), for an HII region aged 3–5 Myr.[4] Given that star-forming regions contain stellar populations that are approximately coeval, star formation can be considered constant over a short period of time, corresponding to the lifetimes of the massive stars, a few to a few tens of Myr.

Table 6.1 lists the monochromatic (single-band) infrared calibrations for the SFR surface density (SFR/area) of regions in nearby galaxies, ordered as a function of wavelength and characteristic region size. The convention for the parameters in Table 6.1 is:

$$\Sigma(SFR) = A\Sigma(\lambda_{\mathrm{IR}})^{\alpha}, \tag{6.1}$$

where $\Sigma(SFR)$ is in units of $\mathrm{M}_{\odot}$ yr^{-1} kpc^{-2} and $\Sigma(\lambda_{\mathrm{IR}})$ is in units of erg s^{-1} kpc^{-2}. The use of surface densities, rather than luminosities, provides the advantage of being able to compare regions of different sizes and at different distances. This, however, comes with the caveat that often the published sizes are dictated by resolution limitations, rather than reflect physical scales. This limitation notwithstanding, Table 6.1 shows a suggestive trend for the slope α to converge to unity for both longer wavelengths and larger region sizes.

This trend is easier to interpret when the relation between SFR and the IR band is fit with a slope of unity (α=1). The scatter in the data is usually large enough, ~0.2–0.3 dex peak–to–peak, that this fit provides a reasonable representation of the behavior of regions and galaxies, and has been widely adopted (e.g. Wu et al., 2005; Zhu et al., 2008; Rieke et al., 2009; Calzetti et al., 2010; Li et al., 2010; Battisti et al., 2015). Table 6.2 lists some recent calibrations between SFR and λ_{IR}, with the format:

$$SFR = C \times 10^{-43} L(\lambda_{\mathrm{IR}}), \tag{6.2}$$

with the SFR in units of $\mathrm{M}_{\odot}$ yr^{-1} and $L(\lambda_{\mathrm{IR}})$ is in units of erg s^{-1}. With a slope of unity, luminosities and surface densities are interchangeable, so we can relate the values of Table 6.2 to those of Table 6.1. The general trend observed in Table 6.2 is for the calibration coefficient C to decrease for increasing region size. However, the trends are different for different bands, as shown in Fig. 6.1.

The trends of Fig. 6.1 are consistent with an increase of the contribution of evolved stellar populations to the IR emission for increasing region size. As the model in Fig. 6.1 indicates (grey stars), the calibration of $L(70)$ as a SFR indicator requires a decreasing calibration constant, in agreement with the mean age of the stellar population increasing from 100 Myr at ~200 parsec size to 10 Gyr for the whole galaxy. In a 10 Gyr old, constant SFR stellar population, the heating from stellar populations older than 100 Myr almost

[4] At about 6 Myr age, a coeval stellar population produces over 10 times less ionizing photons than the same population at 2–3 Myr age (Leitherer et al., 1999).

Table 6.1. *Non-linear infrared SFR indicator calibrations*

Reg. Size (pc)[a]	150–200	300–400	400–600	600–1300	Galaxy
	A	A	A	A	A
	α	α	α	α	α
Surf. Bright.[b]	Ref.[c]	Ref.[c]	Ref.[c]	Ref.[c]	Ref.[c]
$\Sigma(24)$	$(1.82 \pm 0.09) \times 10^{-33}$		$(1.56 \pm 0.22) \times 10^{-35}$		
	0.77 ± 0.01		0.81 ± 0.02		
	1		2		
$\Sigma(70)$	$(1.84 \pm 0.04) \times 10^{-36}$			$(2.50 \pm 0.18) \times 10^{-42}$	$(1.72 \pm 0.17) \times 10^{-40}$
	0.83 ± 0.01			0.97 ± 0.01	0.92 ± 0.01
	1			3	4
$\Sigma(100)$		$(3.34 \pm 0.12) \times 10^{-38}$			
		0.87 ± 0.01			
		1			
$\Sigma(160)$		$(8.04 \pm 0.31) \times 10^{-42}$			$(1.05 \pm 0.11) \times 10^{-45}$
		0.96 ± 0.02			1.05 ± 0.02
		1			4

[a]Diameter of the region used to perform the calibration. "Galaxy" means galaxy-wide calibration. All constants are from observational calibrations, typically performed on local peaks of emission within galaxies, across several hundred regions and/or tens to a few hundred galaxies. This Table is adapted from Li (2014).

[b]Surface brightness in erg s^{-1} kpc^{-2}. Luminosities are given as $\nu L(\nu)$; the relation between $\Sigma(\lambda_{IR})$ and $\Sigma(SFR)$ is given by: $\Sigma(SFR) = A\ \Sigma(\lambda_{IR})^{\alpha}$. $\Sigma(SFR)$ is in units of $M_\odot\ yr^{-1}\ kpc^{-2}$.

[c]References: 1 – Li (2014); Li et al. (2013); 2 – Calzetti et al. (2007); 3 – Li et al. (2010); 4 – Calzetti et al. (2010).

Table 6.2. *Linear infrared SFR indicator calibrations*

Reg. Size (pc)[a]	150–200	300–400	400–600	600–1300	Galaxy
Luminosity[b]	C Ref.[c]	C Ref.[c]	C Ref.[c]	C Ref.[c]	C Ref.[c]
$L(8)$			1.22 ± 0.09 1		1.07 ± 0.20 2
$L(24)$	5.54 ± 0.08 3		2.32 ± 0.09 1		2.04 ± 0.11 4
$L(70)$	1.23 ± 0.02 3			0.94 ± 0.02 5	0.59 ± 0.06 6
$L(100)$		1.12 ± 0.01 3			
$L(160)$		1.67 ± 0.03 3			1.43 ± 0.08 6

[a]Diameter of the region used to perform the calibration. "Galaxy" means galaxy-wide calibration. All constants are from observational calibrations, typically performed on local peaks of emission within galaxies, across several hundred regions and/or tens to a few hundred galaxies. This Table is adapted from Li (2014).
[b]Luminosity in erg s^{-1}. Luminosities are given as $\nu L(\nu)$; the relation between $L(\lambda_{IR})$ and SFR is given by: $SFR = C \times 10^{-43} L(\lambda_{IR})$. SFR is in units of $M_\odot$ yr^{-1}.
[c]References: 1 – Calzetti et al. (2007); 2 – Battisti et al. (2015); 3 – Li (2014); Li et al. (2013); 4 – Rieke et al. (2009); 5 – Li et al. (2010); 6 – Calzetti et al. (2010).

doubles the dust IR (and 70 μm) emission. The calibration constant C increases for shorter star-formation timescales or a decreasing SFR with time, but also for decreasing fractions of dust-absorbed stellar light (Calzetti, 2013). The band-specific behavior also depends on the mean temperature of the dust and its variations within galaxies.

The mean IR spectral energy distribution of a region about 500 parsec in size is shown in Fig. 6.2. This energy distribution is not that of a specific region, but represents the infrared emission of an average 'star-forming region' with a SFR of 1 $M_\odot$ yr^{-1}, as inferred from inverting the calibration constants of Table 6.2. The data have been fitted with the models of Draine and Li (2007). For the model, we use Milky Way dust properties, and the best fit implies significant (~40%) contribution from dust heated by massive stars, together with a diffuse starlight illumination about three times higher than the local Galactic interstellar radiation field. This is typical of regions or galaxies hosting active star formation (Draine et al., 2007). Although the specific details of the fit are likely not accurate, this result validates the calibration constants of Table 6.2. Combining the values at 24, 70, and 160 μm from Fig. 6.2 according to the recipe of Dale and Helou (2002) yields a 8 1000 μm integrated IR luminosity $L(IR) = 2.2 \times 10^{43}$ erg s^{-1} for $SFR = 1$ $M_\odot$ yr^{-1}, only 16% lower than the independent calibration reported in Kennicutt and Evans (2012).

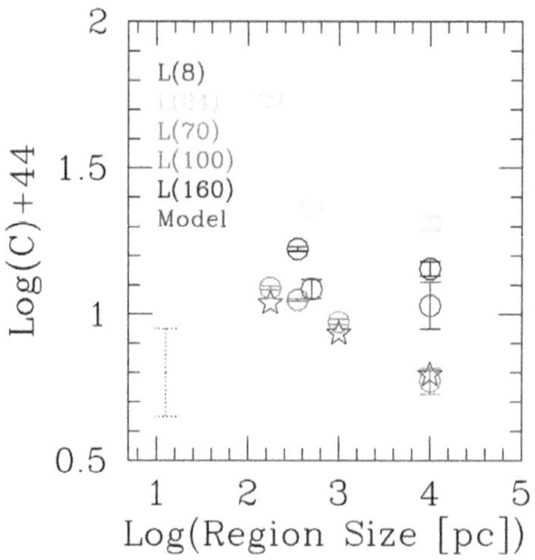

Figure 6.1 The calibration coefficient C (eq. 1.2) as a function of the size of the region over which the SFR indicator has been calibrated. Different IR bands are indicated by different colors, and shown in the Figure. The calibrations for whole galaxies are reported at a fiducial size of 10 kpc. Formal uncertainties are shown with error bars on the data points. The typical peak–to–peak scatter of the data used to derive the calibrations is shown as a vertical dashed bar to the bottom-left of the panel. The scatter in the data is usually significant. A simple model for the 70 μm emission is shown with star symbols, and should be compared with the magenta data points. The stars represent the expected emission at 70 μm for a population of (from small to large sizes): 100 Myr, 1 Gyr, and 10 Gyr. Model assumptions include that half of the IR emission is emitted at 70 μm, and only half of the starlight is reprocessed by dust in the IR. Changing these assumptions will change the vertical location of the models. Color version available online.

An interesting feature of the data in Fig. 6.1 is that the calibrations for $L(8)$, $L(24)$ (beyond 500 pc), and $L(160)$ are fairly constant in value for increasing region size. These are bands more closely associated with heating by the diffuse stellar population, either via thermal heating of large dust grains (160 μm) or stochastic heating of small dust grains (8 and 24 μm); conversely, $L(70)$ is more sensitive to thermal heating of large dust grains by the young, clustered stellar populations (Draine and Li, 2007; Draine et al., 2007; Calzetti et al., 2010). A detailed matching of the calibration constants at different wavelengths with model expectations would require modeling the variations of the IR spectral energy distributions, the fraction of the stellar light absorbed by dust at each characteristic size, and the star-formation history (SFH) of the galaxy; however, the simple model of Fig. 6.1 already suggests that deriving a SFR from the IR emission in sub-galactic regions requires accounting for the contribution from stellar populations that are unrelated to the current star-formation event. This effect had been already noted by many authors since the times of the first extragalactic IR studies with the *IRAS*[5] satellite (for a discussion

[5] *Infrared Astronomical Satellite*; https://irsa.ipac.caltech.edu/IRASdocs/iras.html

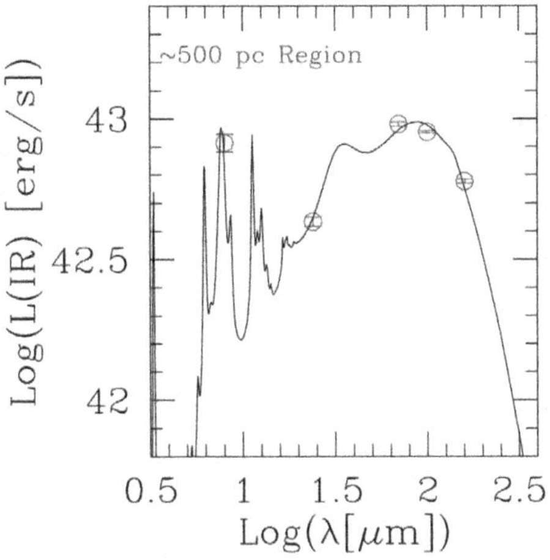

Figure 6.2 The spectral energy distribution of a representative ~500 parsec region with a SFR of 1 $M_\odot$ yr^{-1}. The plotted points (blue circles with error bars) are the inverted *C* values from 6.2, chosen at characteristic diameters ~400–500 pc. The 70 μm calibration constant at this diameter is the interpolation between the two closest calibration constants. A best-fitting dust emission model, from Draine and Li (2007), is overplotted (black line), under the assumption of Milky Way-like dust; it yields about 40% contribution from dust heated by massive stars. The spectral energy distribution does not correspond to any specific HII region, but is the template star-forming region resulting from the SFR calibrations of Table 6.2. The model fitting, while it contains some features (e.g. the "bump" around 30 μm) that are likely to be not real, corresponds to realistic parameters for the dust and starlight intensity and validates the template spectral energy distribution. Color version available online.

and several references, see Kennicutt, 1998; Kennicutt et al., 2009, and Chapter 7), although only the much higher angular resolution of the *Spitzer Space Telescope*[6] and the *Herschel Space Observatory*[7] has enabled an accurate quantification.

Using SFR indicators at wavelengths other than IR brings its own set of challenges. As already mentioned in previous chapters, the UV stellar continuum and hydrogen recombination lines at optical wavelengths (e.g. Hα) require correction for the effects of dust attenuation. The effects of dust are less important at wavelengths longer than the near-IR, and use of hydrogen recombination lines beyond 1 μm to measure SFRs has increased in recent years, due in part to the availability of improved near-IR and radio instruments and facilities. Hydrogen recombination lines and the free-free continuum require presence of ionizing photons to ionize the gas surrounding massive stars. Direct absorption by dust of the Lyman Continuum (LyC) emission, if strong, can in principle invalidate this assumption. However, under most conditions found in galaxies, direct LyC absorption by dust should be a small effect, at the level of 15%–20%, with a more significative impact at high

[6] *Spitzer Space Telescope*; www.spitzer.caltech.edu
[7] *Herschel Space Observatory*; www.cosmos.esa.int/web/herschel/overview

ionization and metallicity values, such those found in Luminous and UltraLuminous IR Galaxies and in the high-density regions of some galaxies (Dopita et al., 2003). For regions of star formation within galaxies, leakage of ionizing photons from the region is likely to be a much larger effect than dust absorption of LyC photons. Leakage is estimated to be around 50%–60%, but could be lower, about 20%–30%, once the correction for differential extinction between star-forming regions and diffuse stellar emission is included (Oey et al., 2007; Pellegrini et al., 2012; Crocker et al., 2013). Local/small-region calibrations do not usually include this correction of about 1/3 missing ionizing photons.

Recombination lines beyond $\sim$1 μm, while offering lower sensitivity to dust attenuation than the optical lines, suffer from increasing dependency on the physical conditions of the emitting nebula. Brα, at 4.05 μm, suffers variations of almost 60% for changes of the electron temperature in the range 5000–20000 K (compared to 15% of Hα), and these variations increase to a factor 2.6 and higher for Radio Recombination Lines (RRLs) that originate from quantum levels n $>$ 20. For RRLs with upper quantum levels n $>$ 80–200 (depending on the gas physical conditions), stimulated emission begins to contribute to the line luminosity, thus complicating calibrations of these lines as SFR indicators. The free-free emission is usually employed to constrain the gas temperature for RRLs and decrease their sensitivity to this parameter. However, determination of the free-free emission luminosity of a galaxy or region requires observations at multiple (sub)-mm and cm wavelengths, in order to separate the contribution of free-free from the tails of both the dust emission at shorter wavelengths ($<$ 2–3 mm) and synchrotron emission beyond 1–3 cm. Star-forming regions within galaxies are dominated by free-free emission at $\sim$1 cm, implying that this emission can be effectively disentangled from dust and synchrotron emission (Murphy et al., 2011). Free-free continuum has a weaker dependency on electron temperature than RRLs, less than a factor 2 for the same range 5000–20000 K, and has been calibrated as a SFR indicator by several authors (Murphy et al., 2011, and references therein).

6.2.2 Multi-Band SFR Indicators

The need to account for both dust attenuated and unattenuated star formation has led to the formulation of SFR indicators that combine an optical or UV tracer with an infrared or radio tracer, both for galaxies (Kennicutt et al., 2009; Hao et al., 2011) and star-forming regions within galaxies (e.g. Calzetti et al., 2007; Liu et al., 2011; Li et al., 2013). The goal of this approach is to produce an "unbiased" SFR indicator, in the case when both stochastic sampling of the IMF and variations in the star-formation history are not relevant. The formulation of such hybrid or multi-band indicators usually refers the calibration to the UV/optical tracer, under the assumption that it is more directly related to the current star formation. UV and the Hα emission line, for instance, directly trace the stellar continuum of massive stars or the gas they ionize, while the IR traces the dust they heat, thus the SFR is measured in a more indirect fashion.

The main calibration effort is to determine the constant to be multiplied to the IR (or radio), so they can be brought to the same luminosity scale as the UV/optical. The main

Table 6.3. *Multi-band SFR indicator calibrations*

Reg. Size (pc)[a]		150–200	300–400	400–600	Galaxy
$L(\lambda_1)$[b]	$L(\lambda_2)$[b]	c_{λ_2} Ref.[c]	c_{λ_2} Ref.[c]	c_{λ_2} Ref.[c]	c_{λ_2} Ref.[c]
$L(\mathrm{H}\alpha)$	$L(24)$	0.095 ± 0.019 1		0.031 ± 0.006 2	0.020 ± 0.006 3
$L(\mathrm{UV})$[b]	$L(24)$			6.0 ± 1.2 4	3.89 ± 0.15 5
$L(\mathrm{H}\alpha)$	$L(70)$		0.021 ± 0.005 1		
$L(\mathrm{H}\alpha)$	$L(100)$		0.022 ± 0.005 1		

[a]Diameter of the region used to perform the calibration. "Galaxy" means galaxy-wide calibration. All constants are from observational calibrations, typically performed on local peaks of emission within galaxies, across several hundred regions and/or tens to a few hundred galaxies. This Table is adapted from Li (2014).
[b]Luminosity in erg s^{-1}. UV and IR luminosities are expressed as $\nu L(\nu)$. L(UV) is the luminosity at 1530 Å(Hao et al., 2011). The relation between SFR and the two luminosities at λ_1 and λ_2 is given by equation 1.3.
[c]References: 1 – Li (2014); Li et al. (2013); 2 – Calzetti et al. (2007); 3 – Kennicutt et al. (2009); 4 – Liu et al. (2011); 5 – Hao et al. (2011).

reasoning behind such calibrations is that while the observed UV/optical is related to the SFR via a power law with exponent >1 (Hao et al., 2011), the observed IR has exponent <1 (Table 6.1). In other words, for increasing SFRs, the UV/optical become insufficient at tracing the true SFR, while the IR compensates for the deficiency. This is consistent with the well-known correlation between SFR and dust attenuation and content of galaxies (Calzetti, 2001, and references therein).

A multi-band SFR indicator is generally indicated as:

$$SFR = C_{\lambda_1}[L(\lambda_1) + c_{\lambda_2} L(\lambda_2)], \tag{6.3}$$

with the SFR in units of $M_\odot\ yr^{-1}$ and $L(\lambda_X)$ is in units of erg s^{-1}. The first coefficient, C_{λ_1} refers to the calibration of $L(\lambda_1)$ as if it were unaffected by dust attenuation. $C_{UV} = 6.5$–4.5×10^{-44} $M_\odot\ yr^{-1}\ erg^{-1}s$ for star-formation durations in the range 10−>100 Myr and $\lambda_{UV} = 1500$ Å. $C_{H\alpha} = 5.4$–5.5×10^{-42} $M_\odot\ erg^{-1}$ for the same star-formation timescales. The constant for the UV calibration changes by ~45% between the shortest and longest duration of star formation considered, while the constant for the Hα changes by 2% only. This is consistent with Hα tracing shorter star-formation timescales than UV, i.e. ~6 Myr versus ~100 Myr (Kennicutt and Evans, 2012).

The values of c_{λ_2} are tabulated in Table 6.3. These constants provide the scaling between the UV/optical and the IR tracers, wrapping into one numbera variety of effects, including the star-formation timescale of the tracer. The mixed indicators that involve the luminosity at 24 μm show, like in Table 6.2 and Fig. 6.1, a decrease in value for larger region sizes, for the same reason discussed in the previous section.

6.3 Star-Formation Rates within Galaxies

The definition, or even the meaning, of a SFR indicator becomes increasingly complicated when the applications involve systems with sporadic and/or low levels of star formation. Examples of such systems are: dwarf star-forming galaxies and random areas within large galaxies.

Local dwarf star-forming galaxies are characterized by long periods of moderate intensity star formation interleaved by brief periods of quiescence, although the variations to this baseline scenario are large (Tolstoy et al., 2009). Among those, the starbursting dwarfs present enhanced star formation over several hundred Myr, with a median value for the duration of 500 Myr, and large fluctuations in intensity over timescales of several tens of Myr (McQuinn et al., 2010) that are spatially distributed in an 'orderly stochastic' fashion (Dohm-Palmer et al., 2002). SFHs have been derived for many dwarf galaxies in the local universe (e.g. Gallart et al., 2005; Annibali et al., 2013; Sacchi et al., 2016; Cignoni et al., 2019, among many others), including the Magellanic Clouds (e.g. Cignoni et al., 2013, 2015). SFHs of dwarfs are less complicated to derive than those of spirals, due to their relatively low stellar densities, low dust content, and small variations in the line-of-sight extinction to the stars, which simplify the measurement and modeling of the stellar populations.

Comparisons of UV-derived SFRs against the SFH-derived SFRs reveal significant discrepancies, especially for SFRs < 0.1 $M_{\odot}$ yr^{-1}, with the SFH-derived SFRs requiring calibration constants that are over 50% higher than the UV-derived SFRs (McQuinn et al., 2015). Without entering into the merits of the detailed reasons for the discrepancy, it is clear that star-forming dwarfs reveal the limitations of standard calibrations in the low-mass and low-SFR regime.

Limitations on the applicability of SFR tracers may be expected also in the case of randomly selected areas within galaxies. These areas are generally referred as: regions, pixels, spaxels, pixellated areas, etc. As they are not specifically centered on, or dominated by, recent star formation, the luminosity at UV, optical, and IR wavelengths may receive significant contribution from older/evolved stellar populations; in addition, the Hα emission may be enhanced by leakage from neighboring star-forming regions.

The UV light in the interam regions of spirals includes contributions from declining star formation and 10%–20% diffuse light; both are consistent with a scenario in which stars mainly form in the spiral arms and stream into the interarm regions over 10–50 Myr timescales, leaking UV light in the process (Crocker et al., 2015). Older stellar populations are still effective at heating the dust in the region, thus enhancing the IR emission and, consequently, requiring a decrease in the calibration constant of IR-based SFR indicators.

As regions within galaxies can present a range of mixtures of newly formed and older stellar populations, the calibration constant can change from region to region.

For multi-band indicators, the calibration factor for the IR band, c_{λ_2}, is correlated with the specific SFR of the region (sSFR=SFR/M_{star}), spanning a factor ~8.5–14 change at 24, 70, and 100 μm over the sSFR range 10^{-11}–$10^{-8.5}$ yr^{-1} (Boquien et al., 2016). The trends for c_{λ_2} are in the direction of increasing values for increasing sSFR: as the star formation in a region increases relative to the existing stellar mass, the IR emission is increasingly contributed by the current star formation and a larger portion of this emission is added to the observed UV to create the SFR indicator. An evaluation of this recipe in the Andromeda Galaxy supports its validity (Lewis et al., 2017).

6.4 Summary

"Recipes" for the determination of SFRs of and within galaxies have enabled many advances in understanding the properties and physics of stellar and galaxy populations. However, like all recipes, they have significant limitations, and are usually only applicable within the luminosity and physical range where they have been calibrated. Outside the validity range, they can lead to under/overestimates of the true SFR up to factors of several. Future efforts will need to concentrate on tying together the SFR calibrations across several different physical regimes by analyzing low-SFR environments such as dwarf galaxies, the interarm regions of spirals, and the outskirts of galaxies.

References

Adamo, A., and Bastian, N. 2018. *The Lifecycle of Clusters in Galaxies*. Astrophysics and Space Science Library, vol. 424. Page 91.

Alonso-Herrero, A., Rieke, G. H., Rieke, M. J. et al. 2006. Near-Infrared and Star-forming Properties of Local Luminous Infrared Galaxies. *Astrophysical Journal*, **650**(2), 835–849.

Annibali, F., Cignoni, M., Tosi, M. et al. 2013. The Star Formation History of the Very Metal-Poor Blue Compact Dwarf I Zw 18 from HST/ACS Data. *Astronomical Journal*, **146**(6), 144.

Asplund, M., Grevesse, N., Sauval, A. J. et al. 2009. The Chemical Composition of the Sun. *Annual Review of Astronomy and Astrophysics*, **47**(1), 481–522.

Battisti, A. J., Calzetti, D., Johnson, B. D. et al. 2015. Continuous Mid-Infrared Star Formation Rate Indicators: Diagnostics for 0 <; z < 3 Star-Forming Galaxies. *Astrophysical Journal*, **800**(2), 143.

Boquien, M., Kennicutt, R., Calzetti, D. et al. 2016. Towards universal hybrid star formation rate estimators. *Astronomy and Astrophysics*, **591**(June), A6.

Calzetti, D. 2001. The Dust Opacity of Star-Forming Galaxies. *Publications of the Astronomical Society of the Pacific*, **113**(790), 1449–1485.

Calzetti, D. 2013. *Star Formation Rate Indicators*. Page 419.

Calzetti, D., Kennicutt, R. C., Engelbracht, C. W. et al. 2007. The Calibration of Mid-Infrared Star Formation Rate Indicators. *Astrophysical Journal*, **666**(2), 870–895.

Calzetti, D., Wu, S. Y., Hong, S. et al. 2010. The Calibration of Monochromatic Far-Infrared Star Formation Rate Indicators. *Astrophysical Journal*, **714**(2), 1256–1279.

Cerviño, M., Valls-Gabaud, D., Luridiana, V. et al. 2002. Confidence Levels of Evolutionary Synthesis Models. II. On Sampling and Poissonian Fluctuations. *Astronomy and Astrophysics*, **381**(Jan), 51–64.

Cignoni, M., Cole, A. A., Tosi, M. et al. 2013. Mean Age Gradient and Asymmetry in the Star Formation History of the Small Magellanic Cloud. *Astrophysical Journal*, **775**(2), 83.

Cignoni, M., Sabbi, E., van der Marel, R. P. et al. 2015. Hubble Tarantula Treasury Project. II. The Star-Formation History of the Starburst Region NGC 2070 in 30 Doradus. *Astrophysical Journal*, **811**(2), 76.

Cignoni, M., Sacchi, E., Tosi, M. et al. 2019. Star Formation Histories of the LEGUS Dwarf Galaxies. III. The Nonbursty Nature of 23 Star-Forming Dwarf Galaxies. *Astrophysical Journal*, **887**(2), 112.

Cook, D. O., Dale, D. A., Johnson, B. et al. 2014. Spitzer Local Volume Legacy (LVL) SEDs and physical properties. *Monthly Notices of the Royal Astronomical Society*, **445**(1), 899–912.

Crocker, A. F., Calzetti, D., Thilker, D. A. et al. 2013. Quantifying Non-Star-Formation-Associated 8 μm Dust Emission in NGC 628. *Astrophysical Journal*, **762**(2), 79.

Crocker, A. F., Chandar, R., Calzetti, D. et al. Origin of the Diffuse, Far Ultraviolet Emission in the Interarm Regions of M101. *Astrophysical Journal*, **808**(1), 76.

Dale, D. A., and Helou, G. 2002. The Infrared Spectral Energy Distribution of Normal Star-Forming Galaxies: Calibration at Far-Infrared and Submillimeter Wavelengths. *Astrophysical Journal*, **576**(1), 159–168.

Dohm-Palmer, R. C., Skillman, E. D., Mateo, M. et al. 2002. Deep Hubble Space Telescope Imaging of Sextans A. I. The Spatially Resolved Recent Star Formation History. *Astronomical Journal*, **123**(2), 813–831.

Dopita, M. A., Groves, B. A., Sutherland, R. S. et al. 2003. Internal Dust Correction Factors for Star Formation Rates Derived for Dusty H II Regions and Starburst Galaxies. *Astrophysical Journal*, **583**(2), 727–731.

Draine, B. T., and Li, Aigen. 2007. Infrared Emission from Interstellar Dust. IV. The Silicate-Graphite-PAH Model in the Post-Spitzer Era. *Astrophysical Journal*, **657**(2), 810–837.

Draine, B. T., Dale, D. A., Bendo, G. et al. 2007. Dust Masses, PAH Abundances, and Starlight Intensities in the SINGS Galaxy Sample. *Astrophysical Journal*, **663**(2), 866–894.

Gallart, C., Zoccali, M., and Aparicio, A. 2005. The Adequacy of Stellar Evolution Models for the Interpretation of the Color-Magnitude Diagrams of Resolved Stellar Populations. *Annual Review of Astronomy and Astrophysics*, **43**(1), 387–434.

Garnett, D. R. 1999 (Jan). Element Abundances in Magellanic Cloud H II Regions from Carbon to Argon. Page 266 of: Chu, Y. H., Suntzeff, N., Hesser, J., and Bohlender, D. (eds), *New Views of the Magellanic Clouds*. IAU Symposium, vol. 190.

Grasha, K., Elmegreen, B. G., Calzetti, D. et al. 2017a. Hierarchical Star Formation in Turbulent Media: Evidence from Young Star Clusters. *Astrophysical Journal*, **842**(1), 25.

Grasha, K., Calzetti, D., Adamo, A. et al. 2017b. The Hierarchical Distribution of the Young Stellar Clusters in Six Local Star-Forming Galaxies. *Astrophysical Journal*, **840**(2), 113.

Hao, C.-N., Kennicutt, R. C., Johnson, B. D. et al. 2011. Dust-Corrected Star Formation Rates of Galaxies. II. Combinations of Ultraviolet and Infrared Tracers. *Astrophysical Journal*, **741**(2), 124.

Kennicutt, Robert C., Jr. 1998. Star Formation in Galaxies Along the Hubble Sequence. *Annual Review of Astronomy and Astrophysics*, **36**(Jan), 189–232.

Kennicutt, Robert C., Jr., Hao, C.-N., Calzetti, et al. 2009. Dust-Corrected Star Formation Rates of Galaxies. I. Combinations of Hα and Infrared Tracers. *Astrophysical Journal*, **703**(2), 1672–1695.

Kennicutt, Robert C., Jr., and Evans, Neal J. 2012. Star Formation in the Milky Way and Nearby Galaxies. *Annual Review of Astronomy and Astrophysics*, **50**(Sep), 531–608.

Kroupa, P. 2001. On the Variation of the Initial Mass Function. *Monthly Notices of the Royal Astronomical Society*, **322**(2), 231–246.

Lada, C. J., Lombardi, M., and Alves, J. F. 2010. On the Star Formation Rates in Molecular Clouds. *Astrophysical Journal*, **724**(1), 687–693.

Lada, C. J., Forbrich, J., Lombardi, M. et al. 2012. Star Formation Rates in Molecular Clouds and the Nature of the Extragalactic Scaling Relations. *Astrophysical Journal*, **745**(2), 190.

Larsen, S. S. 2009. The Mass Function of Young Star Clusters in Spiral Galaxies. *Astronomy and Astrophysics*, **494**(2), 539–551.

Leitherer, C., Schaerer, D., Goldader, J. et al. 1999. Starburst99: Synthesis Models for Galaxies with Active Star Formation. *Astrophysical Journal Supplement Series*, **123**(1), 3–40.

Lewis, A. R., Simones, J. E., Johnson, B. D. et al. 2017. The Panchromatic Hubble Andromeda Treasury. XVII. Examining Obscured Star Formation with Synthetic Ultraviolet Flux Maps in M31. *Astrophysical Journal*, **834**(1), 70.

Li, Y. 2014. *Measures of Star Formation Rates within Galaxies: The Impact of Diffuse Stellar Populations*. Ph.D. thesis, University of Massachusetts – Amherst (U.S.A.). https://scholarworks.umass.edu/dissertations_2/108.

Li, Y., Calzetti, D., Kennicutt, Robert C. et al. 2010. Spitzer 70 μm Emission as a Star Formation Rate Indicator for Sub-Galactic Regions. *Astrophysical Journal*, **725**(1), 677–691.

Li, Y., Crocker, A. F., Calzetti, D. et al. 2013. Star Formation Rates in Resolved Galaxies: Calibrations with Near- and Far-Infrared Data for NGC 5055 and NGC 6946. *Astrophysical Journal*, **768**(2), 180.

Liu, G., Koda, J., Calzetti, D. et al. 2011. The Super-Linear Slope of the Spatially Resolved Star Formation Law in NGC 3521 and NGC 5194 (M51a). *Astrophysical Journal*, **735**(1), 63.

McQuinn, K. B. W., Skillman, E. D., Cannon, J. M. et al. 2010. The Nature of Starbursts. II. The Duration of Starbursts in Dwarf Galaxies. *Astrophysical Journal*, **724**(1), 49–58.

McQuinn, K. B. W., Skillman, E. D., Dolphin, A. E. et al. 2015. Calibrating UV Star Formation Rates for Dwarf Galaxies from STARBIRDS. *Astrophysical Journal*, **808**(2), 109.

Murphy, E. J., Condon, J. J., Schinnerer, E. et al. 2011. Calibrating Extinction-Free Star Formation Rate Diagnostics with 33 GHz Free-Free Emission in NGC 6946. *Astrophysical Journal*, **737**(2), 67.

Oey, M. S., Meurer, G. R., Yelda, S. et al. 2007. The Survey for Ionization in Neutral Gas Galaxies. III. Diffuse, Warm Ionized Medium and Escape of Ionizing Radiation. *Astrophysical Journal*, **661**(2), 801–814.

Pellegrini, E. W., Oey, M. S., Winkler, P. F. et al. 2012. The Optical Depth of H II Regions in the Magellanic Clouds. *Astrophysical Journal*, **755**(1), 40.

Pérez-González, P. G., Kennicutt, Robert C., Jr., Gordon, K. D. et al. 2006. Ultraviolet through Far-Infrared Spatially Resolved Analysis of the Recent Star Formation in M81 (NGC 3031). *Astrophysical Journal*, **648**(2), 987–1006.

Pflamm-Altenburg, J., Weidner, C., and Kroupa, P. 2009. Diverging UV and Hα Fluxes of Star-Forming Galaxies Predicted by the IGIMF Theory. *Monthly Notices of the Royal Astronomical Society*, **395**(1), 394–400.

Relaño, M., Lisenfeld, U., Pérez-González, P. G. et al. 2007. On the Metallicity Dependence of the 24 μm Luminosity as a Star Formation Tracer. *Astrophysical Journal*, **667**(2), L141–L144.

Rieke, G. H., Alonso-Herrero, A., Weiner, B. J. et al. 2009. Determining Star Formation Rates for Infrared Galaxies. *Astrophysical Journal*, **692**(1), 556–573.

Sacchi, E., Annibali, F., Cignoni, M. et al. 2016. Stellar Populations and Star Formation History of the Metal-Poor Dwarf Galaxy DDO 68. *Astrophysical Journal*, **830**(1), 3.

Tolstoy, E., Hill, V., and Tosi, M. 2009. Star-Formation Histories, Abundances, and Kinematics of Dwarf Galaxies in the Local Group. *Annual Review of Astronomy and Astrophysics*, **47**(1), 371–425.

Tremonti, C. A., Heckman, T. M., Kauffmann, G. et al. 2004. The Origin of the Mass-Metallicity Relation: Insights from 53,000 Star-forming Galaxies in the Sloan Digital Sky Survey. *Astrophysical Journal*, **613**(2), 898–913.

Wu, H., Cao, C., Hao, C.-N. et al. 2005. PAH and Mid-Infrared Luminosities as Measures of Star Formation Rate in Spitzer First Look Survey Galaxies. *Astrophysical Journal*, **632**(2), L79–L82.

Zhu, Y.-N., Wu, H., Cao, C. et al. 2008. Correlations between Mid-Infrared, Far-Infrared, Hα, and FUV Luminosities for Spitzer SWIRE Field Galaxies. *Astrophysical Journal*, **686**(1), 155–171.

7

Continuum and Emission-Line Star-Formation Rate Indicators

MÉDÉRIC BOQUIEN AND DANIEL DALE

7.1 Observing Star Formation in Galaxies

Even if they are few in numbers, the young, blue, and massive stars in galaxies play a most critical role in our understanding of the formation and the evolution of galaxies through cosmic times. In effect, not only do they have a strong physical impact on their host galaxy, for instance through supernovae feedback and the production of metals, they also open a unique window that allows us to witness star formation across the universe. What sets massive stars apart in this endeavour is that their short lifetimes, from a few million years to a few hundred million years, ensure that their presence is an unmistakable testimony of recent star formation. In other words, quantifying directly or indirectly their presence is the central tenet behind star-formation rate (SFR) estimators and one of the main pillars upon which rest our models of galaxy evolution.

The most direct way to detect the presence of massive stars is through their photospheric emission escaping from their host galaxy. Doing so at optical or near-infrared wavelengths is difficult as they are generally outshined by the multitude of lower mass stars from current and previous generations. However, their high temperature makes them easier to observe at short wavelengths as even at low levels of star formation they dominate the ultraviolet (UV) energy budget of galaxies. Energetic photons from massive stars are, however, not our only window into star formation. Understanding how they propagate through the Interstellar Medium (ISM) and interact with gas and dust reveals the existence of alternative ways to trace star formation in galaxies.

The main impediment for astronomers to observe the UV radiation from star-forming galaxies is the presence of dust. Dust is especially efficient at absorbing short-wavelength radiation and even a small quantity can be sufficient to capture a large fraction of UV photons. In practice, dust acts as a calorimeter and re-emits most of this energy in the mid- and far-infrared domains. That is, the bolometric emission from dust is a star-formation tracer and increasingly more so when the galaxy is more dusty. Measuring the total emission from dust is a challenge as it dominates the emission of galaxies over a broad wavelength range from a few microns to approximately a millimetre. To make the matter even more complex, dust emission is the combination of multiple components not necessarily in thermal equilibrium (dust grains with different compositions, sizes, temperatures, etc., as well as complex molecules such as Polycyclic Aromatic Hydrocarbons; PAH), giving a rich spectrum with two very prominent features: a modified black body peaking between 100 and 200 microns and various strong PAH bands around 10 microns. The consequence is that the intensity of the emission at any given wavelength is a complex function of the quantity of dust, its composition, its distribution relative to the stellar populations of all ages, the shape of the illuminating radiation field, etc. This means that in principle we cannot easily infer from theory the total dust emission from a single infrared band. A successful empirical approach is to consider that the radiation field from massive stars is overwhelmingly responsible for exciting and heating the dust that emits typically under 100 microns and estimate the dust luminosity from just one of a few bands at such wavelengths. We also refer to Chapter 6 for a detailed discussion of these empirical methods.

As we follow photons from massive stars propagating through their host galaxy, besides escaping or being absorbed by dust, the most energetic of them can also ionise the surrounding gas. The interaction of the neutral and ionised atoms with free electrons gives rise to a complex spectrum: a series of strong emission lines (mainly hydrogen, carbon, nitrogen, and oxygen) at discrete wavelengths as well as continuum emission (free-free and free-bound transitions as well as 2-photon processes). Ionising these elements requires more energetic photons than those used to trace star formation in the UV (the ionisation potential is 13.6 eV for hydrogen and oxygen, 11.3 eV for carbon, and 14.5 eV for nitrogen) and, therefore, ionised gas is only sensitive to the emission from the most massive stars whose lifetimes typically do not exceed ten million years, tracing the most recent star formation. Besides the Lyman series, which suffers from resonant scattering that strongly reduces the probability of Lyman-α photons to escape the galaxy, hydrogen lines are certainly the most popular tracer of ionising stars. They combine the advantage of being prominent in the optical and near-infrared spectrum while depending only weakly on the exact metallic content of the gas in photo-ionised regions or the actual shape of the ionisation field. Metal lines are more complex tracers as they depend more strongly on the local physical conditions. While oxygen and hydrogen share a similar ionisation potential, the intensity of the latter is directly dependent on the excitation conditions. As for singly ionised carbon, besides its dependence on local physical conditions, its lower ionisation potential also makes it somewhat more sensitive to longer timescales. Its main advantage, however, is that it has a very strong fine structure line at 158 microns, which is one of the main coolants of the ISM.

An increasingly promising way to trace ionising photons with the advent of new radio instruments is free-free emission. Continuum emission near Lyman-break wavelengths is heavily dominated first by stellar emission and then by nebular emission. In the millimetre regime, however, free-free becomes the dominant source of emission and can be used to trace star formation. Besides free-free emission, the other major component in the radio spectrum is synchrotron emission. It is generated by cosmic rays accelerated in the remnants of core-collapse supernovae and interacting with a galaxy's magnetic field. The surprisingly strong correlation between the radio continuum and infrared emission, given the unknowns in the diffusion of the cosmic rays and the strength of the magnetic field, shows its capacity as an indirect tracer of star formation.

In this brief overview we have seen that we can trace star formation through a broad variety of observations: photospheric emission from massive stars in the UV, dust emission in the infrared from grains heated and excited by energetic photons, hydrogen and metal recombination lines from the optical to the infrared, and even free-free continuum and synchrotron emission in the radio domain. The first and foremost constraint for astronomers in estimating SFRs is the ability to obtain adequate observations. For instance, distant galaxies may have emission lines shifted beyond the near-infrared, making them inaccessible from the ground, or the object may be too faint for its free-free emission to be detected. The nature of the observed galaxy and the available instruments therefore strongly guide how we can measure star formation. In the context of this chapter we concentrate on detailing how we can use any observation in star-formation tracing bands to measure the SFR as reliably as possible.

7.2 Theoretical Considerations

In practice the diverse zoo of SFR estimators can be classified into two broad categories: theoretical estimators and empirical estimators. The former are more fundamental and can be more direct. They are grounded on models which take into account the evolution of different stellar populations, the radiative transfer through stellar atmospheres, the ionisation and recombination of atoms in the interstellar medium, etc. As such, theoretical estimators are easily parametrised against various physical properties (initial mass function, metallicity, star-formation history, etc.), making them in principle remarkably powerful and adaptable to a broad variety of cases. However, as we will see, in practice they often rely on a set of strong assumptions. To assess the reliability of SFR estimators, it is important to understand these often overlooked assumptions and whether they are valid in a given situation. Barring this, SFR estimators can provide unreliable estimates.

While some estimators such as Hα, UV, or total infrared (TIR) for instance lend themselves well to such a theoretical approach, other estimators (e.g. hybrid estimators, monochromatic infrared bands, metal lines) would require a considerably more complex modelling of the physical processes within galaxies and star-forming regions. A common strategy is to calibrate statistically the luminosity of a representative sample of galaxies (or star-forming regions) in these star-formation tracing bands against theoretical tracers that act as the primary reference. Their quality then depends primarily on the quality of the primary theoretical estimators and on the actual representativity and quality of the calibration sample.

7.2.1 A Brief Overview of the Common Assumptions behind Theoretical SFR Estimators

A naïve approach to measuring the SFR is to consider that the emission in star-formation tracing bands and the SFR are linearly related. After all, if we compare two galaxies with different SFRs, as a first approximation we can expect the number of massive stars and the luminosity in star-formation tracing bands to also be different in the same proportion. The conclusion of this candid reasoning is that a simple proportionality coefficient should be sufficient to convert the luminosity in a star-formation tracing band into an estimate of the SFR. This idea is at the heart of the most popular SFR estimators. Yet it implicitly relies on a set of fairly strong assumptions: an appropriate star-formation history, a well-defined and fully populated initial mass function, a good estimate of the metallicity, and reliable models of star populations.

Star-Formation History

The first major assumption when building SFR estimators is that of the star-formation history (SFH). Indeed, the emission in any star-formation tracing band is not the product (direct or indirect) of a unique stellar population of a single age. Stars of different ages contribute to the overall emission. One of the main difficulties is that very different SFHs can yield the same luminosity in a given star-formation tracing band. For example, a weak but recent episode of star formation can output the same UV luminosity as a stronger but older episode of star formation (see also Chapter 8). This means that the conversion of the luminosity into an SFR requires an assumption on the SFH. The most widespread assumption is that of a constant SFR over the past 100 Myr (or sometimes 1 Gyr). However convenient such a straightforward assumption may be, it is not without important limitations. The SFR can be misjudged if star formation varied somewhat during the past 100 Myr (or 1 Gyr) and this also does not take into account the contribution from stars living longer than 100 Myr (or 1 Gyr). In other words, SFR estimators are dependent on the appropriateness of the assumed SFH.

Initial Mass Function

Star formation tracers rely, directly or indirectly, on the emission from massive, short-lived stars. However, such stars are few in number and they represent only a minor fraction of the total mass of stars formed at any given moment. This means that, to estimate the SFR, we need to be able to estimate the total mass of stars formed from the emission of a small number of them. For this we need to assume an initial mass function (IMF) which describes the distribution of stellar masses at birth. Thus, SFR estimators are sensitive to the shape of the IMF (generally defined as one or several power laws or through a log-normal function), the minimum and maximum stellar masses between which it is defined, and its sampling (e.g. Fig. 7.1). Naturally, adopting an inappropriate IMF has direct consequences on SFR estimations. We also refer to Chapter 2 for an extensive discussion about the impact of the IMF on SFR determinations.

Metallicity

The presence of heavy elements in stars has an influence on their emission in two main ways. First, metals redden the stellar spectra through the presence of numerous absorption

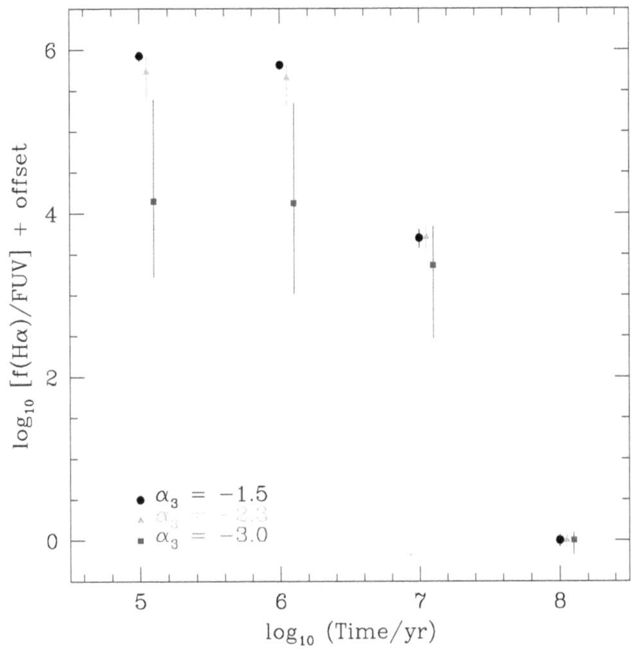

Figure 7.1 Evolution with time of the logarithmic ratio of Hα-to-FUV flux as a function of IMF slope, scaled to match at 10^8 years. Error bars indicate the central 68.3% of stochastically varied simulations as determined by SLUG Krumholz et al. (2015) for a $10^3 M_\odot$ cluster assuming a Kroupa IMF ($\alpha_1 = -0.3, \alpha_2 = -1.3$) and **Starburst99**/Padova stellar tracks. Colour version available online.

lines in their atmospheres that preferentially absorb blue and UV radiation, increasing the photospheric temperature through backscattering and increasing the emission in the near-infrared. The second major effect is the expansion of the stars because of the increased opacity brought by the metals in their envelope. This causes an increase of the radius and a corresponding decrease of its temperature. At the same time, the increased metallicity also affects the evolution of stars, increasing their mass loss through winds for instance. In short, for a given mass, low metallicity stars are bluer.

Single Stellar Populations Models

Modelling the emission of just an individual star in isolation is already a complex matter even when its physical properties are known with certainty. In the end, different approaches to model stars translate into differences in their computed emission spectra, depending on the choice of the evolutionary tracks, stellar atmospheres, winds, etc. In particular, there has been increasing emphasis on understanding the effect of stellar rotation and multiplicity on the emission of stars. The former affects stellar evolution through mixing, which increases the luminosity, the number of ionising photons, and the lifetime of massive stars. The latter affects stellar evolution through mass transfer between stars. Ultimately, different stellar population models include different choices that sometimes have an important effect on the energetic radiation of stars and, therefore, have a correspondingly important effect on star-formation tracers.

7.2.2 Quantitative Impact of the Assumptions behind SFR Estimators

Having laid down these fundamental assumptions behind theoretical SFR estimators, building SFR estimators is relatively straightforward using spectral synthesis models. As an example, we adopt the `CIGALE`[1] code (Burgarella et al., 2005; Noll et al., 2009; Boquien et al., 2019) with the Bruzual and Charlot (2003) single stellar populations to build SFR estimators varying three of the most fundamental assumptions: 1) the IMF from Salpeter (Salpeter, 1955) and from Chabrier (Chabrier, 2003); 2) the metallicity (from $Z = 0.0001$ to $Z = 0.05$); and 3) the SFH (constant over 100 Myr and over 1 Gyr). We consider here some of the most popular SFR estimators:

1. the photospheric radiation from stars in the far- and near-UV bands (FUV and NUV, both assumed to follow the *GALEX*[2] filters) and the U band;
2. the ionising radiation from massive stars as measured from the Hα line, assuming that Lyman continuum photons do not escape; and
3. the Total Infrared (TIR) radiation, assuming that the galaxy is entirely buried in dust (therefore, it is the integral of the stellar+nebular spectrum).

In Table 7.1, we present these estimators for the various aforementioned assumptions. It appears that each of these assumptions has a varying effect on the SFR estimators.

- There is a difference of ~0.2 dex between using the Salpeter and the Chabrier IMF. This is due to the fact that the Salpeter IMF has a smaller fraction of massive stars compared to the Chabrier IMF. The difference between the Chabrier and Kroupa (Kroupa, 2001) IMF is, however, much smaller, differing by no more than 7% (Madau and Dickinson, 2014). As we will see in Section 7.2.2, the IMF can have even more drastic effects on SFR estimators when it is not fully sampled, this issue is also discussed extensively in Chapter 2.
- The SFH can also have an important effect on SFR estimators. This effect varies from being completely negligible for Hα to sometimes exceeding ~0.2 dex in the U band. This is due to the progressive accumulation of stars that contribute to the emission in star-formation tracing bands. The effect is naturally stronger for the TIR and in the U band as they are sensitive to the emission of lower mass stars compared to Hα or UV bands. We will delve in more detail on the effect of the SFH hereafter.

[1] Code Investigating GALaxy Emission; https://CIGALE.lam.fr
[2] Galaxy Evolution Explorer; www.galex.caltech.edu/index.html

- Finally, the stellar metallicity also has a noticeable effect. It most strongly affects Hα with variations as large as ∼ 0.5 dex. This is due to the strong dependence of the ionising flux on the metallicity. We should stress, however, that additional processes such as binary evolution for instance can also cause broad variations in the ionising flux (Stanway et al., 2016) and that by probing only the metallicity, we only sample a part of a much broader parameter space that is considerably influential in estimating the production of Lyman continuum photons.

Table 7.1. *Calibration coefficients k to estimate the SFR with a relation of the form* $\log SFR = \log L + k$, *with the SFR in* $M_\odot\ yr^{-1}$, *and L in W for Hα and TIR and W* Hz^{-1} *for the other tracers*

Z	Hα	FUV	NUV	U	TIR
		100 Myr, Salpeter IMF			
0.0001	−34.388	−20.998	−20.882	−20.791	−36.464
0.0004	−34.274	−20.973	−20.870	−20.789	−36.428
0.004	−34.190	−20.944	−20.868	−20.908	−36.425
0.008	−34.137	−20.937	−20.876	−20.946	−36.413
0.02	−34.054	−20.892	−20.856	−20.892	−36.376
0.05	−33.945	−20.855	−20.850	−20.797	−36.350
		1 Gyr, Salpeter IMF			
0.0001	−34.389	−21.101	−21.032	−21.065	−36.612
0.0004	−34.275	−21.064	−21.004	−21.062	−36.584
0.004	−34.190	−21.013	−20.971	−21.102	−36.576
0.008	−34.137	−20.992	−20.962	−21.108	−36.552
0.02	−34.055	−20.932	−20.924	−21.040	−36.506
0.05	−33.946	−20.877	−20.893	−20.941	−36.476
		100 Myr, Chabrier IMF			
0.0001	−34.610	−21.197	−21.082	−20.992	−36.667
0.0004	−34.496	−21.175	−21.070	−20.989	−36.631
0.004	−34.415	−21.146	−21.070	−21.114	−36.629
0.008	−34.363	−21.141	−21.078	−21.154	−36.618
0.02	−34.279	−21.096	−21.059	−21.098	−36.581
0.05	−34.167	−21.059	−21.052	−20.996	−36.554
		1 Gyr, Chabrier IMF			
0.0001	−34.611	−21.293	−21.220	−21.246	−36.802
0.0004	−34.497	−21.259	−21.194	−21.244	−36.775
0.004	−34.415	−21.210	−21.164	−21.293	−36.768
0.008	−34.363	−21.191	−21.157	−21.302	−36.746
0.02	−34.280	−21.133	−21.122	−21.234	−36.700
0.05	−34.168	−21.080	−21.092	−21.130	−36.669

These variations illustrate how important the assumptions behind SFR estimators can be. Even with such a restricted exploration of parameter space, peak-to-peak differences in SFR can reach ~0.7 dex using Hα, ~0.5 dex using FUV, U, and TIR, and ~0.4 dex using NUV. In some extreme cases errors on SFR estimates can be well over 1 dex.

Star-Formation History

The SFH affects SFR estimators through two main avenues: 1) the short term variations of the SFR and 2) the accumulation of long-lived stars. Because of the finite lifetime of stars, the former induces a delay between the variation of the SFR and the variation of the luminosity in star-formation tracing bands along with a smoothing effect (that is, the variation of the luminosity will be smoother than the variation of the SFR). The latter contaminates directly or indirectly the emission in star-formation tracing bands, leading to an overestimate of the SFR.

Using numerical simulations of main-sequence galaxies at $z = 2$, comparing the true SFR to the SFR estimated from luminosities assuming a constant SFR over 100 Myr, Boquien et al. (2014) showed that short-term variations induce a scatter of ~20%[3] with no systematic biases as episodes of increasing and decreasing SFR average out. However, it has to be noted that single measurements can still be biased by a factor of a few in the case of a rapid variation of the SFR. Taking an extreme example – a step function in the onset or stopping of star formation – the standard SFR estimators would be infinitely biased. In more realistic cases, this can be particularly of concern when measuring SFR at sub-galactic scales[4] and for galaxies undergoing a rapid evolution such as the onset of a starburst or a quenching phenomenon (e.g. Ciesla et al., 2017).

In addition to examining the effect of short-term variations of the SFR, Boquien et al. (2014) also found that the contamination by long-lived stars leads to a systematic bias that can reach a few tens of per cent, with larger biases for longer wavelengths. To limit the bias they recommend assuming a constant SFR over 1 Gyr, while Cerviño et al. (2016) recommend assuming a constant SFR over the age of the galaxy. In any case, knowing what stars contribute which fraction in star-formation tracing bands is important for understanding what SFR estimators really trace.

This prompts the question of the timescales probed by SFR estimators. It is often incorrectly stated that different tracers provide average SFR estimates over different timescales (10 Myr for Hα, 100 Myr for FUV, etc.). In reality, it is not possible to define such an average timescale (Boquien et al., 2014; Cerviño et al., 2016). The reason is that, because of the aforementioned effects, the actual averaging timescale strongly depends on the SFH. In effect, as we can see in Fig. 7.2, stars of different ages can contribute to the emission in star-formation tracing bands.

[3] We should note that this number will be dependent on the actual SFH.
[4] With fewer star forming regions at sub-galactic scale, the SFH is more likely to deviate from the constant SFR assumption.

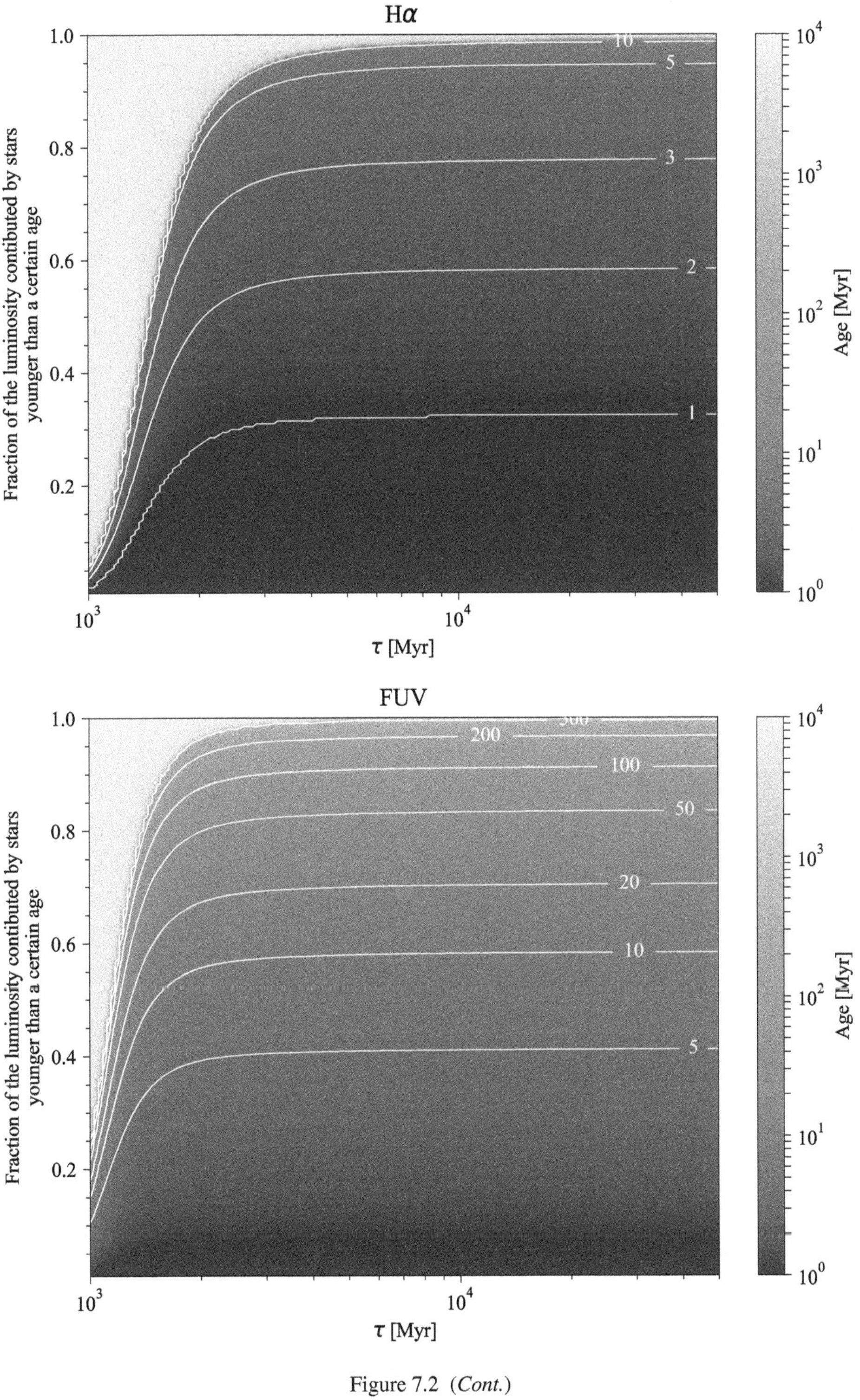

Figure 7.2 (*Cont.*)

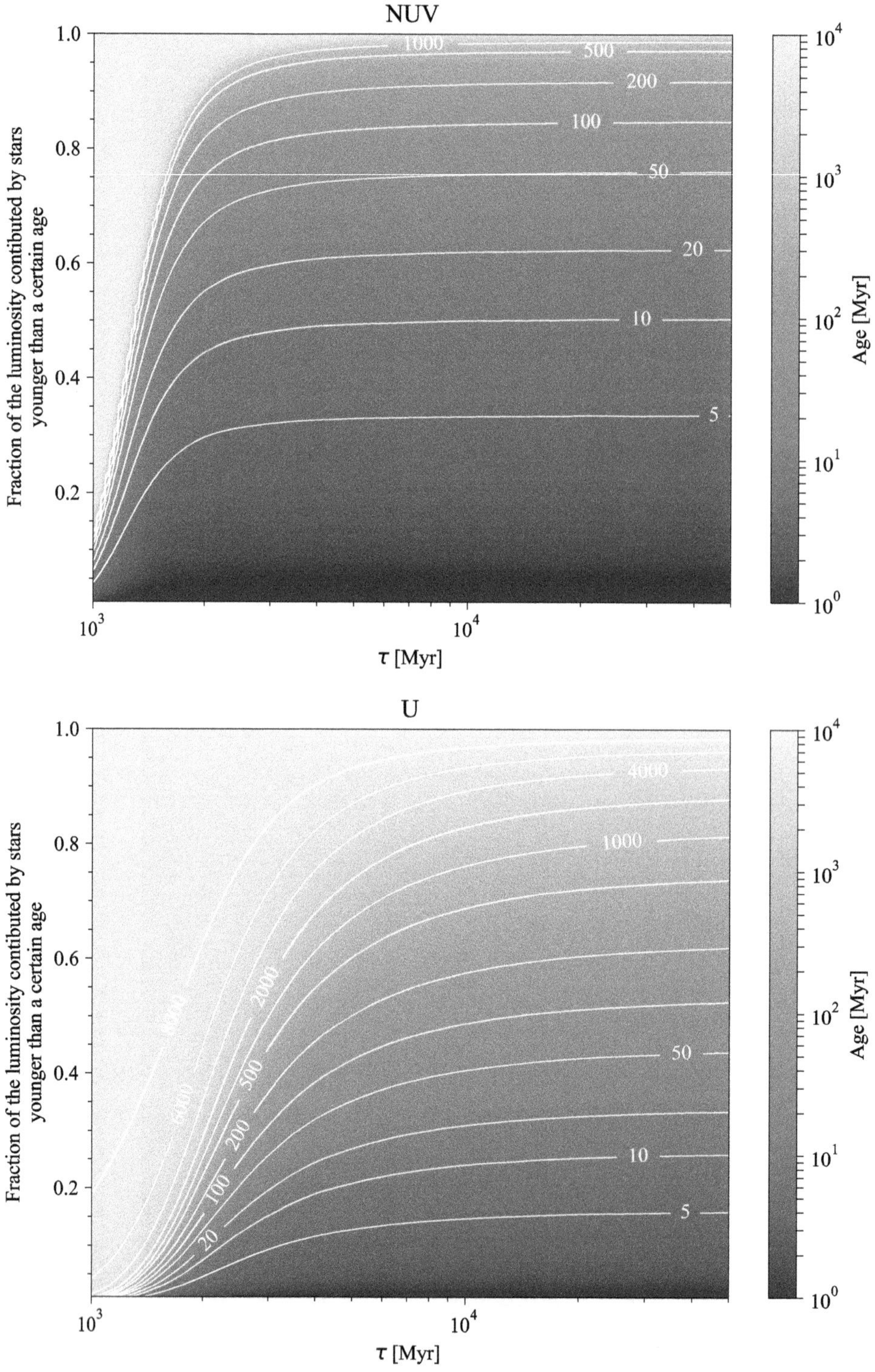

Figure 7.2 (*Cont.*)

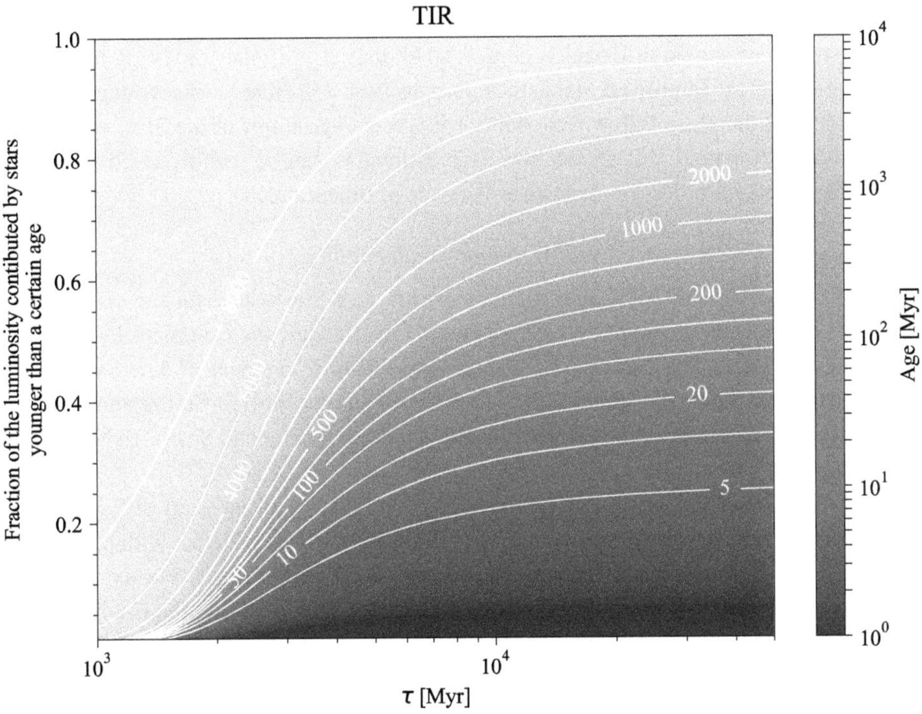

Figure 7.2 (*Cont.*) Cumulative fraction of the emission in star-formation tracing bands contributed by stars younger than a certain age, indicated in colour following the bar to the right, versus the timescale τ for a 10 Gyr old exponentially decaying SFH ($SFR \propto \exp(-t/\tau)$). White lines indicate the fraction of of the total luminosity contributed by stars younger than a given age. For instance the line at 100 Myr gives the fraction of the emission that is contributed by stars no older than 100 Myr as a function of τ. Small values of τ correspond to a near-instantaneous star-formation episode whereas a large τ corresponds to a nearly continuous star-formation. Colour version available online.

For instance, if we consider a nearly constant SFR, ~10% of the FUV luminosity is due to stars older than 100 Myr, leading to an overestimate of the SFR of the same order under the assumption of a constant SFR over 100 Myr. At the same time, almost 60% of the FUV luminosity is contributed by stars no older than 10 Myr. A similar behaviour can be seen for the TIR. With a constant SFR, over ~30% of the emission is due to stars no older than 10 Myr and ~50% is due to stars older than 100 Myr.[5] As we can see in Fig. 7.2,

[5] As mentioned earlier, we assume here that the galaxy is fully attenuated by dust, which is the most extreme case. Partial attenuation will increase the relative contribution of populations of younger ages because the attenuation curve rises towards shorter wavelengths.

these numbers depend strongly on the shape SFH, even for such a simple parametrisation. This illustrates, as shown in Boquien et al. (2014), that even if star formation tracers can be contaminated by long-lived stars, they remain very sensitive to the youngest stellar populations and therefore follow reasonably well recent variations of the SFR, even if in a delayed way. Ultimately, this shows how SFR estimators can be sensitive to the shape of the SFH over several orders of magnitude in terms of timescales.

The IMF Shape and Sampling

As noted earlier, the IMF describes the distribution of stellar masses at the onset of star formation in a localised volume within a galaxy. For decades, the canonical IMF utilised was that described in Salpeter (1955): a single power law $\xi(m) \propto m^{-\alpha}$ where $\alpha = 2.35$. Subsequent descriptions of the stellar IMF have kept a similar form for stellar masses above 0.5–1.0 $M_\odot$ and shallower slopes below this mass regime (Miller and Scalo, 1979; Kroupa, 2001; Chabrier, 2003).

A fully sampled IMF is a foundational underpinning to, essentially, all SFR indicators, but this assumption can be severely compromised in lower mass systems, with particularly important implications for stars missing at the massive end of the IMF. For example, in a survey of ~300 nearby star-forming galaxies, Lee et al. (2009) find that the Hα-based SFR conspicuously underpredicts the FUV-based SFR for the least massive (dwarf) galaxies, i.e. $6 \leq \log_{10}(M_{\rm star}/{\rm M}_\odot) \leq 8$ or, alternatively, $-4 \leq \log_{10}[SFR({\rm H}\alpha){\rm M}_\odot{\rm yr}^{-1}] \leq -2$ (see also Salim et al., 2007; Boselli et al., 2009; Meurer et al., 2009). Lee et al. (2009) show that the semi-empirical Integrated Galactic IMF (Kroupa and Weidner, 2003; Weidner and Kroupa, 2005; Pflamm-Altenburg et al., 2007), where the maximum stellar mass for the IMF depends on the total stellar cluster mass, adequately reproduces the observed trend. However, Weisz et al. (2012) show that a time variable ('bursty') star-formation history does a superior job in matching the data.

Metallicity

As mentioned earlier, the metallicity has an effect on both the stellar atmosphere and stellar evolution itself. To investigate the influence of metallicity on SFR estimators, we plot in Fig. 7.3 the evolution of the emission normalised to $Z = 0.02$ in various star-formation tracers at different metallicities. As could already be inferred from Table 7.1, there are clear variations of SFR tracers with respect to metallicity. An important aspect to note is that there is not just a variation with metallicity, there is also a variation with time. The Hα reaches a constant ratio with respect to $Z = 0.02$ after ~10 Myr and, therefore, a single scaling coefficient allows for a simple conversion in the SFR estimation from one metallicity to another when the SFR is constant for a longer period. This is not the case for the other estimators that keep varying at least until 1 Gyr, and probably beyond. The consequence is that there is no simple way to convert SFR from one metallicity to another without taking into account the SFH. That being said, as can be expected, the largest difference in the SFR estimators occurs for the largest differences on the metallicity. For instance, the difference reaches a factor ~ 3 between $Z = 0.05$ and $Z = 0.0001$ in the case of Hα. For other bands the difference is generally the largest the first few million years, at ages that are not appropriate for SFR estimators. In practice, one generally has a reasonable estimate of the stellar metallicity of recently formed stars and it is unlikely that the bias is larger than 10–20%.

Single Stellar Population Models

Many stellar populations models are available in the literature and we refer to Chapter 3 for a more detailed description of these models. Strikingly, different models can give widely different spectra. A poster-child example is that of the stellar mass inferred from the near-infrared luminosity. The differing impact of thermally pulsating asymptotic giant branch stars can yield notably different spectra using different models even with the same SFH, IMF, and metallicity. Quantitatively, a simple `CIGALE` model shows that assuming an instantaneous burst the Maraston (2005) models can give a luminosity over a factor of 2 higher than the Bruzual and Charlot (2003) models in the *2MASS*[6] Ks band after a few hundred Myr and for a duration of a few hundred Myr. Unfortunately, such tension is not limited to the near-infrared, and different models also yield different predictions for star-formation tracers. As mentioned earlier, there are multiple factors contributing to these differences. Here we quantify two of them: stellar rotation and binarity.

Stellar rotation: To examine the effect of stellar rotation, we adopt the latest `Starburst99`[7] models from Leitherer et al. (2014). They include the stellar evolutionary tracks of Ekström et al. (2012) both without rotation and with a rotation speed of 40% the breakup velocity. The effect of stellar rotation on star-formation tracers is shown in Fig. 7.4.

Considering only quasi-instantaneous bursts, Hα is by far the most affected tracer, with an increase in the luminosity that can attain a factor of $\sim$10. However, the peak occurs after $\sim$6–8 Myr, at a time the Hα luminosity has already dropped by a factor of $\sim$10 to 100 (depending whether there is rotation or not) from its maximum. Nevertheless, for populations aged of a few Myr, the increase in the Hα luminosity is limited to a factor of a few. The UV is less affected: both the FUV and NUV luminosities are increased by a factor of $\sim$1.9 after $\sim$8 Myr for a single stellar population. The peak impact of stellar rotation is slightly lower for the U band and TIR, with peaks a bit after/before 10 Myr.

6 Two Micron All Sky Survey; https://irsa.ipac.caltech.edu/Missions/2mass.html
7 www.stsci.edu/science/starburst99/

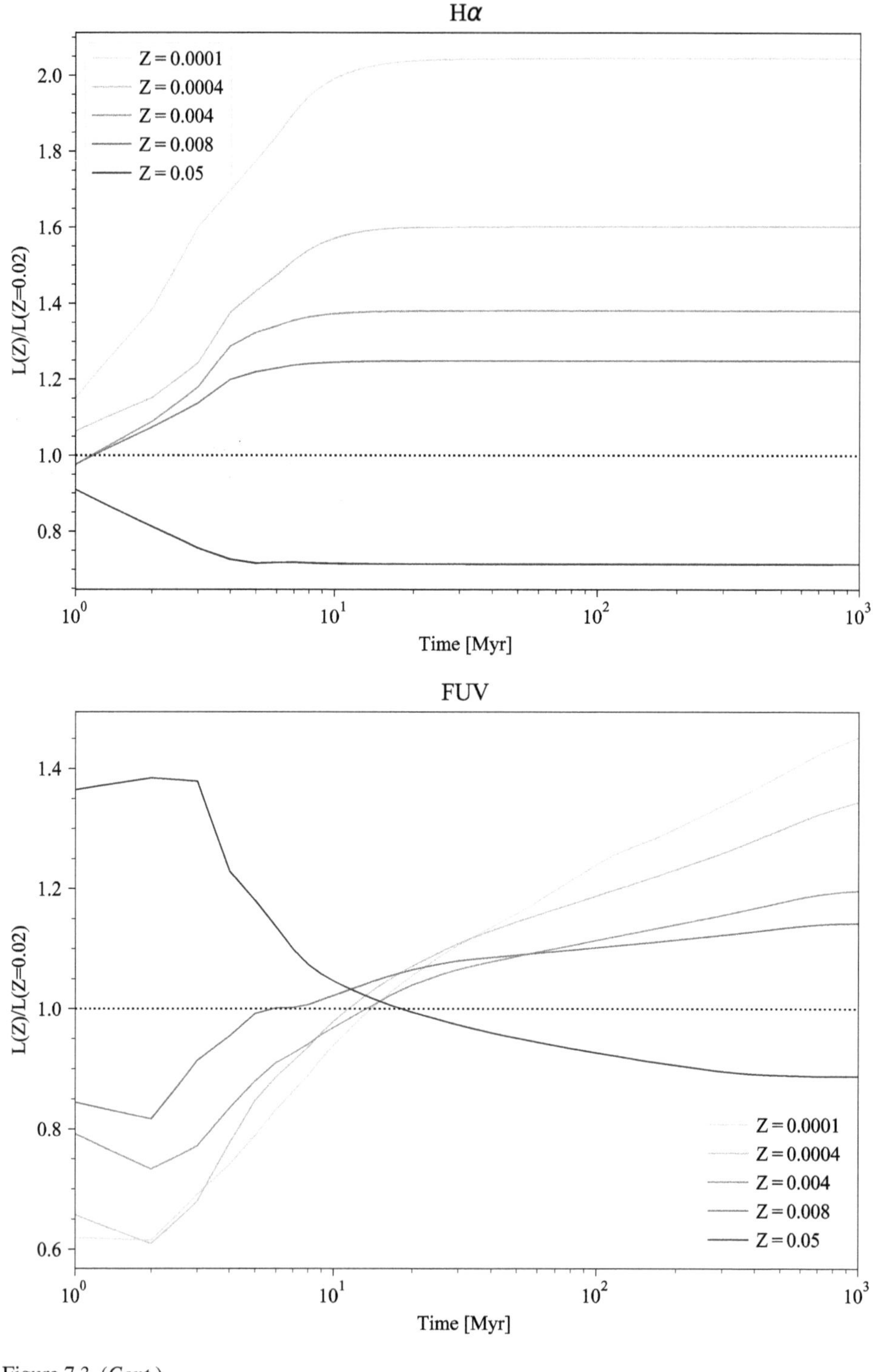

Figure 7.3 (*Cont.*)

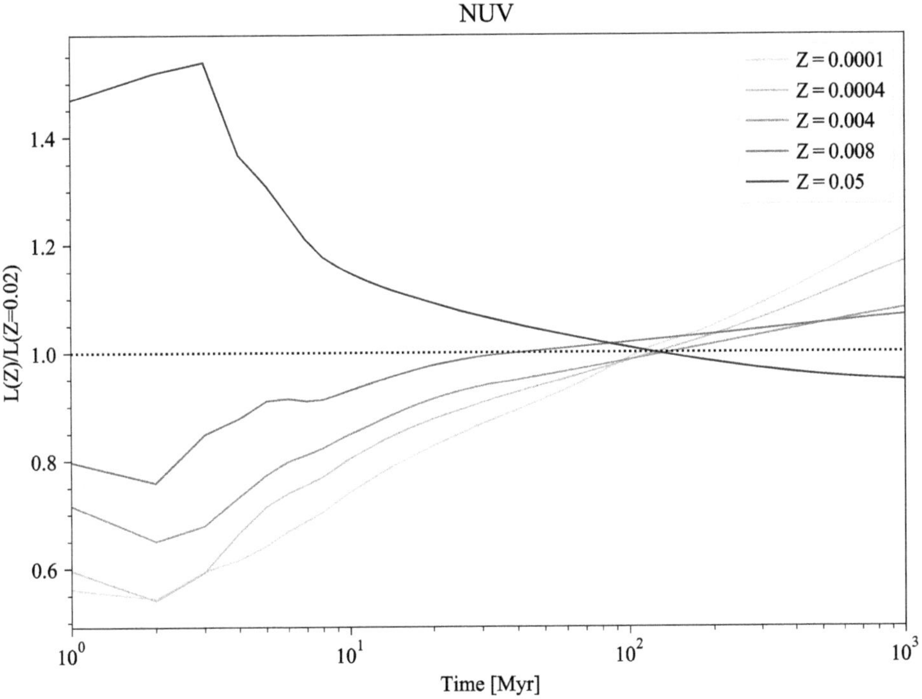

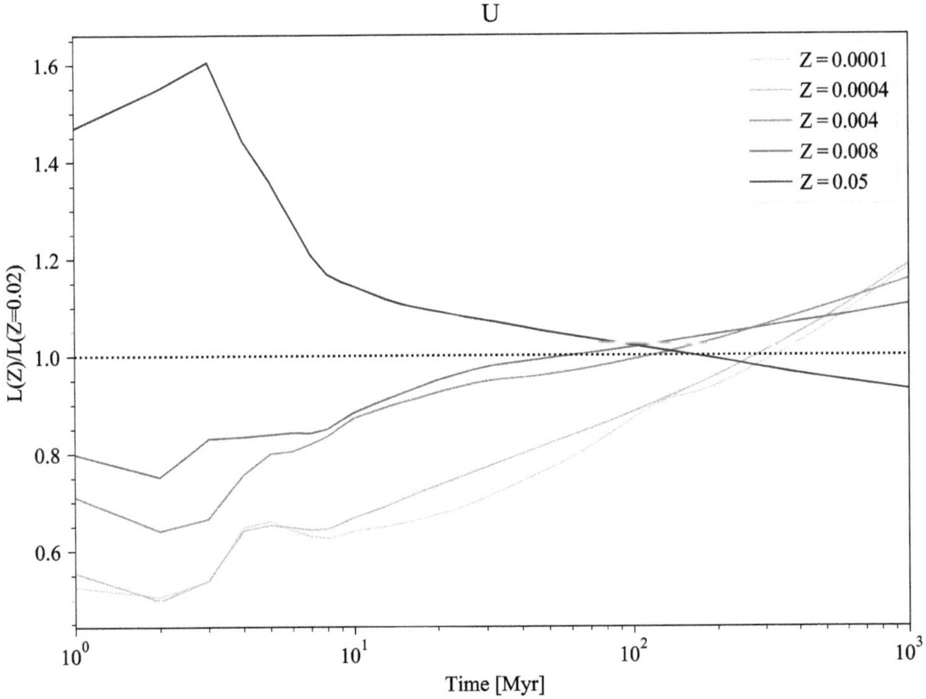

Figure 7.3 (*Cont.*)

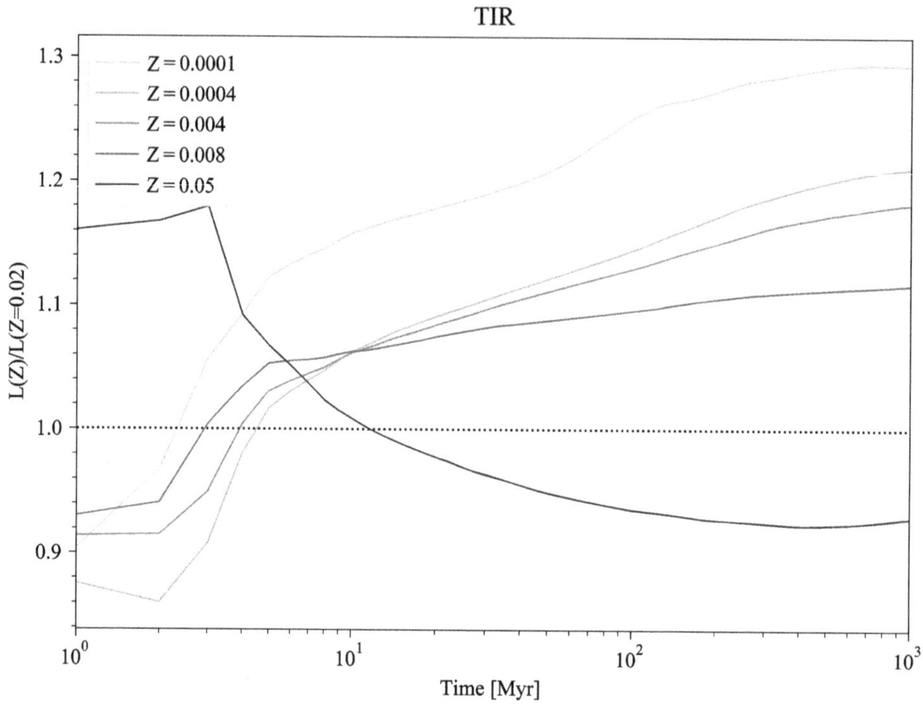

Figure 7.3 (*Cont.*) Evolution with time of the luminosity at different wavelengths for $Z = 0.0001$, $Z = 0.0004$, $Z = 0.004$, $Z = 0.008$, and $Z = 0.05$ normalised to the luminosity at $Z = 0.02$ as a function time. A constant SFR and a Salpeter IMF are assumed. Colour version available online.

While exploring single stellar populations is of interest to evaluate the effect of stellar rotation in isolation, in practice for the purpose of measuring star formation, we instead consider the case of constant star formation. The effect is then much milder and varies slowly with time. The Hα line remains the most affected by stellar rotation, with the luminosity being about twice the luminosity without rotation past the first few Myr. The luminosity excess in the other bands progressively increases to a factor of typically 1.2–1.3.

The exact impact of stellar rotation on SFR estimates is difficult to estimate. Its amplitude is dependent on the star-formation history (independently of the effects developed in the next section) and the models themselves are uncertain as they naturally depend on the actual rotation velocity of the stars. However, if we consider that the Ekström et al. (2012) rotation models are a practical upper limit for rotation velocities and the SFR is constant, the uncertainty should be contained within a factor $\sim$2 for Hα and a factor $\sim$1.3 for other bands.

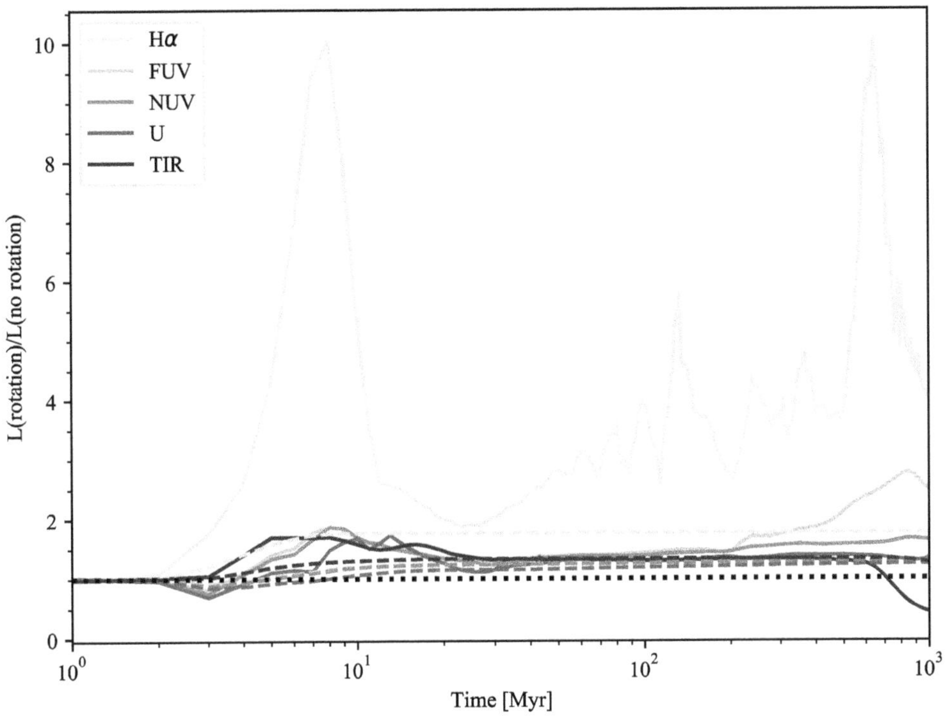

Figure 7.4 Evolution with time of the ratio of the luminosity with and without rotation of a quasi instantaneous burst (an episode of star formation lasting 1 Myr and ceasing abruptly, solid lines) and for a constant SFR (dashed lines). We assume a solar metallicity. We have adopted the single stellar populations of `Starburst99` (Leitherer et al., 2014) following the Ekström et al. (2012) stellar evolutionary tracks. The luminosities in Hα (blue), FUV (orange), NUV (green), U (red), and TIR (violet) have been computed using `CIGALE`. Colour version available online.

Binarity: To examine the effect of binarity, we adopt the `BPASS`[8] models of Eldridge and Stanway (2009) version 2.1 Eldridge et al. (2017) (see also Chapter 3). In Fig. 7.5, we compare the evolution of the ratio of the luminosity with and without binary evolution in star-formation tracing bands for a single stellar population and a constant SFR at $Z = 0.02$.

We see that the effects of binary evolution can be strong for single stellar populations, in particular at shorter wavelengths. However, the effect is stronger after the luminosity in star-formation tracing bands has already dropped to much lower levels than a zero-age population, limiting the practical effect. In the case of a constant SFR over 100 Myr, the effect remains fairly mild, typically contained within ~20%.

[8] Binary Population and Spectral Synthesis

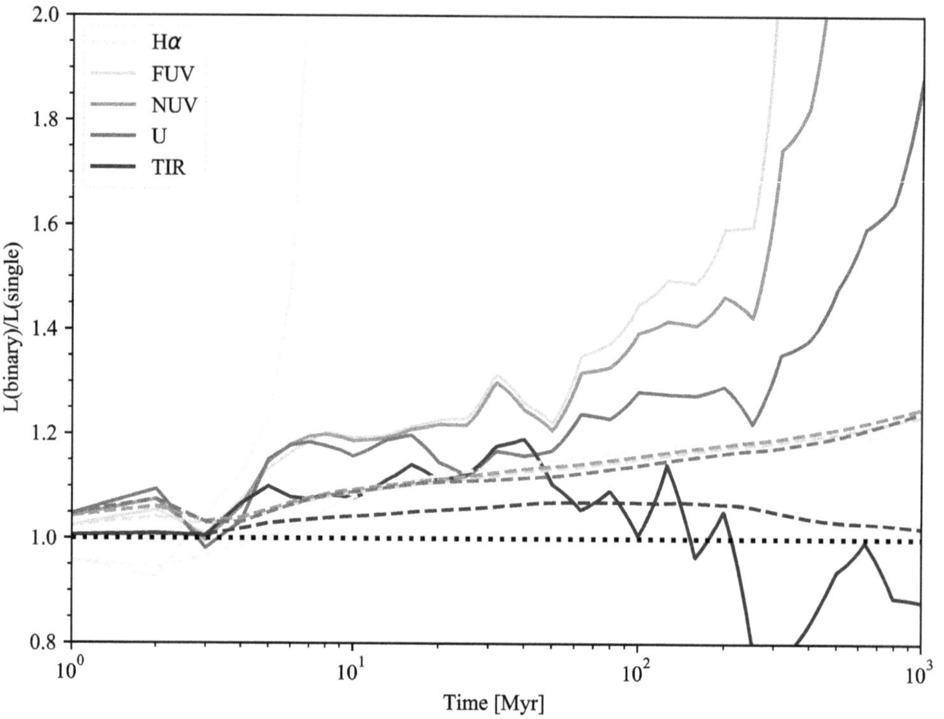

Figure 7.5 Evolution with time of the ratio of the luminosity with and without binary evolution of a single stellar population (solid lines) and for a constant SFR (dashed lines) at solar metallicity. We have adopted the BPASS models of Eldridge and Stanway (2009) version 2.1 (Eldridge et al., 2017) and a metallicity $Z = 0.02$. The luminosities in Hα (blue), FUV (orange), NUV (green), U (red), and TIR (violet) have been computed using `CIGALE`. Colour version available online.

7.3 Observational Constraints

7.3.1 Ultraviolet Continuum

The UV continuum is commonly used to gauge the star-formation rate in galaxies (e.g. Buat et al., 2005; Daddi et al., 2007). This popularity relies on the fact that the UV continuum, and in particular the far-UV continuum at $\lambda < 2000$ Å, mainly traces the emission from massive young stars; any robust SED-based measure of the star-formation history of a galaxy critically relies on information from the UV (e.g. Boquien et al., 2009). Moreover, in a study of 60,000 low redshift galaxies, Salim et al. (2007) show that the UV provides more reliable star-formation rates for galaxies with weak Hα or appreciable AGN contributions, characteristics that hold true for about half of their sample. However, as noted above and alluded to again below, UV light is acutely susceptible to absorption and scattering by interstellar dust grains. Even in normal star-forming galaxies where the attenuation by dust is moderate, the typical attenuation level ($A_V \sim 1$ mag $\rightarrow A_{UV} \sim 3$ mag) corresponds to observations actually detecting less than 10% of the intrinsic UV emission (see also

Chapters 4 and 6). Fortunately, UV studies of the star formation in the interesting outer disk regions of nearby galaxies generally suffer from less extinction (Thilker et al., 2007). An additional important constraint on UV astronomy is that short-wavelength observations are best carried out by space-based (or at least balloon- or rocket-based) facilities, adding to the complexity and preciousness of galaxy UV databases (Gil de Paz et al., 2007; Lee et al., 2011).

7.3.2 Emission Lines

With hydrogen being the most abundant element in the universe, it is not surprising that hydrogen emission lines are commonly utilised to trace the SFR in galaxies. In particular, hydrogen emission lines are targeted for observations of massive star formation arising from within HII regions, where the energy of far-UV starlight exceeds the 13.6 eV ionisation potential of HI. Lyman–α is intrinsically the strongest hydrogen emission line and is commonly utilised to identify actively star-forming galaxies in the distant universe – for these galaxies its rest-frame wavelength of 0.1216 μm is conveniently observed to be redshifted into the optical regime. However, Lyman–α typically suffers from extreme attenuation by dust and absorption by hydrogen, and for this emission line it is notoriously difficult to accurately constrain its escape fraction (e.g. Atek et al., 2008; Hayes et al., 2011).

Going to longer wavelengths, infrared hydrogen recombination lines (e.g. Paschen–α at 1.875 μm and Brackett–α at 4.051 μm) are much less extinguished but intrinsically faint. The optical Balmer-α (Hα) line at 0.6563 μm, on the other hand, presents a happy compromise: relatively unextinguished but also quite bright intrinsically, resulting in a comparatively bright *observed* line frequently adopted as an SFR indicator. One significant drawback to SFR(Hα) is that it is compromised for dwarf galaxies, perhaps because of IMF variations that result in stochastically rather than fully sampled, and thus the presence of the massive stars that power HII regions is not guaranteed (Lee et al., 2009; Fumagalli et al., 2011, and see Chapter 2).

Other emission lines are also employed to quantify a galaxy's SFR, particularly when the use of hydrogen recombination lines is infeasible or unavailable. For example, the [OII] forbidden line at 0.3727 μm has been used as a SFR indicator, and its blue rest-frame wavelength conveniently allows it to be a probe of the intermediate redshift universe with near-infrared instrumentation (Kennicutt, 1998; Kewley et al., 2004; Ly et al., 2007). In addition, the advent of sensitive long-wavelength spectroscopy has led to the study of far-infrared fine structure lines as SFR indicators (e.g. Boselli et al., 2002; De Looze et al., 2011, 2014; Sargsyan et al., 2012; Herrera-Camus et al., 2015). The far-infrared has the natural advantage of suffering from minimal extinction, but among the diversity of far-infrared lines the [CII] 158 μm line presents a particularly compelling case due to its intrinsic brightness – approximately 0.1%–1% of the total infrared luminosity for most galaxies (e.g. Díaz-Santos et al., 2013; Smith et al., 2017). Indeed, the [CII] 158 μm line has been used to characterise star-forming galaxies at high redshift (e.g. Stacey et al., 2010; Aravena et al., 2016).

7.3.3 Free-Free and Synchrotron Radio Emission

Galaxy emission at radio wavelengths of course suffers from essentially zero attenuation by interstellar dust grains, though establishing robust SFR indicators in this regime can be observationally challenging. Moreover, the observations must be disentangled into their 'thermal' free-free (Bremsstrahlung) and 'non-thermal' synchrotron components; the latter tends to dominate at wavelengths longer than 1 cm (frequencies lower than 30 GHz) and the former tends to dominate at wavelengths shorter than 1 cm (frequencies higher than 30 GHz) until the Rayleigh-Jeans tail of the dust emission starts to become appreciable around 0.3 cm (see Condon, 1992). Synchrotron emission can be employed, in large part because of a well known one-to-one scaling between 1.4 GHz emission and the total infrared emission (de Jong et al., 1985; Helou et al., 1985; Yun et al., 2001; Appleton et al., 2004; Ivison et al., 2010; Delhaize et al., 2017).

Murphy et al. (2011) employ 26–40 GHz emission to calibrate the star-formation rate for 10 star-forming regions in NGC 6946, but care is taken to account for the presence of anomalous mm emission potentially from spinning dust (see also Mezger and Henderson, 1967; Turner and Ho, 1983; Klein and Graeve, 1986; Turner et al., 1998).

7.3.4 Dust Emission

Interstellar dust grains capture and reprocess into infrared radiation a significant portion of the UV/optical light from stars, about 50 per cent overall based on studies of the cosmic background (e.g. Dole et al., 2006). This absorption is carried out by interstellar 'dust' spanning all sizes from planar PAH just a few Angstroms across to solid carbonaceous and silicate grains as large as a micron (Hollenbach and Tielens, 1999; Draine, 2003; Draine and Li, 2007).

The infrared dust emission can be calibrated as a star-formation rate indicator, in either a bolometric sense or in a monochromatic fashion, that uses common filters centreed at rest-frame wavelengths of 8 μm, 12 μm, 22/24/25 μm, and 70 μm (Bell, 2003; Wu et al., 2005; Calzetti et al., 2007; Rieke et al., 2009; Boquien et al., 2010; Hao et al., 2011; Murphy et al., 2011; Jarrett et al., 2013; Li et al., 2013; Cluver et al., 2017, see also Chapter 6). But, like with all SFR indicators, there are caveats to the usage of the infrared dust continuum. For example, not all UV/optical light emanating from star-forming regions will be captured by interstellar dust grains, an issue that is particularly problematic for many dwarf galaxies with relatively porous interstellar media. Another major caveat is that a substantial fraction of the infrared emission may arise from heating by older stellar populations ($\geq$100 Myr) that are not directly related to the current star-formation rate. The exact proportion depends on a particular galaxy's star-formation history and the relative distribution of interstellar dust and stars. Indicative numbers for the percentage of infrared emission that arises from dust heating by older stellar populations are $\sim$50% of the spatially integrated 70 μm emission for NGC 5055 and NGC 6946 (Li et al., 2013) and one-third of the bolometric infrared emission for a typical spiral galaxy (Bell, 2003; Kennicutt and Evans, 2012).

Another caveat for employing infrared radiation as a star-formation rate indicator is that low-metallicity regions such as those commonly found throughout dwarf irregular galaxies and the outskirts of normal star-forming galaxies exhibit depressed 8 μm PAH emission (Madden et al., 2006; Draine et al., 2007; Calzetti et al., 2007; Smith et al., 2007; Engelbracht et al., 2008; Sandstrom et al., 2012). Hence, when sampling a range of ISM environments, it is more reliable to utilise monochromatic indicators that trace warm dust at longer wavelengths that avoid PAH emission features (20 μm $< \lambda <$ 70 μm), or to utilise the bolometric infrared emission with a correction for dust heated by older stars.

7.4 Summary

Measuring the physical properties of galaxies in general and their star-formation rate in particular is key to understanding their formation and evolution. The fundamental idea behind star-formation rate estimators is to measure the luminosity of galaxies using certain conspicuous and readily observable emission lines or continuum spectral regions for which the emission is heavily dominated by the emission from young stellar populations, either directly (e.g. FUV, NUV, or U bands) or indirectly (e.g. TIR, Hα, or the radio continuum). While simple in appearance, the conversion from the luminosity in a star-formation rate tracer into a reliable SFR depends on the system's metal abundance, its star-formation history, the shape and sampling of its initial mass function, and correcting for extinction. Additional complications arise concerning stellar rotation and stellar binarity. Measurements of the metallicity and extinction of a galaxy or star-forming region are limited by their observational (and theoretical) uncertainties, and the other caveats listed above all require assumptions to be made. Calibrations of the various star-formation rate indicators can vary by 0.5 dex or more depending on the choices made for these parameters, and thus a multi-pronged, multi-wavelength approach is preferred where feasible.

References

Appleton, P. N., Fadda, D. T., Marleau, F. R. et al. 2004. The Far- and Mid-Infrared/Radio Correlations in the Spitzer Extragalactic First Look Survey. *Astrophysical Journal Supplement Series*, **154**(Sept.), 147–150.

Aravena, M., Decarli, R., Walter, F. et al. 2016. The ALMA Spectroscopic Survey in the Hubble Ultra Deep Field: Search for [CII] Line and Dust Emission in 6. *Astrophysical Journal*, **833**(Dec.), 71.

Atek, H., Kunth, D., Hayes, M. et al. 2008. On the Detectability of Lyα Emission in Star Forming Galaxies. The Role of Dust. *Astronomy and Astrophysics*, **488**(Sept.), 491–509.

Bell, E. F. 2003. Estimating Star Formation Rates from Infrared and Radio Luminosities: The Origin of the Radio-Infrared Correlation. *Astrophysical Journal*, **586**(Apr.), 794–813.

Boquien, M., Bendo, G., Calzetti, D. et al. J. 2010. Total Infrared Luminosity Estimation of Resolved and Unresolved Galaxies. *Astrophysical Journal*, **713**(Apr.), 626–636.

Boquien, M., Buat, V., and Perret, V. 2014. Impact of Star Formation History on the Measurement of Star Formation Rates. *Astronomy and Astrophysics*, **571**(Nov.), A72.

Boquien, M., Burgarella, D., Roehlly, Y. et al. 2019. CIGALE: A Python Code Investigating GALaxy Emission. *Astronomy and Astrophysics*, **622**(Feb.), A103.

Boquien, M., Duc, P.-A., Wu, Y. et al. 2009. Collisional Debris as Laboratories to Study Star Formation. *Astronomical Journal*, **137**(June), 4561–4576.

Boselli, A., Gavazzi, G., Lequeux, J. et al. 2002. [CII] at 158 mu m as a Star Formation Tracer in Late-Type Galaxies. *Astronomy and Astrophysics*, **385**(Apr.), 454–463.

Boselli, A., Boissier, S., Cortese, L. et al. 2009. High-Mass Star Formation in Normal Late-Type Galaxies: Observational Constraints to the Initial Mass Function. *Astrophysical Journal*, **706**(Dec.), 1527–1544.

Bruzual, G., and Charlot, S. 2003. Stellar Population Synthesis at the Resolution of 2003. *Monthly Notices of the Royal Astronomical Society*, **344**(Oct.), 1000–1028.

Buat, V., Iglesias-Páramo, J., Seibert, M. et al. 2005. Dust Attenuation in the Nearby Universe: A Comparison between Galaxies Selected in the Ultraviolet and in the Far-Infrared. *Astrophysical Journal*, **619**(Jan.), L51–L54.

Burgarella, D., Buat, V., and Iglesias-Páramo, J. 2005. Star Formation and Dust Attenuation Properties in Galaxies from a Statistical Ultraviolet-to-Far-Infrared Analysis. *Monthly Notices of the Royal Astronomical Society*, **360**(July), 1413–1425.

Calzetti, D., Kennicutt, R. C., Engelbracht, C. W. et al. 2007. The Calibration of Mid-Infrared Star Formation Rate Indicators. *Astrophysical Journal*, **666**(Sept.), 870–895.

Cerviño, M., Bongiovanni, A., and Hidalgo, S. 2016. Recent SFR Calibrations and the Constant SFR Approximation. *Astronomy and Astrophysics*, **589**(May), A108.

Chabrier, G. 2003. Galactic Stellar and Substellar Initial Mass Function. *Publications of the Astronomical Society of the Pacific*, **115**(July), 763–795.

Ciesla, L., Elbaz, D., and Fensch, J. 2017. The SFR-M_* Main Sequence Archetypal Star-Formation History and Analytical Models. *Astronomy and Astrophysics*, **608**(Dec.), A41.

Cluver, M. E., Jarrett, T. H., Dale, D. A. et al. 2017. Calibrating Star Formation in WISE Using Total Infrared Luminosity. *Astrophysical Journal*, **850**(Nov.), 68.

Condon, J. J. 1992. Radio Emission from Normal Galaxies. *Annual Review of Astronomy and Astrophysics*, **30**, 575–611.

Daddi, E., Dickinson, M., Morrison, G. et al. 2007. Multiwavelength Study of Massive Galaxies at z˜2. I. Star Formation and Galaxy Growth. *Astrophysical Journal*, **670**(Nov.), 156–172.

de Jong, T., Klein, U., Wielebinski, R. et al. 1985. Radio Continuum and Far-Infrared Emission from Spiral Galaxies – A Close Correlation. *Astronomy and Astrophysics*, **147**(June), L6–L9.

De Looze, I., Baes, M., Bendo, G. J. et al. 2011. The Reliability of [C II] as an Indicator of the Star Formation Rate. *Monthly Notices of the Royal Astronomical Society*, **416**(Oct.), 2712–2724.

De Looze, I., Cormier, D., Lebouteiller, V. et al. 2014. The Applicability of Far-Infrared Fine-Structure Lines as Star Formation Rate Tracers over Wide Ranges of Metallicities and Galaxy Types. *Astronomy and Astrophysics*, **568**(Aug.), A62.

Delhaize, J., Smolčić, V., Delvecchio, I. et al. 2017. The VLA-COSMOS 3 GHz Large Project: The Infrared-Radio Correlation of Star-Forming Galaxies and AGN to $z \sim 6$. *Astronomy and Astrophysics*, **602**(June), A4.

Díaz-Santos, T., Armus, L., Charmandaris, V. et al. 2013. Explaining the [C II]157.7 μm Deficit in Luminous Infrared Galaxies – First Results from a Herschel/PACS Study of the GOALS Sample. *Astrophysical Journal*, **774**(Sept.), 68.

Dole, H., Lagache, G., Puget, J.-L. et al. 2006. The Cosmic Infrared Background Resolved by Spitzer. Contributions of Mid-Infrared Galaxies to the Far-Infrared Background. *Astronomy and Astrophysics*, **451**(May), 417–429.

Draine, B. T. 2003. Interstellar Dust Grains. *Annual Review of Astronomy and Astrophysics*, **41**, 241–289.

Draine, B. T., and Li, A. 2007. Infrared Emission from Interstellar Dust. IV. The Silicate-Graphite-PAH Model in the Post-Spitzer Era. *Astrophysical Journal*, **657**(Mar.), 810–837.

Draine, B. T., Dale, D. A., Bendo, G. et al. 2007. Dust Masses, PAH Abundances, and Starlight Intensities in the SINGS Galaxy Sample. *Astrophysical Journal*, **663**(July), 866–894.

Ekström, S., Georgy, C., Eggenberger, P. et al. 2012. Grids of Stellar Models with Rotation. I. Models from 0.8 to 120 $M_\odot$ at Solar Metallicity ($Z = 0.014$). *Astronomy and Astrophysics*, **537**(Jan.), A146.

Eldridge, J. J., and Stanway, E. R. 2009. Spectral Population Synthesis including Massive Binaries. *Monthly Notices of the Royal Astronomical Society*, **400**(Dec.), 1019–1028.

Eldridge, J. J., Stanway, E. R., Xiao, L. et al. 2017. Binary Population and Spectral Synthesis Version 2.1: Construction, Observational Verification, and New Results. *Publications of the Astronomical Society of Australia*, **34**(Nov.), e058.

Engelbracht, C. W., Rieke, G. H., Gordon, K. D. et al. 2008. Metallicity Effects on Dust Properties in Starbursting Galaxies. *Astrophysical Journal*, **678**(May), 804–827.

Fumagalli, M., da Silva, R. L., and Krumholz, M. R. 2011. Stochastic Star Formation and a (Nearly) Uniform Stellar Initial Mass Function. *Astrophysical Journal*, **741**(Nov.), L26.

Gil de Paz, A., Boissier, S., Madore, B. F. et al. 2007. The GALEX Ultraviolet Atlas of Nearby Galaxies. *Astrophysical Journal Supplement Series*, **173**(Dec.), 185–255.

Hao, C.-N., Kennicutt, R. C., Johnson, B. D. et al. 2011. Dust-Corrected Star Formation Rates of Galaxies. II. Combinations of Ultraviolet and Infrared Tracers. *Astrophysical Journal*, **741**(Nov.), 124.

Hayes, M., Schaerer, D., Östlin, G. et al. 2011. On the Redshift Evolution of the Lyα Escape Fraction and the Dust Content of Galaxies. *Astrophysical Journal*, **730**(Mar.), 8.

Helou, G., Soifer, B. T., and Rowan-Robinson, M. 1985. Thermal Infrared and Nonthermal Radio – Remarkable Correlation in Disks of Galaxies. *Astrophysical Journal*, **298**(Nov.), L7–L11.

Herrera-Camus, R., Bolatto, A. D., Wolfire, M. G. et al. 2015. [C II] 158 μm Emission as a Star Formation Tracer. *Astrophysical Journal*, **800**(Feb.), 1.

Hollenbach, D. J., and Tielens, A. G. G. M. 1999. Photodissociation Regions in the Interstellar Medium of Galaxies. *Reviews of Modern Physics*, **71**(Jan.), 173–230.

Ivison, R. J., Magnelli, B., Ibar, E. et al. 2010. The Far-Infrared/Radio Correlation as Probed by Herschel. *Astronomy and Astrophysics*, **518**(July), L31.

Jarrett, T. H., Masci, F., Tsai, C. W. et al. 2013. Extending the Nearby Galaxy Heritage with WISE: First Results from the WISE Enhanced Resolution Galaxy Atlas. *Astronomical Journal*, **145**(Jan.), 6.

Kennicutt, Jr., R. C., and Evans, N. J. 2012. Star Formation in the Milky Way and Nearby Galaxies. *Annual Review of Astronomy and Astrophysics*, **50**(Sept.), 531–608.

Kennicutt, Jr., R. C. 1998. Star Formation in Galaxies along the Hubble Sequence. *Annual Review of Astronomy and Astrophysics*, **36**, 189–232.

Kewley, L. J., Geller, M. J., and Jansen, R. A. 2004. [O II] as a Star Formation Rate Indicator. *Astronomical Journal*, **127**(Apr.), 2002–2030.

Klein, U., and Graeve, R. 1986. Multi-Frequency Radio Continuum Observations of Dwarf Irregular Galaxies. *Astronomy and Astrophysics*, **161**(June), 155–168.

Kroupa, P. 2001. On the Variation of the Initial Mass Function. *Monthly Notices of the Royal Astronomical Society*, **322**(Apr.), 231–246.

Kroupa, P., and Weidner, C. 2003. Galactic-Field Initial Mass Functions of Massive Stars. *Astrophysical Journal*, **598**(Dec.), 1076–1078.

Krumholz, M. R., Fumagalli, M., da Silva, R. L. et al. 2015. SLUG – Stochastically Lighting Up Galaxies. III. A Suite of Tools for Simulated Photometry, Spectroscopy, and Bayesian Inference with Stochastic Stellar Populations. *Monthly Notices of the Royal Astronomical Society*, **452**(Sept.), 1447–1467.

Lee, J. C., Gil de Paz, A., Tremonti, C. et al. 2009. Comparison of Hα and UV Star Formation Rates in the Local Volume: Systematic Discrepancies for Dwarf Galaxies. *Astrophysical Journal*, **706**(Nov.), 599–613.

Lee, J. C., Gil de Paz, A., Kennicutt, Jr., R. C. et al. 2011. A GALEX Ultraviolet Imaging Survey of Galaxies in the Local Volume. *Astrophysical Journal Supplement Series*, **192**(Jan.), 6.

Leitherer, C., Ekström, S., Meynet, G. et al. 2014. The Effects of Stellar Rotation. II. A Comprehensive Set of Starburst99 Models. *Astrophysical Journal Supplement Series*, **212**(May), 14.

Li, Y., Crocker, A. F., Calzetti, D. et al. 2013. Star Formation Rates in Resolved Galaxies: Calibrations with Near- and Far-Infrared Data for NGC 5055 and NGC 6946. *Astrophysical Journal*, **768**(May), 180.

Ly, C., Malkan, M. A., Kashikawa, N. et al. 2007. The Luminosity Function and Star Formation Rate between Redshifts of 0.07 and 1.47 for Narrowband Emitters in the Subaru Deep Field. *Astrophysical Journal*, **657**(Mar.), 738–759.

Madau, P., and Dickinson, M. 2014. Cosmic Star-Formation History. *Annual Review of Astronomy and Astrophysics*, **52**(Aug.), 415–486.

Madden, S. C., Galliano, F., Jones, A. P. et al. 2006. ISM Properties in Low-Metallicity Environments. I. Mid-Infrared Spectra of Dwarf Galaxies. *Astronomy and Astrophysics*, **446**(Feb.), 877–896.

Maraston, C. 2005. Evolutionary Population Synthesis: Models, Analysis of the Ingredients and Application to High-Z Galaxies. *Monthly Notices of the Royal Astronomical Society*, **362**(Sept.), 799–825.

Meurer, G. R., Wong, O. I., Kim, J. H. et al. 2009. Evidence for a Nonuniform Initial Mass Function in the Local Universe. *Astrophysical Journal*, **695**(Apr.), 765–780.

Mezger, P. G., and Henderson, A. P. 1967. Galactic H II Regions. I. Observations of their Continuum Radiation at the Frequency 5 GHz. *Astrophysical Journal*, **147**(Feb.), 471.

Miller, G. E., and Scalo, J. M. 1979. The Initial Mass Function and Stellar Birthrate in the Solar Neighborhood. *Astrophysical Journal Supplement Series*, **41**(Nov.), 513–547.

Murphy, E. J., Condon, J. J., Schinnerer, E. et al. 2011. Calibrating Extinction-Free Star Formation Rate Diagnostics with 33 GHz Free-Free Emission in NGC 6946. *Astrophysical Journal*, **737**(Aug.), 67.

Noll, S., Burgarella, D., Giovannoli, E. et al. 2009. Analysis of Galaxy Spectral Energy Distributions from Far-UV to Far-IR with CIGALE: Studying a SINGS Test Sample. *Astronomy and Astrophysics*, **507**(Dec.), 1793–1813.

Pflamm-Altenburg, J., Weidner, C., and Kroupa, P. 2007. Converting Hα Luminosities into Star Formation Rates. *Astrophysical Journal*, **671**(Dec.), 1550–1558.

Rieke, G. H., Alonso-Herrero, A., Weiner, B. J. et al. 2009. Determining Star Formation Rates for Infrared Galaxies. *Astrophysical Journal*, **692**(Feb.), 556–573.

Salim, S., Rich, R. M., Charlot, S. et al. 2007. UV Star Formation Rates in the Local Universe. *Astrophysical Journal Supplement Series*, **173**(Dec.), 267–292.

Salpeter, E. E. 1955. The Luminosity Function and Stellar Evolution. *Astrophysical Journal*, **121**(Jan.), 161.

Sandstrom, K. M., Bolatto, A. D., Bot, C. et al. 2012. The Spitzer Spectroscopic Survey of the Small Magellanic Cloud (S^4MC): Probing the Physical State of Polycyclic Aromatic Hydrocarbons in a Low-Metallicity Environment. *Astrophysical Journal*, **744**(Jan.), 20.

Sargsyan, L., Lebouteiller, V., Weedman, D. et al. 2012. [C II] 158 μm Luminosities and Star Formation Rate in Dusty Starbursts and Active Galactic Nuclei. *Astrophysical Journal*, **755**(Aug.), 171.

Smith, J. D. T., Croxall, K., Draine, B. et al. 2017. The Spatially Resolved [CII] Cooling Line Deficit in Galaxies. *Astrophysical Journal*, **834**(Jan.), 5.

Smith, J. D. T., Draine, B. T., Dale, D. A. et al. 2007. The Mid-Infrared Spectrum of Star-Forming Galaxies: Global Properties of Polycyclic Aromatic Hydrocarbon Emission. *Astrophysical Journal*, **656**(Feb.), 770–791.

Stacey, G. J., Hailey-Dunsheath, S., Ferkinhoff, C. et al. 2010. A 158 μm [C II] Line Survey of Galaxies at z ~ 1-2: An Indicator of Star Formation in the Early Universe. *Astrophysical Journal*, **724**(Dec.), 957–974.

Stanway, E. R., Eldridge, J. J., and Becker, G. D. 2016. Stellar Population Effects on the Inferred Photon Density at Reionization. *Monthly Notices of the Royal Astronomical Society*, **456**(Feb.), 485–499.

Thilker, D. A., Boissier, S., Bianchi, L. et al. 2007. Ultraviolet and Infrared Diagnostics of Star Formation and Dust in NGC 7331. *Astrophysical Journal Supplement Series*, **173**(Dec.), 572–596.

Turner, J. L., and Ho, P. T. P. 1983. VLA Observations of Massive Star Formation in Spiral Nuclei. *Astrophysical Journal*, **268**(May), L79–L83.

Turner, J. L., Ho, P. T. P., and Beck, S. C. 1998. The Radio Properties of NGC 5253 and Its Unusual H II Regions. *Astronomical Journal*, **116**(Sept.), 1212–1220.

Weidner, C., and Kroupa, P. 2005. The Variation of Integrated Star Initial Mass Functions among Galaxies. *Astrophysical Journal*, **625**(June), 754–762.

Weisz, D. R., Johnson, B. D., Johnson, L. C. et al. 2012. Modeling the Effects of Star Formation Histories on Hα and Ultraviolet Fluxes in Nearby Dwarf Galaxies. *Astrophysical Journal*, **744**(Jan.), 44.

Wu, H., Cao, C., Hao, C.-N. et al. 2005. PAH and Mid-Infrared Luminosities as Measures of Star Formation Rate in Spitzer First Look Survey Galaxies. *Astrophysical Journal*, **632**(Oct.), L79–L82.

Yun, M. S., Reddy, N. A., and Condon, J. J. 2001. Radio Properties of Infrared-selected Galaxies in the IRAS 2 Jy Sample. *Astrophysical Journal*, **554**(June), 803–822.

8

Star-Formation Rates from Spectral Energy Distributions of Galaxies

DENIS BURGARELLA

8.1 Introduction

The star-formation rate (SFR) is one of the fundamental parameters for the evolution of galaxies but also the evolution of the universe (e.g. Burgarella et al., 2013; Madau and Dickinson, 2014). By integrating the cosmic SFR density, we infer the amount of stellar mass formed as a function of cosmic times, i.e. the cosmic stellar mass density (e.g. Davidzon et al., 2017). These parameters are, therefore, keys to open the door to the understanding of galaxies and their genesis.

To estimate the SFR of galaxies, we need to translate some information encoded into the electromagnetic spectrum into SFR. The emission of galaxies is not limited to the usual optical range but extends to a much wider spectral range with, most notably, the emissions in the X-ray, ultraviolet (UV), infrared (IR), and radio ranges. Their detection and study brings a lot of information because the physical phenomena at the origin of the various emissions are different. Each of them, separately or combined, allows to analyse a different facet of the physics of galaxies: gas, stars, dust, active galactic nuclei (AGN) that are necessary to model the spectral energy distributions (SEDs) and derive the relevant information to constrain the star-formation history (SFH) and the SFR of individual galaxies and, more globally, of the universe. Here, I will present the tools that are available to the astrophysicians to decipher the message sent by local, distant, or even the most remote galaxies observed in the universe. More specifically, we will try to understand how the spectral observed SEDs of galaxies are compared to modelled SEDs and how we use these comparisons to derive and estimate the physical parameters of the galaxies.

Like any other astrophysical object containing baryonic matter, galaxies emit light over the entire electromagnetic spectrum. Understanding this emission means that we can get a handle on the associated physical processes. It has become clear that addressing the question of the formation and the evolution of galaxies in a cosmological context implies that we must understand their emission over the broadest electromagnetic spectrum (Domínguez-Sánchez et al., 2014; Catalán-Torrecilla et al., 2015; Wang et al., 2016; Brown et al., 2017). For instance, UV and far-infrared (FIR) data are mandatory if we wish to estimate the

SFR of galaxies and the cosmic SFR density. The FIR traces dust-absorbed star formation and its contribution to the total star-formation density can reach up to 90% below redshift $z < 3-4$ (Takeuchi et al., 2005; Burgarella et al., 2013). Still, the amount of star formation can be up to 50% at very low redshifts and it might reach higher levels in the early universe. Modern astrophysicians prefer using multi-wavelength data consistently as the emission in the various spectral ranges can be physically connected. However, the drawback of this more informative approach is that we do need more information in terms of data. So, we need to handle them by using more powerful computers and smart codes that are able to run in a reasonable amount of time and deal with a wealth of data and a huge number of models. A statistical approach is mandatory.

8.2 Why Spectral Energy Distributions to Estimate the Star-Formation Rate?

In a given SED, we can find emission from a variety of components: gas, stars with a range of ages, temperatures, luminosities, dust grains with a range of compositions, sizes, etc. that modify the distribution of energy. To these, we can add the contribution from an AGN. Finally, the inter-galactic medium (IGM) and the redshift must also be taken into account, if we study galaxies at $z > 0$. Even though all these processes are not directly related to the star formation, they contribute to the total galaxy SED. It is, therefore, necessary to identify which component is present in a galaxy to correctly model its spectrum, and derive the best SFR value among other quantities.

The various physical processes that are at the origin of SEDs (see Fig. 8.1):

- *Stars*: In a very naive way, the SFR counts the number of stars formed in 1 year, on average over some period (see also Chapter 6). The stellar emission is obviously related to the number of stars present in a galaxy. So, we expect a relation between the flux directly measured from stars in the UV-optical-near-IR (NIR) range, and the SFR. Single Stellar Population (SSP) (e.g. Leitherer et al., 1999; Bruzual and Charlot, 2003; Maraston, 2005; Eldridge and Stanway, 2016, and see Chapter 3) are used to model the stellar unreddened emission.
- *Dust*: Dust grains modify the shape of the SED over the entire wavelength range. As extensively described in Chapter 4, they absorb energy from the galaxy in the UV and re-emit it in the IR. The way energy is absorbed in UV is characterised by the dust attenuation law (e.g. Calzetti et al., 2000; Buat et al., 2012). The impact of dust grains on the SED strongly depends on their characteristics (chemical composition, size), but also on their geometric distribution at the scale of the galaxy. We refer to Chapter 9 and Chapter 4 for a more detailed description of this impact. Estimating what is the amount of dust attenuation in a galaxy is crucial because a valid and safe SFR estimate must account for both the UV and the IR contributions, especially for dusty galaxies for which up to 90% of the SFR is located in the IR. The energy balance can be used to constrain the total dust emission. This means that the energy absorbed in UV controls the IR

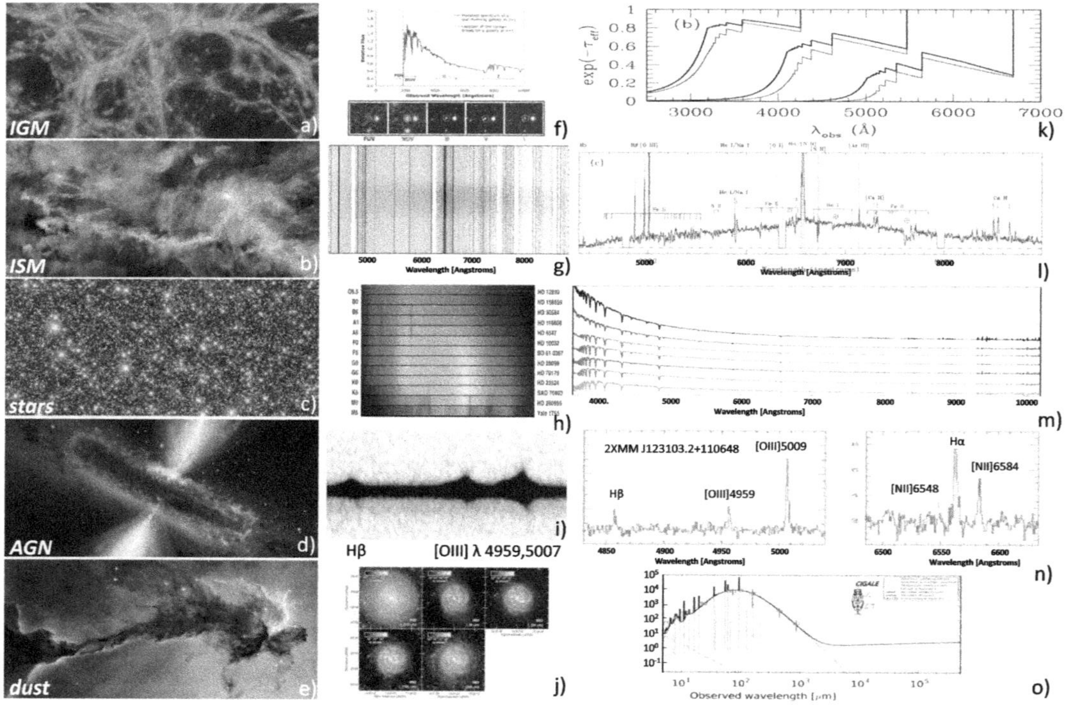

Figure 8.1 Most physical processes at play in a galaxy contribute to the shape of the emitted SED. The left-hand part of the figure shows observed or art illustrations of the various components. The central and right-hand parts present representative emissions for each of them. As detailed in the text, all of them (inter-galactic medium, stars, gas, active galactic nuclei, and dust) impact on the final emitted galaxy spectra. If we wish to understand galaxies that are multi-facet objects, we need to be able to model each and every physical process shown here. That is what SED fitting tries to do. Credits: a) from Schaye et al. (2015) by permission of Oxford University Press on behalf of the Royal Astronomical Society; b) ESA/Herschel/PACS, SPIRE/Gould Belt survey Key Programme/Palmeirim et al. (2013); c) NASA, ESA, and T. Brown (STScI); d) ESA/NASA, the AVO project, and Paolo Padovani; e) NASA, ESA, and the Hubble Heritage Team (STScI/AURA); g) and l) from Smith et al. (2018) by permission of Oxford University Press on behalf of the Royal Astronomical Society; h) NOAO/AURA/NSF; i) from Villar-Martín et al. (2011) by permission of Oxford University Press on behalf of the Royal Astronomical Society; j) from Jones et al. (2015) by permission of Oxford University Press on behalf of the Royal Astronomical Society; k) from Meiksin (2006) by permission of Oxford University Press on behalf of the Royal Astronomical Society; m) from Kesseli et al. (2017) © AAS, reproduced with permission; n) from Ho et al. (2012) © AAS, reproduced with permission. Colour version available online.

luminosity. Codes based on the energy balance (e.g. `CIGALE`,[1] `MAGPHYS`,[2] `PROSPECTOR`,[3] `Bagpipes`[4]) successfully reproduce observed SEDs. The shape of the dust emission is either modelled with simple functions (usually modified black bodies, Casey et al. (2012)), or with templates (Draine and Li, 2007; Dale et al., 2014).

- *Gas*: Stars are formed from gas. Gas left over in the star-formation process can be ionised by young and hot stars. The emission lines from this gas, therefore, provide us with another useful tracer of star formation. The hydrogen lines such as Hα and other hydrogen lines or even other lines such as [OII]3727 could be used to estimate the SFR. But, other physical parameters like the chemical abundances should be assumed for these metallic lines. We refer to Chapter 7 for a detailed description of these star-formation estimators. Again, the effect of dust must be accounted for if we wish to estimate the total SFR. Moreover, some of these lines can be very strong and can contributed a non-negligible percentage of the flux density in some broad bands. It is therefore necessary to evaluate this contribution to correctly fit the observed SEDs with models.
- *AGN*: AGNs can be detected in the spectrum of galaxies, but do not always significantly contribute to the flux density. We need to account for their contributions to evaluate how much of the SED is related to star formation and how much to the AGN. Again, different models or templates can be used (e.g. Fritz et al., 2006; Siebenmorgen and Krügel, 2007; Dale et al., 2014).
- *IGM and Redshifting*: For galaxies at high redshifts ($z > 2-3$), the IGM absorbs light in the UV (Madau et al., 1996; Meiksin, 2006) and must be accounted for in the modelling. Of course, it is also mandatory to redshift the models (or un-redshift the observations).

Figure 8.2 shows the observed Spectral Energy Distribution (SED) of the nearby star-bursting galaxy Messier 82 (M82) as modelled by the SED-fitting code `CIGALE` (Burgarella et al., 2005; Noll et al., 2009; Boquien et al., 2019). We can see the effects of these various physical processes in the SED and the range where they emit light. For instance, stars initially emit most of their flux in the rest-frame UV and optical, while dust grains predominantly radiate in the IR, sub-millimeter (sub-mm), and radio. No AGN is present is this modelled SED for M82, though. However, both the wavelength range and the strength of the emission can vary, depending on the type of galaxy and on the components (see Fig. 8.3): young massive stars will show a strong UV flux whereas older stars will mainly contribute to the near-IR. The emission of an AGN in a Seyfert 2 galaxy can strongly modify the SED in the mid-IR but hardly in the FIR (Hatziminaoglou et al., 2010). Emission lines corresponding to nebular emission, on the other hand, can be observed over most of the wavelength range.

[1] Code Investigating GALaxy Emission; https://cigale.lam.fr
[2] Multi-wavelength Analysis of Galaxy Physical Properties; http://astronomy.swinburne.edu.au/~ecunha/ecunha/MAGPHYS.html
[3] Python code for Stellar Population Inference from Spectra and SEDs; https://prospect.readthedocs.io/en/latest/
[4] Bayesian Analysis of Galaxies for Physical Inference and Parameter EStimation; https://bagpipes.readthedocs.io/en/latest/

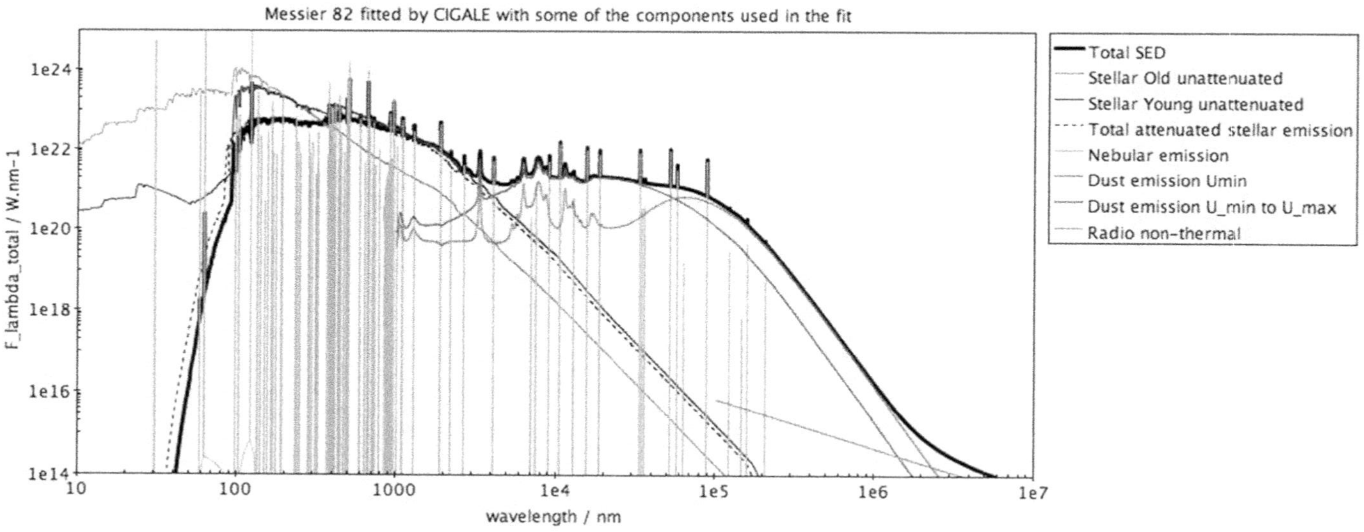

Figure 8.2 Messier 82 is a dusty starbursting galaxy in the local universe, that is often used as a representative of this type of objects because of the excellent wavelength coverage. Here, we have fitted Messier 82 with the `CIGALE SED` fitting code. All the major emission components in this galaxy are drawn to give an idea of how and where they contribute to the SED. Colour version available online.

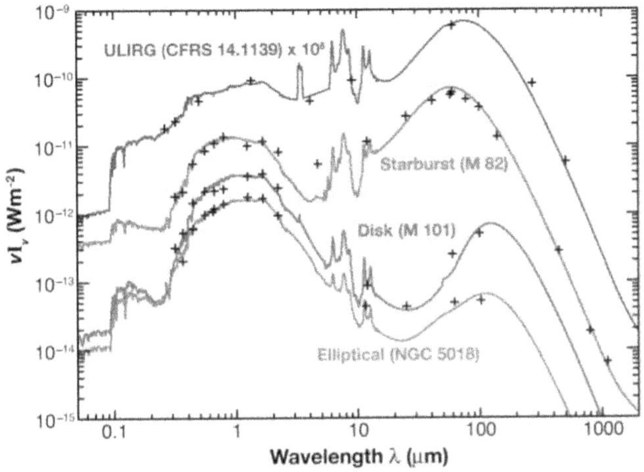

Figure 8.3 Variety of SED from Lagache et al. (2005). Galaxies present various spectral energy distributions because their components show a large range of properties. For instance, an ultra luminous IR galaxy is a very bright and dusty galaxy that can be detected even in the distant universe at large redshifts. On the other hand, Messier 82 is more normal whilst still the vast majority of its emission is found in the rest-frame far-IR. Messier 101 is a typical disky galaxy and finally, NGC 5018 is an elliptical galaxy where there is almost no present star formation. Because of this variety the stellar ages are different, as are the amount of dust grains and gas. Thus, the final spectrum of these objects are clearly distinct. SED fitting translates this variety into astrophysical information. From Galliano, 2004, PhD thesis, reproduced with the agreement of the author. Colour version available online.

8.3 What Information Can Be Extracted from the Spectral Energy Distribution?

Estimating the SFR and related physical parameters via SED fitting provides the advantage of making use of the information that is spread over a large part of the electromagnetic spectrum (Walcher et al., 2011; Conroy, 2013, for reviews). However, beyond the difficulty of securing photometric or spectroscopic information over the widest possible wavelength range, it is important to know how these SFR values compare with those estimated from other methods.

It is hardly possible to compare all the SFR methods over the wide variety of objects. But an example is shown here where we compare SED-based SFRs to hybrid SFRs from the *SDSS*[5]–DR7[6] version of MPA/JHU[7] catalogue. The latter make use of various indicators

[5] Sloan Digital Sky Survey; www.sdss.org
[6] SDSS 7th Data Release; https://classic.sdss.org/dr7/
[7] Max-Planck Institut für Astrophysik/Johns Hopkins University

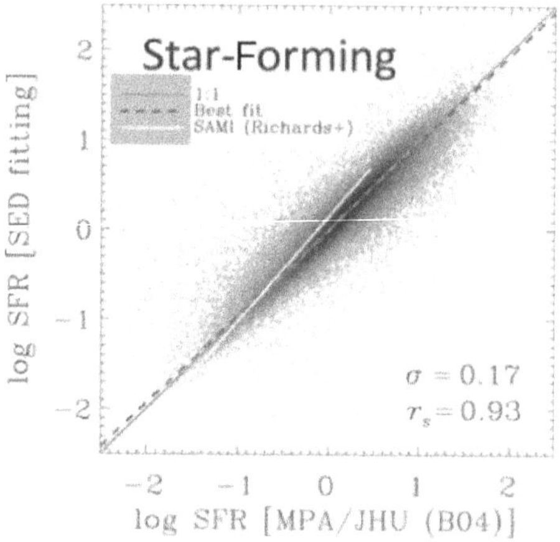

Figure 8.4 Comparison of SFRs of star-forming galaxies derived from SED fitting to SFRs from the DR7 MPA/JHU catalogue (Brinchmann et al., 2004). The latter is an hybrid SFR derived from emission lines, D4000 and SED fitting. No systematic differences or significant non-linearities are observed, which is probably due to the fact that both methods make use of multi-wavelength data. This is in contrast with the relation derived from integral-field spectroscopy Hα measurements from SAMI Galaxy Survey (Richards et al., 2016), shown as the white solid line (SAMI vs MPA/JHU). Here, the difference might be related to the fact that Hα-based SFRs sample younger stellar populations than SFR-based ones but are still unexplained (Salim et al., 2016). From Salim et al. (2016), © AAS. Reproduced with permission. Colour version available online.

that include emission lines, D4000 and SED fitting. And this is not likely to present a strong bias. We use the results from a very large and statistical analysis over 700000 galaxies as presented in Salim et al. (2016). The data used in this analysis are from several telescope surveys that cover very large areas of the sky: *GALEX*[8] in UV, *SDSS* in optical, and *WISE*[9] in IR. The representativeness of the very large sample in the DR7 version of MPA/JHU catalogue provides SFRs that are often used as a reference for comparisons in other papers. As we can see, SED-based SFRs favourably compare to the hybrid SFRs. We conclude that the SFRs estimated from SED fitting can be considered as reliable.

Theoretically, the observed SED contains the information related to the nature and history of the galaxy. However, practically, it is not always easy to get back to the physical parameters behind the galaxy SED.

We can easily understand that the quality of the observed SED is at the origin of one of these limitations. If data of good quality are available only for part of the UV to FIR wavelength range, it will be difficult to safely extract any information. For instance, estimating the total SFR of dusty galaxies requires good quality data in the wavelength range corresponding to dust emission at tens to hundreds of Kelvin, that is in the rest-frame

[8] Galaxy Evolution Explorer; www.galex.caltech.edu/index.html
[9] Wide-field Infrared Survey Explorer; http://wise2.ipac.caltech.edu/docs/release/allsky/

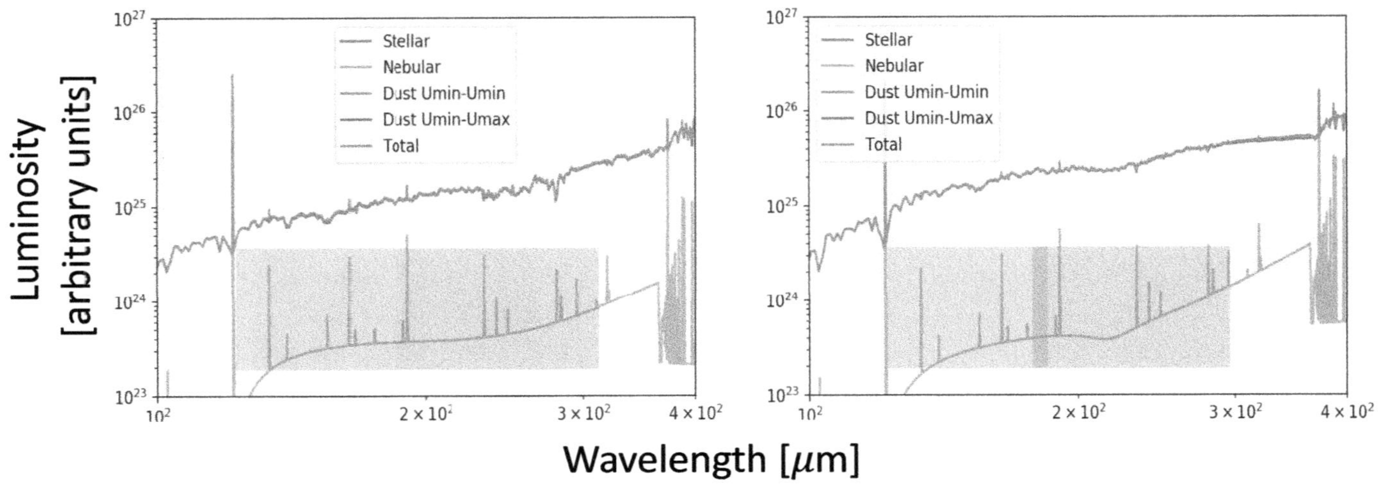

Figure 8.5 Age-dust degeneracy computed with `CIGALE`. The two colour bands approximately illustrate the position of the *GALEX* FUV and NUV filters. The flux densities inside the two bands are quite similar while the physical parameters are very different (see text). Colour version available online.

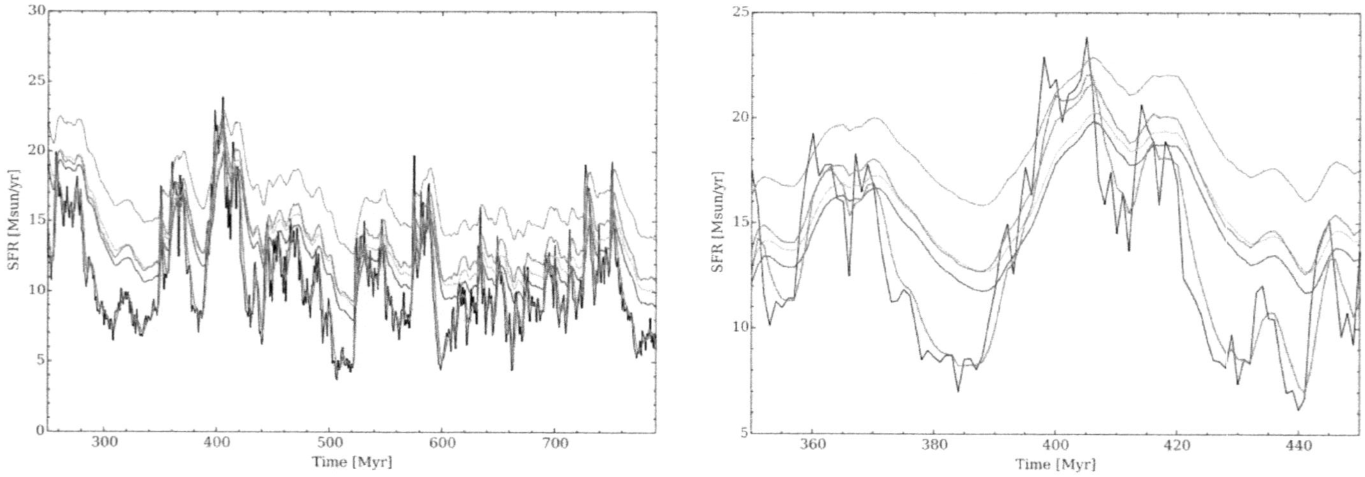

Figure 8.6 Comparison between SFRs from hydrodynamical simulations and SFR estimations using single photometric bands. The x-axis is the time in Myr and the y-axis is the input SFR and the corresponding output SFR using single bands. *Left panel:* shows the comparison of SFRs for one of the simulations. True SFR is in black; SFR estimated from: Lyman continuum, FUV, NUV, U, and IR are respectively in magenta, blue, cyan, green, and red. We can see that true SFR and SFR estimated from Lyman continuum are very close but the other SFR estimators are more or less smoothed and exhibit an offset in time. *Right panel:* same plot by zoomed between 350 Myr and 450 Myr. Credit: Boquien et al. (2014), reproduced with permission © ESO. Colour version available online.

mid-IR to FIR. However, one of the issues is that the amount of data and, therefore, the amount of information available is often quite limited. It is not rare that the dust emission is not well sampled, or even that we only have an upper limit from which we must constrain the total dust emission. Collecting deep FIR data can be a strong challenge for intrinsically faint nearby objects or for bright but distant objects. Without reliable IR data, crucial parameters such as the dust mass, the dust temperature and, of course, the SFR cannot be safely inferred.

Besides, even when good data are available, we might face degeneracies when using broad-band data. For instance, the reddening of the UV-optical spectrum can be due to dust and/or to the age of the stellar populations as illustrated in Fig. 8.5. In this figure the left-hand panel shows the spectrum of an old (800 Myr) stellar population with almost no dust ($A_V = 0.24$) simulated with the `CIGALE` code in the rest-frame UV (100–400 nm) and the colour bands represent *GALEX* far-UV (FUV) and near-UV (NUV) bands. The right-hand panel presents the spectrum of a young (100 Myr) and dusty ($A_V = 1.95$) stellar population. The FUV-NUV colours of the two models are very similar and the degeneracy is strong. It is known that, with the help of FIR data, we can remove this degeneracy by using additional information on the dust attenuation which means that we can, to some extent, hope to decouple the effects of dust attenuation from that of the age on the UV-optical spectrum (Takagi et al., 1999).

Within the context of this chapter, our main concern will be to estimate the SFR and related parameters like the dust attenuation (i.e. the amount of energy emitted by the stellar population that is absorbed by dust grains and re-emitted in the IR), the SFH, and the stellar mass, which represents the end-product of star formation in galaxies.

The imprint of star-formation rate variations on the emission of a galaxy at different wavelengths has been studied by Boquien et al. (2014) who combined hydrodynamical simulations of $1 < z < 2$ galaxies with the `CIGALE` modelling code. Each simulation results in a star-formation history that produces a galaxy with a spectrum. Using these modelled spectra, they estimated the SFR from classical estimators (see Chapter 7) and compared them to the true SFR known from the simulations. An example is shown in Fig. 8.6. They found that except for the Lyman continuum, classical SFR estimators calibrated over 100 Myr overestimate the SFR by 25% in the FUV to 65% in the U band because of the sizeable contribution of long-lived stars to the luminosity in these star-formation tracing bands. Fitting simultaneously all the available data allows to better account for the different stellar populations present in a galaxy.

8.4 How to Estimate the Quality of the Fits and the Physical Parameters Associated to the Star Formation?

To fit an SED, we must compare the observed data to models and to select the best-fit model. The goodness of fit of a model describes how well it fits a set of observations. But, how do we assess whether a given model represents an observed dataset?

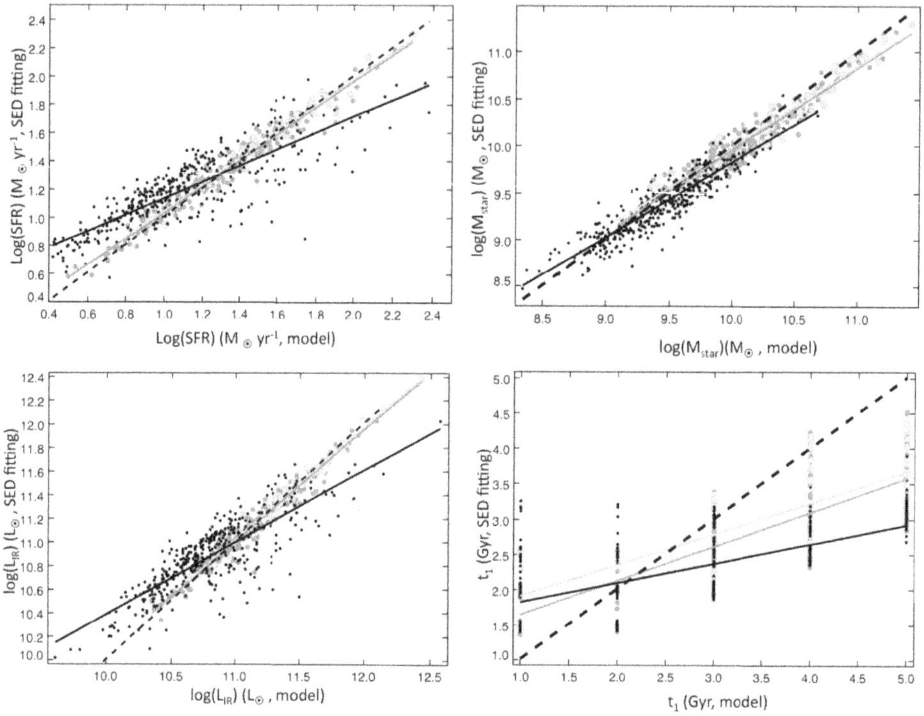

Figure 8.7 Comparison of parameters estimated by SED fitting for a catalogue of artificial galaxies. The initial values of the parameters are on the x-axis and the values estimated after SED fitting on the y-axis. The galaxies without IR data are plotted as black dots, the galaxies with only a MIPS detection with green filled circles and those also detected with PACS with large red filled circles, the regression lines for each subsample are also plotted as solid lines and the 1:1 relation as a dashed line. From the left top to the right bottom the parameters considered are the SFR, M_{star}, L_{IR}, and the age of the stellar populations estimated by **CIGALE**. This type of plot clearly helps in estimating the validity of the estimation for each parameter. For instance, in this specific case (but, it is a general trend for most SED fittings), the age of the stellar population is not accurately estimated. Credit: Buat et al. (2012), reproduced with permission © ESO. Colour version available online.

The simplest method is to use the so-called chi-squared test χ^2. This χ^2 is computed as the sum of the squared difference between the observed ($S_{\rm obs,i}$) and the modelled ($S_{\rm mod,i}$) measurements, divided by the uncertainties ($\sigma_{\rm i}$) associated to the measurements:

$$\chi^2 = \sum_{\rm i} [(S_{\rm obs,i} - m S_{\rm mod,i})/\sigma_{\rm i}]^2$$

where m is a normalisation parameter and i denotes each individual band observations.

The model that minimises χ^2 provides the best-fit model along with all the associated physical parameters.

However, there might be other models that also fit almost as well the data. The best-fit model provides only one of the possible answers. For a better analysis, we should also account for these models in a statistically robust way. This is why we use a Bayesian approach where not only the best model but the entire population of models is used to build the probability distribution function (PDF). The discrete PDF of a discrete random variable X provides the probabilities Pr(X = x) for all possible values of x (e.g. Noll et al., 2009; da Cunha et al., 2008). The code estimates each physical property and the associated uncertainty as the likelihood-weighted mean and standard deviation of the models. This review will not detail the mathematics behind the SED fitting method (for a short introduction see e.g. Kauffmann et al., 2003).

Of course, it is wise when performing any SED modelling to estimate the uncertainties on the physical parameters. However, beyond these uncertainties, it is even wiser to control the level of degeneracies by carrying out tests on a mock catalogue built to be not-too-far from the observed data and for which we have a full knowledge (as best as we can from a model) of the parameters we are trying to estimate (e.g. Buat et al., 2012). These tests allow us to compare the input and output values of the physical parameters and quantitatively evaluate our ability to estimate them (see Fig. 8.7).

8.5 What Kind of Star-Formation History?

Various analytical SFHs can be used when fitting observed SEDs to estimate SFRs. For instance, Salpeter (1959) wrote:

> The most drastic and naive form of this assumption is simply to assume that population-II stars were formed 'instantaneously' at time t = 0, whereas the population-I stars in the solar neighborhood have been forming (between t = 0 and the present time, t = t_0) at a constant rate per cubic parsec (independent of the gas density).

While being very simple, we know that this constant SFH is not physical and that we need something more complex. Schmidt (1959) considers a stellar formation rate per cubic parsec which is proportional to ρ^n, where ρ is the gas density and n is some constant. Note that this so-called Schmidt–Kennicutt law implies continuous accretion of gas from the environment into the discs of spiral galaxies. While Schmidt (1959) considers the effects of different numerical values for n, he only accounts for the population-I systems contained in a cylinder with axis in the z-direction.

Since stars are built from gas, the SFR is related to the amount of gas converted in stars in a given galaxy: $\psi(t) = -\frac{df_{\rm gas}}{dt}$. As we saw, the Schmidt–Kennicutt law assumes that the SFR varies with a power n of the density of interstellar gas, $\psi(t) = \psi_0 f_{\rm gas}^{\rm n}$ [$M_\odot$ yr^{-1}], with $1 \leq n \leq 2$.

The exponentially declining SFRs have been used to model galaxies. However, when trying to understand high-redshift galaxies, declining SFHs are not realistic. While rising SFHs are more realistic for some time, they cannot keep on rising very long before using up all the available gas. We recommend a delayed SFH ($SFR(t) = t/\tau^2 e^{-t/\tau}$ Sandage, 1986) that addresses this issue (see Fig. 8.8).

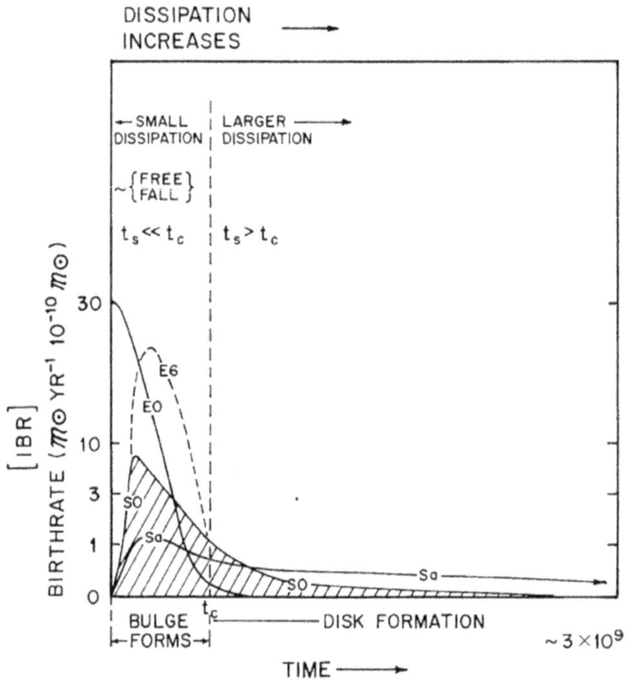

Figure 8.8 Variation of the SFR with time for galaxies of type E, S0, and Sa. The vertical dashed line shows the limit where a bulge (left) and a disk (right) will form. This type of star-formation history presents the advantage of being more realistic and more adapted both to the high redshift and low redshift. Credit: Sandage (1986), reproduced with permission © ESO.

Ciesla et al. (2017) modified the delayed SFH to allow it to account for both high and low SFRs, in the most recent ages. They added flexibility in the recent SFH that models an enhancement or decline of the SFR as: $SFR(t) \propto (t\ e^{-t/\tau})$ when t $\leq t_0$ and $SFR(t) \propto r_{SFR} \times SFR(t = t_0)$, where t $\geq t_0$, t_0 is the time where the time-dependence of SFR changes, and r_{SFR} is the ratio $\mathrm{SFR}_{t \leq t_0}$ / $\mathrm{SFR}_{t \geq t_0}$. Using this SFH, the SED fitting is now able to model the emission of galaxies above (starburst galaxies) and below (quenched galaxies) the main sequence of star-forming galaxies, the so-called relation found between the SFR and the stellar mass of galaxies (e.g. Brinchmann et al., 2004; Noeske et al., 2007). Such a SFH has been tested on a mock-galaxy sample and on a sample of real *GOODS*-South[10] galaxies with redshifts $1.5 < z < 2.5$. There are no artificial limitations to the large SFR values and galaxies are present both at high and low SFRs. There is no age structure along the main sequence. Finally, the youngest stellar populations are found at low stellar masses whereas the oldest galaxies are found at high stellar masses. So, the behaviour is close to what is expected for a population of galaxies evolving on the MS with

[10] The Great Observatories Origins Deep Survey; www.stsci.edu/science/goods/

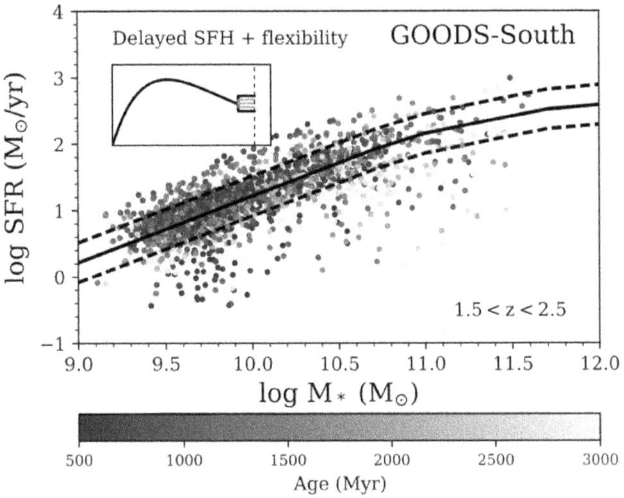

Figure 8.9 Main Sequence relation from data in the *GOODS*-South field. Galaxies are in the range $1.5 < z < 2.5$ obtained using the delayed SFH with an additional flexibility in the recent SFH, colour-coded by the age obtained from the fitting procedure. The black line indicates the position of the MS obtained by Schreiber et al. (2015) at $z = 2$. These results were first tested on a mock galaxy sample using the delayed +recent flexibility SFR (not shown here, fig. 16-topf Ciesla et al. (2017)), and on the *GOODS*-South observed galaxies presented here. More details are given in Ciesla et al. (2017) but we see here that the general behaviour observed here resembles to what is expected for such a sample in terms of SFR, age, and stellar mass distribution: no bias is observed. Credit: Ciesla et al. (2017), reproduced with permission © ESO. Colour version available online.

some random episodes of enhanced and decreased star formation. The strong advantage of this SFH is that it allows either an enhancement or a decrease of the SFR. Such a SFH is versatile to model different types of galaxies. It provides very good estimates of the SFR of normal, starbursting, and rapidly quenched galaxies as shown in Fig. 8.9, where galaxies both above and below the main sequence can be fitted.

8.6 Codes and Ingredients

There are two broad classes of efforts to model SEDs:

- *Radiative Transfer Modelling*: The emission of the stellar light is scattered, absorbed, and re-emitted at longer wavelengths by dust grains present in the environment around stars. That is what we refer to when we mention the dust attenuation. The spectrum emerging from this circumstellar processing provides the astronomer with the only available information about the embedded stellar populations.

  Radiative Transfer codes model the transport of radiation in these dusty circumstellar environments by making assumptions on important parameters such as the chemical composition, the star-formation history, and the dust/star geometry of galaxies. These codes solve the radiation transport equation coupled self-consistently with the equation of motion for the outflow of gas and dust grains. They include the properties for the most common types of astronomical dust and support various analytical forms for the density distribution in galaxies (see Chapter 9).

Among the main phases in building the modelled SEDs, the specification of geometry is an important one. It means that the user must specify the distributions of stars and dust, both at different scales in the model. This is a fundamental phase before actually running the radiative transfer calculations to derive the radiation fields in galaxies that will permit to evaluate the temperature distribution of grains of different sizes and composition as a function of position in the galaxy. Finally, by integrating over all positions in a given galaxy, we obtain the modelled SED (e.g. Silva, 2009; Baes et al., 2011; Popescu et al., 2011; Efstathiou et al., 2013).

- *Physically motivated and fast SED modelling using discrete emission components and energy balance*: These physically motivated codes use a different philosophy. They model the effect of dust on the stellar emission and the corresponding dust and gas emission under the assumption of an energy balance. Adding an optional AGN emission allows for a better modelling of all observed galaxies. Depending on which code is used and which philosophy is adopted, several options for stellar populations, dust attenuation law, dust emission processes, etc. are available. Usually, most of the components are either directly modelled inside the code or published models/templates are included and offered to the user.

  Several codes use the energy balance to model SEDs and fit observed ones (e.g. `MAGPHYS`: da Cunha et al. (2008); `CIGALE`: Boquien et al. (2019)).

In order to better understand how this type of SED fitting code works we will now present and discuss the workflow of `CIGALE`.[11] This code features a wide range of analytical star-formation histories that can be used (constant, exponentially increasing or declining, delayed, periodic, etc.), but it is also possible to use a table describing the time evolution of the star-formation history that might be output from semi-analytical models.

Once the star-formation history is selected we choose a Simple Stellar Population (SSP) model (Leitherer et al., 1999; Bruzual and Charlot, 2003; Maraston, 2005; Eldridge and Stanway, 2016). Once the SSP is selected, it is combined with the star-formation history to obtain the dust-free emission of the complex stellar population. In the presence of dust, we have to account for the dust attenuation. `CIGALE` offers the possibility to use parametric dust attenuation laws where the basis is either power laws (Charlot and Fall, 2000) or the Calzetti law. However, since we know that the latter is not valid for all galaxies, we can modify the slope and/or add a bump at 217.5 nm.

[11] SED fitting analysis can take very long for a very large number of models and/or very large sample of objects. To make runs possible in a reasonable amount of time, `CIGALE` is parallelised and can run on multi-core computers. `CIGALE` is also very modular see (Fig. 8.10).

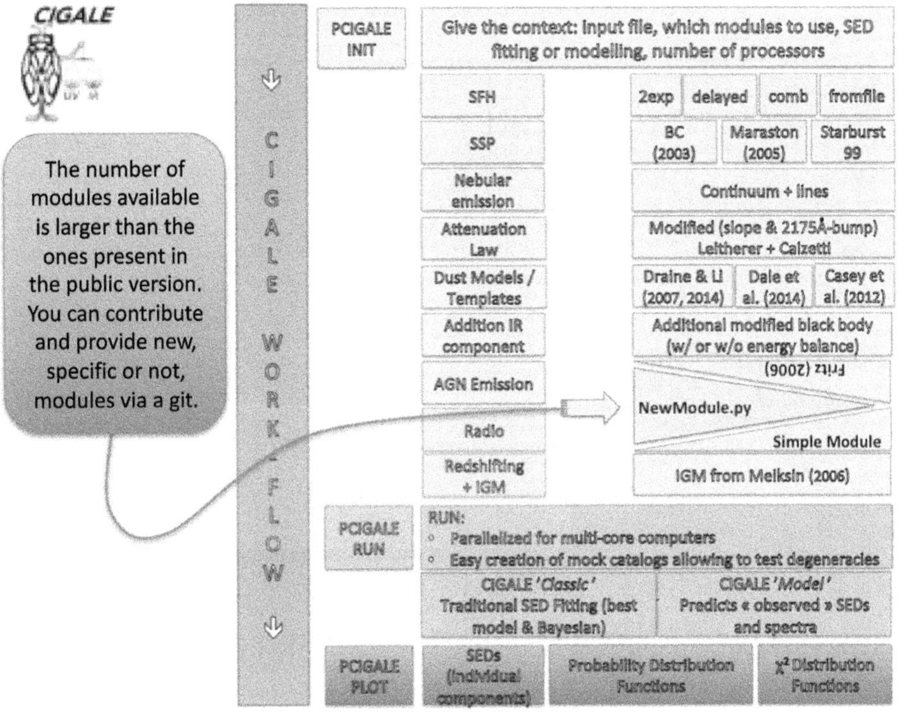

Figure 8.10 The modularity of CIGALE is illustrated in this flowchart. Any new module can be added to modify the code or introduce a new functionality. Colour version available online.

As in any energy balance model, it is assumed that all the energy processed by dust is re-emitted in the infrared. Then, the shape of the infrared emission is provided by one of several proposed templates (Draine and Li, 2007; Casey et al., 2012; Dale et al., 2014).

The nebular emission (continuum and lines) can also computed from the UV to the NIR. Given the number of Lyman continuum photons, CIGALE first computes the Hβ line luminosity and then the other lines using metallicity and radiation-field intensity dependent templates that provide the ratio between individual lines and Hβ.

Emission from an AGN can be finally added using, for example the Fritz et al. (2006) models. For a detailed description of modelled SEDs using an AGN model, we refer to Ciesla et al. (2015).

Finally, the modelled SEDs are redshifted and the absorption from the IGM added to the spectra (Meiksin, 2006).

Models can also be built without any observed data. After fitting, plotting the best model that matches the observed SED (see Figs 8.2 and 8.7) allows to check the quality of the fit. However, beyond this best model CIGALE provides the mean value and uncertainty for each analysed parameter as well as its associated probability density function.

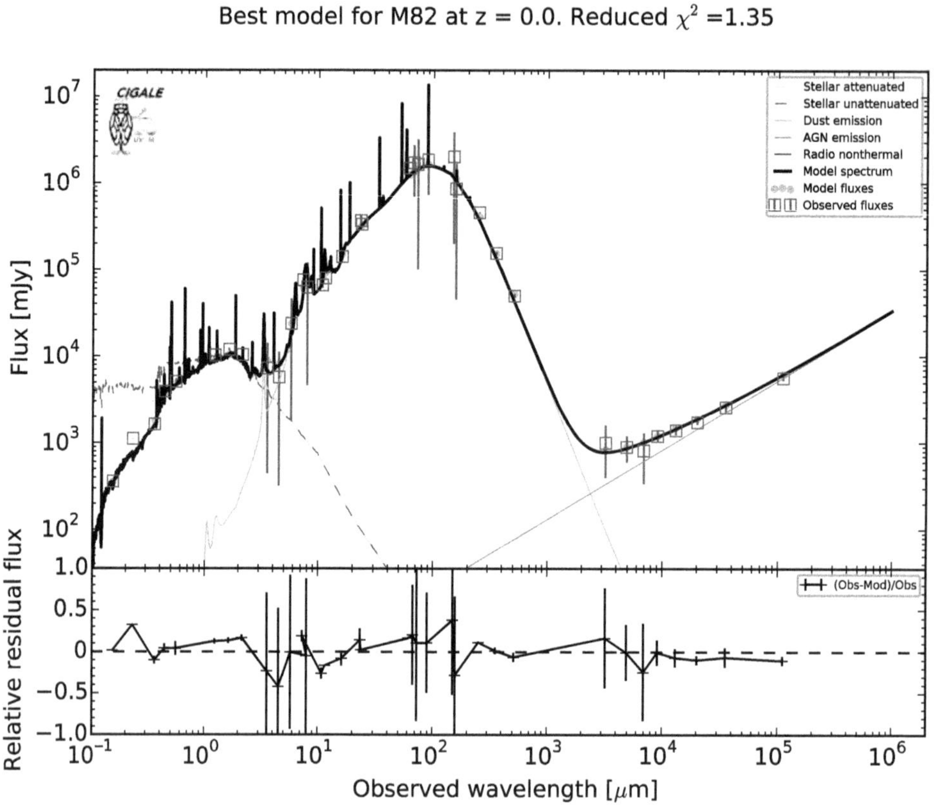

Figure 8.11 Using the various modules presented above, CIGALE fit the observed SED of M82 as shown here. The lower panel shows the fit residuals. Colour version available online.

Figure 8.11 shows an example of a fit performed by CIGALE on the starbursting galaxy Messier 82.

References

Baes, M., Verstappen, J., De Looze, I. et al. 2011. Efficient Three-Dimensional NLTE Dust Radiative Transfer with SKIRT. *Astrophysical Journal Supplement Series*, **196**(Oct.), 22.

Boquien, M., Buat, V., and Perret, V. 2014. Impact of Star Formation History on the Measurement of Star Formation Rates. *Astrophysical Journal*, **571**(Nov.), A72.

Boquien, M., Burgarella, D., Roehlly, Y. et al. 2019. CIGALE: A Python Code Investigating GALaxy Emission. *Astronomy and Astrophysics*, **622**(Feb.), A103.

Brinchmann, J., Charlot, S., White, S. D. M. et al. 2004. The Physical Properties of Star-Forming Galaxies in the Low-Redshift Universe. *Monthly Notices of the Royal Astronomical Society*, **351**(July), 1151–1179.

Brown, M. J. I., Moustakas, J., Kennicutt, R. C. et al. 2017. Calibration of Ultraviolet, Mid-infrared, and Radio Star Formation Rate Indicators. *Astrophysical Journal*, **847**(Oct.), 136.

Bruzual, G., and Charlot, S. 2003. Stellar Population Synthesis at the Resolution of 2003. *Monthly Notices of the Royal Astronomical Society*, **344**(Oct.), 1000–1028.

Buat, V., Noll, S., Burgarella, D. et al. 2012. GOODS-Herschel: Dust Attenuation Properties of UV Selected High Redshift Galaxies. *Astronomy and Astrophysics*, **545**(Sept.), A141.

Burgarella, D., Buat, V., and Iglesias-Páramo, J. 2005. Star Formation and Dust Attenuation Properties in Galaxies from a Statistical Ultraviolet-to-Far-Infrared Analysis. *Monthly Notices of the Royal Astronomical Society*, **360**(July), 1413–1425.

Burgarella, D., Buat, V., Gruppioni, C. et al. 2013. Herschel PEP/HerMES: The Redshift Evolution ($0 < z < 4$) of Dust Attenuation and of the Total (UV+IR) Star Formation Rate Density. *Astronomy and Astrophysics*, **554**(June), A70.

Calzetti, D., Armus, L., Bohlin, R. C. et al. 2000. The Dust Content and Opacity of Actively Star-Forming Galaxies. *Astrophysical Journal*, **533**(Apr.), 682–695.

Casey, C. M., Berta, S., Béthermin, M. et al. 2012. A Population of $z > 2$ Far-Infrared Herschel-SPIRE-Selected Starbursts. *Astrophysical Journal*, **761**(Dec.), 139.

Catalán-Torrecilla, C., Gil de Paz, A., Castillo-Morales, A. et al. 2015. Star Formation in the Local Universe from the CALIFA Sample. I. Calibrating the SFR Using Integral Field Spectroscopy Data. *Astronomy and Astrophysics*, **584**(Dec.), A87.

Charlot, S., and Fall, S. M. 2000. A Simple Model for the Absorption of Starlight by Dust in Galaxies. *Astrophysical Journal*, **539**(Aug.), 718–731.

Ciesla, L., Charmandaris, V., Georgakakis, A. et al. 2015. Constraining the Properties of AGN Host Galaxies with Spectral Energy Distribution Modelling. *Astronomy and Astrophysics*, **576**(Apr.), A10.

Ciesla, L., Elbaz, D., and Fensch, J. 2017. The SFR-M_* Main Sequence Archetypal Star-Formation History and Analytical Models. *Astronomy and Astrophysics*, **608**(Dec.), A41.

Conroy, C. 2013. Modeling the Panchromatic Spectral Energy Distributions of Galaxies. *Annual Review of Astronomy and Astrophysics*, **51**(Aug.), 393–455.

da Cunha, E., Charlot, S., and Elbaz, D. 2008. A Simple Model to Interpret the Ultraviolet, Optical and Infrared Emission from Galaxies. *Monthly Notices of the Royal Astronomical Society*, **388**(Aug.), 1595–1617.

Dale, D. A., Helou, G., Magdis, G. E. et al. 2014. A Two-Parameter Model for the Infrared/Submillimeter/Radio Spectral Energy Distributions of Galaxies and Active Galactic Nuclei. *Astrophysical Journal*, **784**(Mar.), 83.

Davidzon, I., Ilbert, O., Laigle, C. et al. 2017. The COSMOS2015 Galaxy Stellar Mass Function. Thirteen Billion Years of Stellar Mass Assembly in Ten Snapshots. *Astronomy and Astrophysics*, **605**(Sept.), A70.

Domínguez-Sánchez, H., Bongiovanni, A., Lara-López, M. A. et al. 2014. Herschel Far-IR Counterparts of SDSS Galaxies: Analysis of Commonly Used Star Formation Rate Estimates. *Monthly Notices of the Royal Astronomical Society*, **441**(June), 2–23.

Draine, B. T., and Li, A. 2007. Infrared Emission from Interstellar Dust. IV. The Silicate-Graphite-PAH Model in the Post-Spitzer Era. *Astrophysical Journal*, **657**(Mar.), 810–837.

Efstathiou, A., Christopher, N., Verma, A. et al. 2013. Active Galactic Nucleus Torus Models and the Puzzling Infrared Spectrum of IRAS F10214+4724. *Monthly Notices of the Royal Astronomical Society*, **436**(Dec.), 1873–1882.

Eldridge, J. J., and Stanway, E. R. 2016. BPASS Predictions for Binary Black Hole Mergers. *Monthly Notices of the Royal Astronomical Society*, **462**(Nov.), 3302–3313.

Fritz, J., Franceschini, A., and Hatziminaoglou, E. 2006. Revisiting the Infrared Spectra of Active Galactic Nuclei with a New Torus Emission Model. *Monthly Notices of the Royal Astronomical Society*, **366**(Mar.), 767–786.

Hatziminaoglou, E., Omont, A., Stevens, J. A. et al. 2010. HerMES: Far Infrared Properties of Known AGN in the HerMES Fields. *Astronomy and Astrophysics*, **518**(July), L33.

Ho, L. C., Kim, M., and Terashima, Y. 2012. The Low-Mass, Highly Accreting Black Hole Associated with the Active Galactic Nucleus 2XMM J123103.2+110648. *Astrophysical Journal Letters*, **759**(1), L16.

Jones, A. G., Bendo, G. J., Baes, M. et al. 2015. The Relationship between Polycyclic Aromatic Hydrocarbon Emission and Far-Infrared Dust Emission from NGC 2403 and M83. *Monthly Notices of the Royal Astronomical Society*, **448**(1), 168–187.

Kauffmann, G., Heckman, T. M., White, S. D. M. et al. 2003. Stellar Masses and Star Formation Histories for 10^5 Galaxies from the Sloan Digital Sky Survey. *Monthly Notices of the Royal Astronomical Society*, **341**(May), 33–53.

Kesseli, A. Y., West, A. A., Veyette, M. et al. 2017. An Empirical Template Library of Stellar Spectra for a Wide Range of Spectral Classes, Luminosity Classes, and Metallicities Using SDSS BOSS Spectra. *Astrophysical Journal Supplement Series*, **230**(2), 16.

Lagache, G., Puget, J.-L., and Dole, H. 2005. Dusty Infrared Galaxies: Sources of the Cosmic Infrared Background. *Annual Review of Astronomy and Astrophysics*, **43**(Sept.), 727–768.

Leitherer, C., Schaerer, D., Goldader, J. D. et al. 1999. Starburst99: Synthesis Models for Galaxies with Active Star Formation. *Astrophysical Journal Supplement Series*, **123**(July), 3–40.

Madau, P., and Dickinson, M. 2014. Cosmic Star-Formation History. *Annual Review of Astronomy and Astrophysics*, **52**(Aug.), 415–486.

Madau, P., Ferguson, H. C., Dickinson, M. E. et al. 1996. High-Redshift Galaxies in the Hubble Deep Field: Colour Selection and Star Formation History to z˜4. *Monthly Notices of the Royal Astronomical Society*, **283**(Dec.), 1388–1404.

Maraston, C. 2005. Evolutionary Population Synthesis: Models, Analysis of the Ingredients and Application to High-z Galaxies. *Monthly Notices of the Royal Astronomical Society*, **362**(Sept.), 799–825.

Meiksin, A. 2006. Colour Corrections for High-Redshift Objects Due to Intergalactic Attenuation. *Monthly Notices of the Royal Astronomical Society*, **365**(Jan.), 807–812.

Noeske, K. G., Weiner, B. J., Faber, S. M. et al. 2007. Star Formation in AEGIS Field Galaxies since z=1.1: The Dominance of Gradually Declining Star Formation, and the Main Sequence of Star-Forming Galaxies. *Astrophysical Journal Letters*, **660**(1), L43–L46.

Noll, S., Burgarella, D., Giovannoli, E. et al. 2009. Analysis of Galaxy Spectral Energy Distributions from Far-UV to Far-IR with CIGALE: Studying a SINGS Test Sample. *Astronomy and Astrophysics*, **507**(Dec.), 1793–1813.

Palmeirim, P., André, Ph., Kirk, J. et al. 2013. Herschel View of the Taurus B211/3 Filament and Striations: Evidence of Filamentary Growth? *Astrophysical Journal*, **550**(Feb), A38.

Popescu, C. C., Tuffs, R. J., Dopita, M. A. et al. 2011. Modelling the Spectral Energy Distribution of Galaxies. V. The Dust and PAH Emission SEDs of Disk Galaxies. *Astronomy and Astrophysics*, **527**(Mar.), A109.

Richards, S. N., Bryant, J. J., Croom, S. M. et al. 2016. The SAMI Galaxy Survey: Can We Trust Aperture Corrections to Predict Star Formation? *Monthly Notices of the Royal Astronomical Society*, **455**(Jan.), 2826–2838.

Salim, S., Lee, J. C., Janowiecki, S. et al. 2016. GALEX - SDSS - WISE Legacy Catalog (GSWLC): Star Formation Rates, Stellar Masses, and Dust Attenuations of 700,000 Low-redshift Galaxies. *Astrophysical Journal Supplement Series*, **227**(Nov.), 2.

Salpeter, E. E. 1959. The Rate of Star Formation in the Galaxy. *Astrophysical Journal*, **129**(May), 608.

Sandage, A. 1986. Star Formation Rates, Galaxy Morphology, and the Hubble Sequence. *Astronomy and Astrophysics*, **161**(June), 89–101.

Schaye, J., Crain, R. A., Bower, R. G. et al. 2015. The EAGLE Project: Simulating the Evolution and Assembly of Galaxies and Their Environments. *Monthly Notices of the Royal Astronomical Society*, **446**(1), 521–554.

Schmidt, M. 1959. The Rate of Star Formation. *Astrophysical Journal*, **129**(Mar.), 243.

Schreiber, C., Pannella, M., Elbaz, D. et al. 2015. The Herschel View of the Dominant Mode of Galaxy Growth from z = 4 to the Present Day. *Astronomy and Astrophysics*, **575**(Mar), A74.

Siebenmorgen, R., and Krügel, E. 2007. Dust in Starburst Nuclei and ULIRGs. SED Models for Observers. *Astronomy and Astrophysics*, **461**(Jan.), 445–453.

Silva, L. 2009 (Nov.). Multi-Wavelength Modelling of Dusty Galaxies. GRASIL and Applications. Pages 83–93 of: *Revista Mexicana de Astronomia y Astrofisica Conference Series*. Revista Mexicana de Astronomia y Astrofisica Conference Series, vol. 37.

Smith, N., Andrews, J. E., Rest, A. et al. 2018. Light Echoes from the Plateau in Eta Carinae's Great Eruption Reveal a Two-Stage Shock-Powered Event. *Monthly Notices of the Royal Astronomical Society*, **480**(2), 1466–1498.

Takagi, T., Arimoto, N., and Vansevičius, V. 1999. Age and Dust Degeneracy for Starburst Galaxies Solved? *Astrophysical Journal*, **523**(1), 107–113.

Takeuchi, T. T., Buat, V., and Burgarella, D. 2005. The Evolution of the Ultraviolet and Infrared Luminosity Densities in the Universe at $0 < z < 1$. *Astronomy and Astrophysics*, **440**(Sept.), L17–L20.

Villar-Martín, M., Tadhunter, C., Humphrey, A. et al. 2011. Interactions, Star Formation and Extended Nebulae in SDSS Type 2 Quasars at $z \sim 0.3$-0.6. *Monthly Notices of the Royal Astronomical Society*, **416**(1), 262–278.

Walcher, J., Groves, B., Budavári, T. et al. 2011. Fitting the Integrated Spectral Energy Distributions of Galaxies. *Astrophysical Journal Supplement Series*, **331**(Jan.), 1–52.

Wang, L., Norberg, P., Gunawardhana, M. L. P. et al. 2016. GAMA/H-ATLAS: Common Star Formation Rate Indicators and their Dependence on Galaxy Physical Parameters. *Monthly Notices of the Royal Astronomical Society*, **461**(Sept.), 1898–1916.

9

Modelling the Spectral Energy Distribution of Star-Forming Galaxies with Radiative Transfer Methods

CRISTINA POPESCU

9.1 Introduction

Understanding star-formation in galaxies, and of the Universe as a whole, is strongly connected to our ability of accurately measuring the rate and spatial distribution of star-formation from available observations of galaxies. Moreover, knowledge of the output and three dimensional distribution of all constituents of galaxies, stars of all ages, gas, dust, and cosmic rays, is a pre-requisite for understanding the process of star formation along the cosmic time, and ultimately the formation and evolution of galaxies. However, what we observe is the signature of the various galactic components, in terms of mainly photons of various energies, although multi-messenger detections are also developing at an accelerated speed. Thus, we measure the spatial and spectral energy distribution (SED) of galaxies which, ideally, should be modelled and converted into intrinsic properties of galaxies. From a pure mathematical point of view, this is equivalent with solving an inverse problem in a highly inhomogeneous medium, involving highly anisotropic processes that operate at multiple scales. Because of the complexity of self-consistently solving the inverse problem for galactic systems, extracting physical parameters of galaxies has usually been reliant on either phenomenological models or various correlations between astrophysical quantities of interest. For the specific case of star-formation rates (SFR), the so-called star-formation rate indicators have been widely used (see Chapters 6 and 7) or phenomenological SED models (see Chapter 8). Nonetheless, it is still the self-consistent treatment of the inverse problem that should ultimately lead to the 'true' solution, as long as the problem is well posed and degeneracies are avoided.

From the outset we should clarify the meaning of 'self-consistent', as this term has been widely used in the literature, not always to express the same 'consistency'. In this chapter, we understand a self-consistent model to be one that involves a radiative-transfer calculation that can explicitly follow the interaction between stellar photons and, in this case, dust particles and make predictions for all emission mechanisms involved. More generally, radiative-transfer calculations could include non-stellar photons and interactions with gas particles/cosmic rays. Since interactions between photons and particles are usually followed by a re-distribution of their energy (e.g. ultraviolet (UV)/optical photons are absorbed by dust and re-emitted in the infrared (IR)/submm) and a redistribution of their direction of propagation (through scattering), tracing the whole energy flow and the anisotropy of the problem requires modelling of the spectral energy distributions spanning a broad range in wavelengths, from the ultraviolet to the IR/submm (or even broader ranges if non-stellar sources are included) and modelling the imaging/spatial distributions. Thus, we speak about radiative-transfer modelling of the panchromatic spectral energy distributions of galaxies, and, when information is available, of the (panchromatic) spatial energy distribution as well.

But, what is the link between measurements of SFRs and self-consistent modelling of the SEDs of galaxies? Since the SFR is a measure, on average, of the stars formed in one year, it is clear that there is a link between this quantity and the spectral flux density emitted by these stars, a link that is usually quantified by stellar population models. A radiative-transfer model accurately calculates the stellar SEDs emitted by the newly-formed stars. It does this by both calculating the effect of dust attenuation throughout the galaxy, which otherwise would hinder the measurement of flux emitted by the young stars, and by providing a three dimensional picture of the stellar emission of these stars. Moreover, the radiative-transfer model produces a solution for the 3D distributions of all stellar components of a galaxy (stars of all ages and from different morphological components, like disks, bulges, and bars) and of the dust distribution, giving us a detailed understanding of the make-up of a galaxy, both of its stellar content and of the interstellar medium structure. A radiative-transfer model thus provides more than a simple measurement of the global (spatially integrated) SFR.

There are, in fact, two approaches to the self-consistent modelling of the SED of galaxies and the derivation of their SFRs. The first approach is to use multi-wavelength observations as a starting point and build models that can be directly applied to the panchromatic observations to decode their information. These are the SED model tools which perform a direct translation between observed quantities and physical quantities, for example to derive intrinsic distributions of stellar and dust emissivities in galaxies and intrinsic physical parameters, including SFRs. Then the physical quantities can be compared with predicted physical quantities derived from theory. This process has been called decoding observed panchromatic SED (Popescu and Tuffs, 2010), or solving the inverse problem.

A second approach is to start with a theory, in other words to assume 'one knows how galaxies form', and make predictions for some physical quantities, for example to use as a starting point a simulation of a galaxy instead of an observation. Subsequently a model that can deal with the dust physics and the transfer of radiation in the given simulation needs to be applied to obtain simulated panchromatic images of a galaxy for comparison with observations (e.g. Silva et al., 1998). The applied model is again a SED model tool, except that this works in the opposite direction from the previous approach. This process was called encoding predicted physical quantities in Popescu and Tuffs (2010) and can be also used to model the SEDs of galaxies and predict their SFRs. In this chapter we will only consider the former, decoding approach.

9.2 The Propagation of Starlight in Star-Forming Galaxies

Galaxies, in particular star-forming galaxies, contain dust grains, which, because they absorb and scatter stellar photons, partially or wholly prevent a direct measurement of the spatial and spectral distribution of the sources of the UV/optical stellar photons. This dust pervades all components of the interstellar medium (ISM), ranging from the diffuse ionised and neutral medium filling most of the volume of the gaseous disk, through embedded neutral and molecular clouds of intermediate sizes and densities, down to the dense cloud cores on sub-parsec scales which are the sites of formation of stars. The ubiquity and high abundance (relative to the available pool of interstellar metals) of grains is a fundamental result of the solid state being the favoured repository for refractory elements in all but the coronal component of the ISM, and affects the very nature of galaxies. In particular, dust plays a major role in determining the thermodynamic balance of the ISM, through photoelectric heating and inelastic interactions with gas species, influencing the propensity of galaxies to accrete, cool and condense gas into stars (see e.g. Popescu and Tuffs, 2010). In view of the physical role played by dust particles in the process of star formation it can be regarded as a minor perversity of nature that, by virtue of their strong interaction cross section with stellar photons, the very same particles strongly inhibit and distort our view of the resulting stellar populations, thus preventing a straightforward confrontation of theories for star formation in galaxies with observations. The propagation of light depends in a complex way on the relative distribution of stellar emissivity and dust opacity, with structures ranging in scale from parsecs to kiloparsecs. However, the stellar light that is absorbed by dust is re-radiated in the IR/submm range, with spectral characteristics directly related to the heating by interstellar radiation fields (ISRFs). Thus, images of the amplitude and colour of the IR/submm emission can potentially place strong constraints on the distribution of the UV/optical ISRF within galaxies, even when stellar populations are obscured. Moreover, if, as is almost always the case, galaxies are optically thin in the IR/submm, observed images can provide a more direct constraint on the spatial distribution of the emissivity, and thereby the ISRF, in these bands.

When modelling star-forming galaxies it is important to keep in mind that these are inhomogeneous systems, containing highly obscured massive star-formation regions, as well as more extended large-scale distributions of stars and dust. The large-scale distribution of diffuse dust plays a major role in mediating the propagation of photons in galaxy disks and dominates the total bolometric output of dust emission. The discovery of this diffuse component was one of the highlights of the *Infrared Space Observatory*[1] (see Tuffs and Popescu, 2005; Sauvage et al., 2005, for reviews on the ISO science legacy on normal nearby galaxies) and was later confirmed by subsequent infrared space missions, with detail investigations being carried by the *Spitzer Space Telescope (Spitzer)*,[2] the *Infrared Imaging Satellite AKARI*,[3] and the *Herschel Space Observatory*.[4]

[1] *Infrared Space Observatory (ISO)*; www.cosmos.esa.int/web/iso
[2] *Spitzer Space Telescope*; www.spitzer.caltech.edu
[3] https://global.jaxa.jp/projects/sas/astro_f/
[4] *Herschel Space Observatory*; www.cosmos.esa.int/web/herschel/overview

9.3 Main Ingredients

The calculation of the SED of a galaxy and the measurement of its SFR requires several key ingredients, including a dust model, a radiative-transfer model, a prescription of the intrinsic SEDs of the stellar populations, and a prescription of the geometries for dust and stars.

9.3.1 Dust Models

A dust model gives a prescription for the optical properties, the chemical composition, and the grain size distribution of grains. Because the properties of dust grains in the diffuse component are expected to be different from those of the star-forming clouds, different treatments are needed for the two components. Thus, the parameters of the models describing the properties of dust in the diffuse large scale component are usually constrained to simultaneously fit the extinction and emission properties of the diffuse cirrus in the Milky Way, as well as the abundance constraints. However, as shown by Zubko et al. (2004), there is no unique dust model that can simultaneously fit the main observational constraints on such models.

The absorptivities and emissivities of the different grains are usually calculated under the assumption of spherical shapes for the dust particles, for which the Mie theory gives exact analytic solutions. Although empirically anchored through laboratory measurements in the UV/optical/near-IR/far-IR range, the dust efficiencies are quite uncertain in the submm. So far, it has been proven difficult to disentangle in an unambiguous way the effects of dust properties from those of large-scale geometry for stars and dust (Popescu et al., 2000; Xilouris, 2005; Dasyra et al., 2005; Mosenkov et al., 2016; Popescu et al., 2017).

While in the diffuse ISM the properties of dust are reasonably well constrained over a large range of wavelengths (except perhaps in the submm), in the star-forming clouds the properties of dust are largely unknown, as, according to local conditions, various ices can form deep into the clouds, giving rise to a complex chemistry. Because of this, radiative-transfer SED models use empirical/model templates for the star-forming clouds, as introduced in Silva et al. (1998) and Popescu et al. (2000), avoiding this way to introduce an explicit treatment of the optical properties of grains in these dense environments. This approach has been adopted and followed by subsequent work in Popescu et al. (2011), De Looze et al. (2014), Popescu et al. (2017), and Viaene et al. (2017).

9.3.2 Radiative-Transfer Codes

The main tool that allows the construction of a self-consistent model SED of a galaxy is a radiative-transfer code, that can deal with the interaction between photons and dust particles, if the modelling is for the stellar and dust emission part of the electromagnetic spectrum. More generally, a radiative-transfer code could also deal with the interaction between photons and gas particles, as well as with the interaction between cosmic rays and photons. Here we only restrict ourself to the dust radiative-transfer codes.

There are two main methods that are used in radiative-transfer codes: ray-tracing and Monte-Carlo methods (see reviews of Kylafis and Xilouris, 2005; Steinacker et al., 2013). The Monte-Carlo method is the most popular, as it has reduced memory requirements, since there is no need to store the intensity as a function of the angular direction at each spatial position. By itself the Monte-Carlo approach is not efficient in producing images, but, when combined with ray-tracing procedures (e.g. peel-off methods; Yusef-Zadeh et al., 1984) and other acceleration techniques to maximise the use of photon particles (Steinacker et al., 2013), they do become the favourite methods for imaging models. On the other hand, the Monte-Carlo codes are not efficient for radiation fields calculations due to the inherent noise in the results. The ray-tracing methods are therefore the most natural techniques for deriving noise-free and accurate values of radiation fields in galaxies.

Monte-Carlo codes used in modelling SEDs of galaxies are given by `TRADING`[5] (Bianchi, 2008; MacLachlan et al., 2011), `SKIRT`[6] (Baes et al., 2003, 2005, 2011; Camps and Baes, 2015), and `Hyperion`[7] (Robitaille, 2011). Previous Monte-Carlo codes were developed by Witt (1977) and Wood (1997). Ray-tracing codes were constructed by Kylafis and Bahcall (1987) and Natale et al. (2014, 2015, 2017; `DartRay`[8]).

9.3.3 The Intrinsic SED of the Stellar Populations

The wavelength dependence of the intrinsic SED of the stellar populations can be either fixed from population synthesis models or can be empirically derived in the optimisation process of fitting radiative-transfer models to observations. Sometimes, a combination of the two is needed. Thus, the empirical determination of the intrinsic SED is usually considered for the old stellar populations and is optimised in the optical/near-IR range, while theoretical prescriptions are desirable for the highly obscured young stellar populations mainly emitting in the UV range, in particular for edge-on galaxies. The optimisation approach that derives the intrinsic SEDs of old stars from observations was pioneered by Xilouris et al. (1997, 1998, 1999), and was adopted in subsequent work by Popescu et al. (2000), Misiriotis et al. (2001), Bianchi (2008), Bianchi and Xilouris (2011), De Looze et al. (2012a,b), de Geyter et al. (2013, 2014, 2015); Mosenkov et al. (2016).

9.4 Geometries for Stars and Dust

As mentioned before, star-forming galaxies can be described as systems that contain both a large-scale distribution of stars and dust, extending on kiloparsec scales, as well as a component of star-forming clouds operating on parsec scales, where the massive stars inside the clouds locally heat the components of dust associated with the parent molecular cloud. This concept of describing galaxies was introduced by Silva et al. (1998) and Popescu

[5] Transfer of RAdiation through Dust In Galaxies
[6] Stellar Kinematics Including Radiative Transfer; www.skirt.ugent.be/root/_landing.html
[7] www.hyperion-rt.org
[8] www.star.uclan.ac.uk/dartray_doc/

et al. (2000) and has been now adopted as the standard way of prescribing the geometry of stars and dust in radiative-transfer models of galaxies. This concept was also adopted by phenomenological models like MAGPHYS[9] (da Cunha et al., 2008; da Cunha and Charlot, 2011) since, even though they cannot explicitly deal with the geometry of a system, they can adopt separate templates for the diffuse and star forming cloud components.

As with the wavelength dependence of the intrinsic SED of the stellar populations, the geometry of the old stars that are seen through relatively optically thin lines of sight are obtained through an optimisation process that compares optical/near-IR imaging observations with synthetic images produced via radiative-transfer calculations. This approach was introduced by Xilouris et al. (1997, 1998, 1999) and was followed by the different groups working in the field (see references cited in Section 9.3.3). In this process the geometry of the associated diffuse dust is also jointly derived.

The geometry of the young stellar populations is derived differently in edge-on and face-on galaxies. In edge-on galaxies most of the diffuse ultraviolet radiation coming from the young stars in the disk will be obscured, with the little that is seen coming from the outer disk or from the halo. Because of this, theoretical and physical considerations are used to prescribe their geometry, which is then further optimised from the comparison of the dust-emission maps with the corresponding models. In this way, the geometry of all stellar and dust components can be derived. This approach was introduced by Popescu et al. (2000). On the other hand, in face-on galaxies it is possible to derive the radial distribution of the young stellar populations by directly fitting the UV images, similarly with the method applied to the old stellar populations in the optical/near-IR.

In general the emissivity of the young and old stellar populations in the disk-like structures is described by exponential functions in radial directions and sech^2 or exponential functions in vertical direction (Bahcall and Soneira, 1980, 1984; Kylafis and Bahcall, 1987; Misiriotis et al., 2000), with outer and sometimes inner truncations. The emissivity of the bulge is described by functions that in projection produce the Sersic distribution, although King distributions (Silva et al., 1987) or Hubble distributions (Xilouris et al., 1999) have been also used in the past.

For the dust distribution similar double exponential functions or exponential plus sech^2 functions are considered (Kylafis and Bahcall, 1987), as for the stellar emissivity in the disk.

In addition more detailed 3D modelling requires the introduction of bar-like structures, peanut-boxy bulges, spiral arms, and rings. For the spiral arms there are analytic functions describing general trends found in the Milky Way or other nearby galaxies (Wainscoat et al., 1992; Cohen, 1994; Misiriotis et al., 2000; Robitaille et al., 2012; Schechtman-Rook et al., 2012; Natale et al., 2014). The same situation applies for the modelling of rings (Wainscoat et al., 1992; Robitaille et al., 2012). For the peanut-boxy bulges there are no analytic functions describing their spatial distribution, although they

[9] Multi-wavelength Analysis of Galaxy Physical Properties; http://astronomy.swinburne.edu.au/~ecunha/ecunha/MAGPHYS.html

can be produced in numerical simulations. More recently De Looze et al. (2014) and Viaene et al. (2017) modelled the detailed 3D morphology of spirals, by using various approximations. For example, Viaene et al. (2017) considered the projected surface brightness in the far-UV GALEX[10] (FUV) band to be a good approximation of the radial distribution of young stars. Thus, a uniform dust attenuation across the disk was assumed in this work.

The second geometrical component needed to describe star-forming galaxies arises from birth-clouds of massive stars. Because these clouds are spatially correlated with their progeny on parsec scales, they are illuminated by a strong UV-dominated radiation field of intensity 10 to 100 times that in the diffuse ISM. This gives rise to a localised component of emission from grains in thermal equilibrium with these intense radiation fields, which, despite the tiny filling factor of the star-forming regions in spiral galaxies, can nevertheless exceed the entire diffuse IR emission of a galaxy at intermediate wavelengths (ca. 20–ca. 60 μm). It is, therefore, particularly important to incorporate a clumpy component of dust associated with the opaque parent molecular clouds of massive stars. These clumps have been usually assumed to have the same spatial distribution as the young stellar disk. Furthermore, it has been assumed that the properties of these clumps do not systematically depend on their radial location within the galaxy. In reality, we expect star formation complexes in more pressurised regions (such as the inner disk) to be more compact and, therefore, have warmer far-IR colours than their counterparts in low pressure regions (such as the outer disk), as modelled by Dopita et al. (2005).

9.5 Calculating the SED of Galaxies

The first step in the calculation is the determination of the extinction and emission properties of the population of grains assumed in the dust model, as these are needed in the radiative-transfer calculations and in the calculations of the dust temperatures. If we assume grains to be spherical and have radius a, then the absorption cross-section of grains of composition i, $C_{\rm abs,i}$, is obtained by integrating the absorption efficiencies $Q_{\rm abs\,i}$ over the grain size distribution $n(a)$:

$$C_{\rm abs,\,i}(\lambda) = \int_{a_{\rm min}}^{a_{\rm max}} \pi\, a^2\, n(a)\, Q_{\rm abs,\,i}(a,\lambda)\, da \tag{9.1}$$

where $C_{\rm abs,\,i}$ is given in units of $\rm cm^2\,H^{-1}$, $a_{\rm min}$ is the minimum grain size and $a_{\rm max}$ is the maximum grain size.

Similarly, the scattering cross-section of grains of composition i, $C_{\rm sca,\,i}$, is obtained by integrating the scattering efficiencies $Q_{\rm sca,\,i}$ over the grain size distribution $n(a)$:

$$C_{\rm sca,\,i}(\lambda) = \int_{a_{\rm min}}^{a_{\rm max}} \pi\, a^2\, n(a)\, Q_{\rm sca,\,i}(a,\lambda)\, da. \tag{9.2}$$

[10] Galaxy Evolution Explorer; www.galex.caltech.edu/index.html

Then, by summing over the grain composition i we obtain the total absorption and scattering cross-sections, C_{abs} and C_{sca}:

$$C_{abs}(\lambda) = \sum_i C_{abs,\, i}(\lambda), \tag{9.3}$$

$$C_{sca}(\lambda) = \sum_i C_{sca,\, i}(\lambda). \tag{9.4}$$

The extinction cross-section C_{ext} is the sum of the absorption and scattering cross-sections:

$$C_{ext}(\lambda) = C_{abs}(\lambda) + C_{sca}(\lambda), \tag{9.5}$$

and the albedo is defined as

$$albedo(\lambda) = C_{sca}(\lambda)/C_{ext}(\lambda). \tag{9.6}$$

The averaged anisotropy of the scattering phase function g needed in the radiative-transfer calculation is obtained in a similar manner to Eqs 9.1–9.4.

$$g_i(\lambda) = \int_{a_{min}}^{a_{max}} \pi\, a^2\, n(a)\, Q_{sca,\, i}(a, \lambda)\, Q_{phase,\, i}(a, \lambda)\, da \tag{9.7}$$

$$g(\lambda) = \sum_i g_i(\lambda) \tag{9.8}$$

where $Q_{phase,\, i}$ is the anisotropy efficiency.

The images of the model galaxy as would be seen from an outside observer as well as the radiation fields illuminating the dust are then calculated with a radiative-transfer code using the albedo and anisotropy of the scattering phase functions as defined by Eqs 9.6 and 9.8. By equating the rate of absorption of energy to the rate of emission of energy the temperature of the dust grains is obtained:

$$\pi\, a^2\, c \int Q_{abs,\, i}(a, \lambda)\, u_\lambda\, d\lambda = 4\pi\, (\pi\, a^2) \int Q_{abs,\, i}(a, \lambda) \times \int B_\lambda(a, T)\, P_i(a, T)\, dT\, d\lambda \tag{9.9}$$

where c is the speed of light, B_λ is the Planck function, u_λ is the energy density of the radiation fields, and $P_i(a, T)$ is the probability distribution of temperature. This formula accounts for the fact that a large majority of grains in the interstellar medium have cooling timescales that are shorter than the typical time interval between impacts of UV/optical photons, resulting in dust particles not reaching equilibrium with the radiation fields. The process is known as 'stochastic heating', meaning that grains exhibit large probability distribution of temperature instead of radiating at an equilibrium temperature. When the grains are large and the radiation fields are strong, then the temperature distributions become narrow, eventually getting closer to a delta function when equilibrium is reached. Radiative-transfer calculations for transiently heated grains were first performed by Siebenmorgen et al. (1992) and applied to model the SED of the starburst galaxy M82 in Krüegel and Siebenmorgen (1994). Models of spiral galaxies that include explicit calculations of stochastically heated grains were introduced by

Popescu et al. (2000) and are now a standard feature of available radiative-transfer codes like the SKIRT, TRADING, and DartRay codes, with benchmark testing being done in Camps et al. (2015).

If the temperature distribution of a grain is known, the brightness $I_{\lambda,\,\mathrm{i}}$ of a grain of radius a and composition i is then given by:

$$I_{\lambda,\,\mathrm{i}}(a) = Q_{\mathrm{abs,\,i}}(a,\lambda) \int B_{\lambda}(a,T)\, P_{\mathrm{i}}(a,T)\, dT \tag{9.10}$$

where $I_{\lambda,\,\mathrm{i}}$ is in units of $[\mathrm{W\,m^{-2}sr^{-1}\AA^{-1}}]$.

The IR emissivity per H atom of grains of a given composition i, $j^{\mathrm{H}}_{\lambda,\,\mathrm{i}}$ is calculated by integrating over the grain size distribution the IR brightnesses $I_{\lambda,\,\mathrm{i}}$:

$$j^{\mathrm{H}}_{\lambda,\,\mathrm{i}} = \int_{a_{\min}}^{a_{\max}} \pi\, a^2\, n(a)\, I_{\lambda,\,\mathrm{i}}(a)\, da \tag{9.11}$$

where $j^{\mathrm{H}}_{\lambda,\,\mathrm{i}}$ is in units of $\mathrm{W\,sr^{-1}\AA^{-1}H^{-1}}$.

The total IR emissivity per H atom, j^{H}_{λ} is obtained by summing over the grain composition i

$$j^{\mathrm{H}}_{\lambda} = \sum_{\mathrm{i}} j^{\mathrm{H}}_{\lambda,\,\mathrm{i}} \tag{9.12}$$

and can be scaled to the volume density of dust to obtain the volume luminosity density of IR emissivity and images of dust emission, as seen from an outside observer.

Thus, by running the radiative-transfer codes for a trial distribution of stellar emissivity and dust distribution, one can obtain both dust attenuated images, as well as dust-emission images, as seen from an outside observer. In the case of our Milky Way the same radiative-transfer codes can be used to produce images of the Galaxy as seen from inside, in particular from the position of the Sun (Popescu et al., 2017).

9.6 Applications of Radiative-Transfer Modelling: Fitting the SEDs of Galaxies and Measuring Their SFRs

As mentioned in the introduction, the SFR determination is only one of the many outputs of a radiative-transfer SED model. Amongst others, a radiative-transfer model is designed to provide a comprehensive picture of all stellar and non-stellar constituents of a galaxy and their spatial distributions. This means that such a model has the power to predict the panchromatic images of a galaxy, in addition to the global (spatially integrated) SED. Thus, the measurement of SFR comes jointly with the determination of the geometry of the system. As we will see here, in detailed modelling of individual galaxies the geometry is derived in an iterative process within the fit to panchromatic images, while in statistical studies the relative geometry between emitters and absorbers is fixed to statistical trends found from studies of individual galaxies. However, in both cases the geometry is essential for a correct determination of the overall solution and, therefore, of the star-formation rate.

9.6.1 Nearby Individual Galaxies

The first radiative-transfer models of galaxies have been applied to edge-on spirals, since in this orientation it is possible to see the vertical stratification of the different stellar populations and dust distribution. It was the pioneering work of Kylafis and Bahcall (1987) that opened the field of detailed radiative-transfer modelling of galaxy images. These models were not designed for deriving star-formation rates, since they were only modelling optical images, and their scientific motivation was very different. Nonetheless, they can be regarding as the foundation of an emerging field of research. A next step was established in the series of papers of Xilouris et al. (1997, 1998, 1999), who fitted optical images with model images produced with radiative-transfer calculations, where the geometry of the stars and dust were fitting parameters. Again, these models were not designed to measure star-formation rates, but represented a major step in our understanding of the geometry of galaxies, which is a prerequisite for the accurate determination of almost any intrinsic property of a galaxy, including the SFR and its spatial distribution. The fundamental result was that the distribution of old stars relative to that of the dust follows some reproductible trends. These results were later confirmed for larger samples in de Geyter et al. (2014) using the `FitSKIRT` algorithm (de Geyter et al., 2013).

The next major step was the modelling of the whole spectral energy distribution, from the UV to the far-IR/submm, which enabled young stellar populations to be modelled, and star-formation rate determinations to be made self-consistently with the SED fitting. This was achieved for NGC891 in Popescu et al. (2000) (see also Popescu et al., 2011 for an updated model and Fig. 9.1). By taking the solution for the large-scale distribution of old stars and dust from Xilouris et al. (1999) and supplementing it with physical assumptions for the distribution of young stars, Popescu et al. found that there was an energy-balance problem in the sense that the submm fluxes were underestimated by a factor of approximately 3 in the model. This energy-balance problem was confirmed by all later studies of edge-on galaxies, including Bianchi (2008), Baes et al. (2010), de Geyter et al. (2015), Mosenkov et al. (2016), and Mosenkov et al. (2017).

To solve this problem, several solutions have been proposed or adopted (Popescu et al., 2000; Dasyra et al., 2005; Bianchi, 2008; Baes, 2010; Saftly et al., 2015), including the existence of additional large-sale diffuse or clumpy components and of modified dust grain properties. Although no consensus has been yet reached on this, the modelling of edge-on galaxies has provided the path to modelling the multi-wavelength images and SEDs of spiral galaxies, including a formalism for calculating the contribution of the different stellar populations in heating the dust as a function of IR wavelength (Popescu et al., 2000; De Looze et al., 2014; Natale et al., 2015) and the determination of star-formation rates.

In the context of edge-on galaxies one typical example is our own Milky Way, which can be modelled using a similar methodology as for other edge-on systems, obviously adapted to the inner view of a galaxy. Although determinations of the SFR of our Galaxy and of its stellar populations are usually based on measurements of individual stars, it is still the case that the large majority of them, in particular the young stars, are completely obscured by

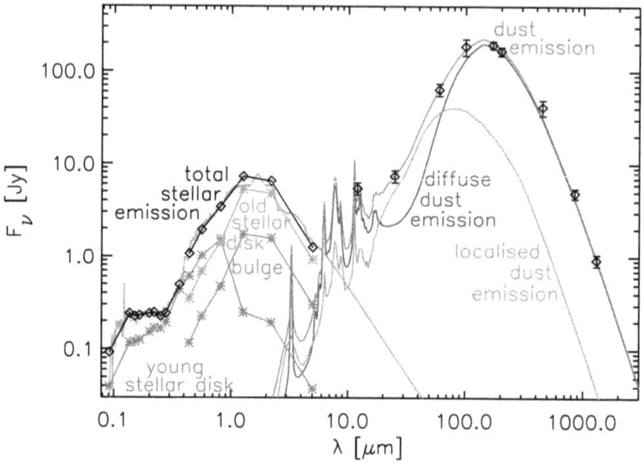

Figure 9.1 The observed and modelled SED of the edge-on spiral galaxy NGC891. The SFR derived from the SED modelling is 2.88 $M_{\odot}$ yr^{-1}. Credit: Popescu et al. (2011), reproduced with permission © ESO. Colour version available online.

dust from the vantage point of our Sun. Because of this, a panchromatic radiative-transfer modelling approach of the all-sky emission is a powerful tool in determining the star-formation rate of our Galaxy and, moreover, of its whole stellar structure, including that at the Galactic centre and the anti-centre. Previous radiative-transfer models of the Milky Way were derived only for restricted wavelength ranges or regions on the sky (Misiriotis et al., 2006; Robitaille et al., 2012). With the advent of the Planck all-sky imaging observations it became possible to perform SED modelling of the Milky Way (Popescu et al., 2017) similar to that of external galaxies. The radiative-transfer derived SFR of our Galaxy was found to be 1.2 $M_{\odot}$ yr^{-1}.

In face-on galaxies the main uncertainty is related to the vertical distribution of stars, since otherwise more detail is available from direct imaging. In particular, the UV emission from the disc can be easily detected and used to constrain the problem, unlike in the edge-on view. The detailed structure seen in face-on systems would make, in principle, non-axisymmetric models ideal for fitting panchromatic images. However, a direct inversion of the non-axisymmetric structure would require a solution for the stellar emissivity SED and dust opacity in each independent spatial element sampled by the observations, resulting in a non-linear optimisation problem in the order of a million independent variables. This is a very challenging problem, considering the resources needed to perform a large number of radiation transfer calculations in any iterative convergence scheme, which to date has not been solved.

To date, two different approaches have been developed for modelling face-on galaxies, each involving some approximations. One approach is to use non-axisymmetric radiative transfer models (De Looze et al., 2014; Viaene et al., 2017; Williams et al., 2019), but without attempting to directly invert the imaging data to predict non-axisymmetric structure, due to the prohibited computational time. Instead, these models use fixed spatial templates for the geometry of stars and dust, and only fit the spatially integrated SEDs. Thus, De Looze et al. (2014) produced a non-axisymmetric model of the face-on galaxy M51 which was able to account for the overall observed SED, providing a determination of the SFR of M51 of 3 $M_{\odot}\ yr^{-1}$. The Andromeda galaxy was also modelled in a relatively similar manner by Viaene et al. (2017) (see Fig. 9.2), providing a determination of the SFR between 0.2 and 0.4 $M_{\odot}\ yr^{-1}$.

The second approach is to use axisymmetric models (Thirlwall et al., 2020) and directly solve the inverse problem for the imaging data, albeit under the approximation of axisymmetry. This approach has the disadvantage that it cannot account for the arm-interarm contrast, but has the advantage that it can optimise for the radial distributions of stars and dust in the system. Interestingly, M33 has been modelled with both non-axisymmetric (Williams et al., 2019) and axi-symmetric models (Thirlwall et al., 2020), giving very different solutions. Williams et al. (2019) used a fixed spatial template for the geometry of stars and dust and was unable to fit the spatially integrated SED unless modified dust grain properties were introduced. On the other hand, Thirlwall et al. (2020) fitted the large scale geometry of M33 and produced a model that accounted for the integrated SED of M33 using Milky Way type dust, with a derived SFR of 0.28 $M_{\odot}\ yr^{-1}$. The results from Thirlwall et al. (2020) do not constitute a proof that the dust in M33 has Milky Way type properties, but they do provide a consistency check that existing grain models can account for the observed dust emission and attenuation in M33.

A unique feature of radiative-transfer models that fit the geometry of face-on systems is their power to predict local properties of galaxies, including star-formation rates as a function of position in a galaxy. A recent example is the finding that, when looking at the different morphological components of a galaxy from the inner to the outer disks, a much steeper relation is defined in the star-formation rate surface density vs stellar-mass surface density space than that expected from the main sequence of star-forming galaxies (Thirlwall et al., 2020). This new relation was called a 'structurally resolved main sequence'. Although this relation has only been established for one galaxy (M33) for the moment, its generality, if proven to exist, may point towards stellar feedback mechanisms regulating the growth of galaxy disks that shape the slope of this relation.

9.6.2 Statistical Samples

Detailed radiative-transfer models that can account for the detail multi-wavelength images of a galaxy can only be applied to model a small number of objects, due to the large computational time involved in the optimisation process. If the goal is to understand global parameters of galaxies derived from large statistical samples, then the use of libraries of model panchromatic SEDs is the only viable solution. Libraries obtained with radiative-transfer calculations have been initially produced for generic geometries, with the purpose of only accounting for the spatially integrated emission. Thus, models for starburst galaxies were developed in Rowan-Robinson and Efstathiou (1993),

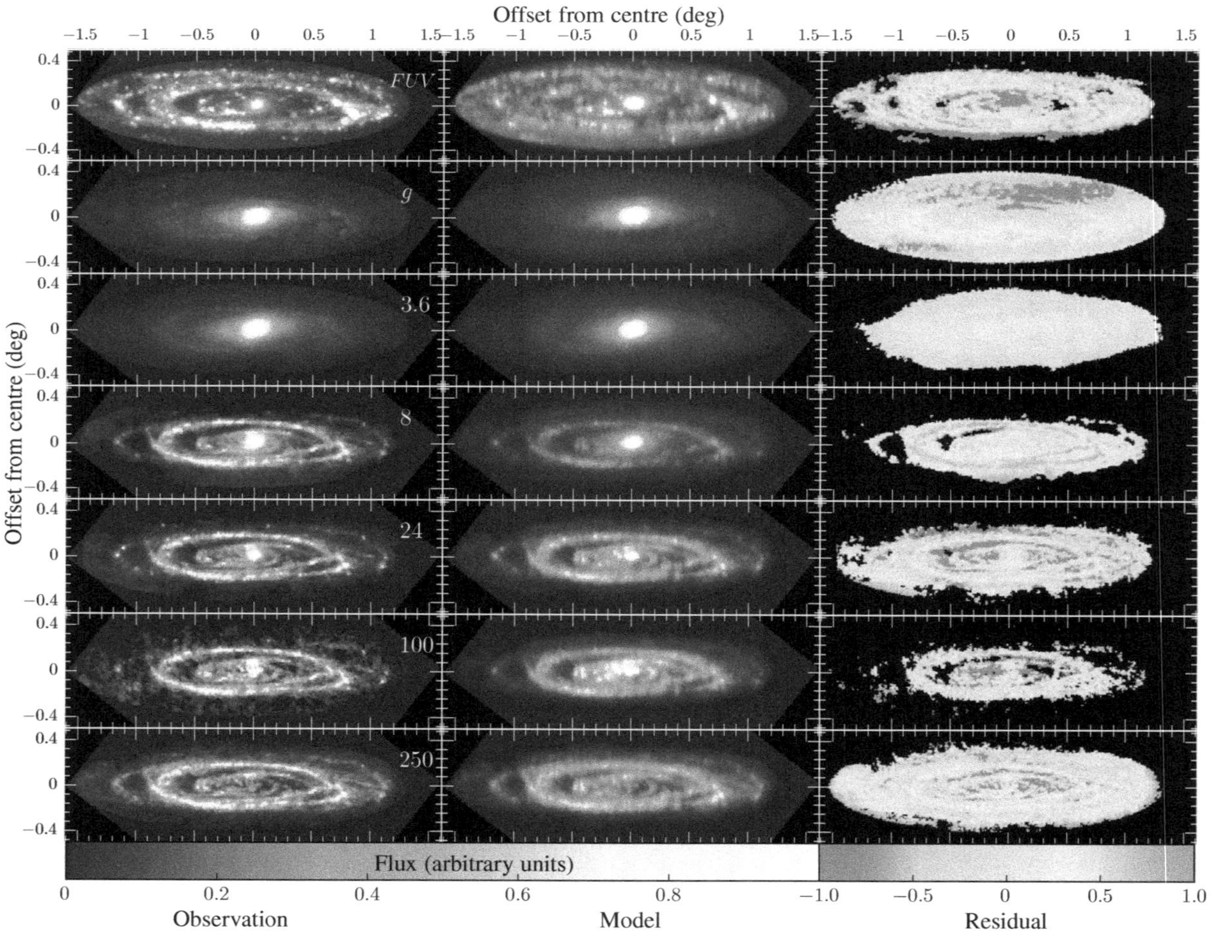

Figure 9.2 Examples of model fits of the Andromeda galaxy to observations in selected wavebands. The derived SFR was found to be between 0.2 and 0.4 $M_\odot$ yr^{-1}. Credit: Viaene et al. (2017), reproduced with permission © ESO. Colour version available online.

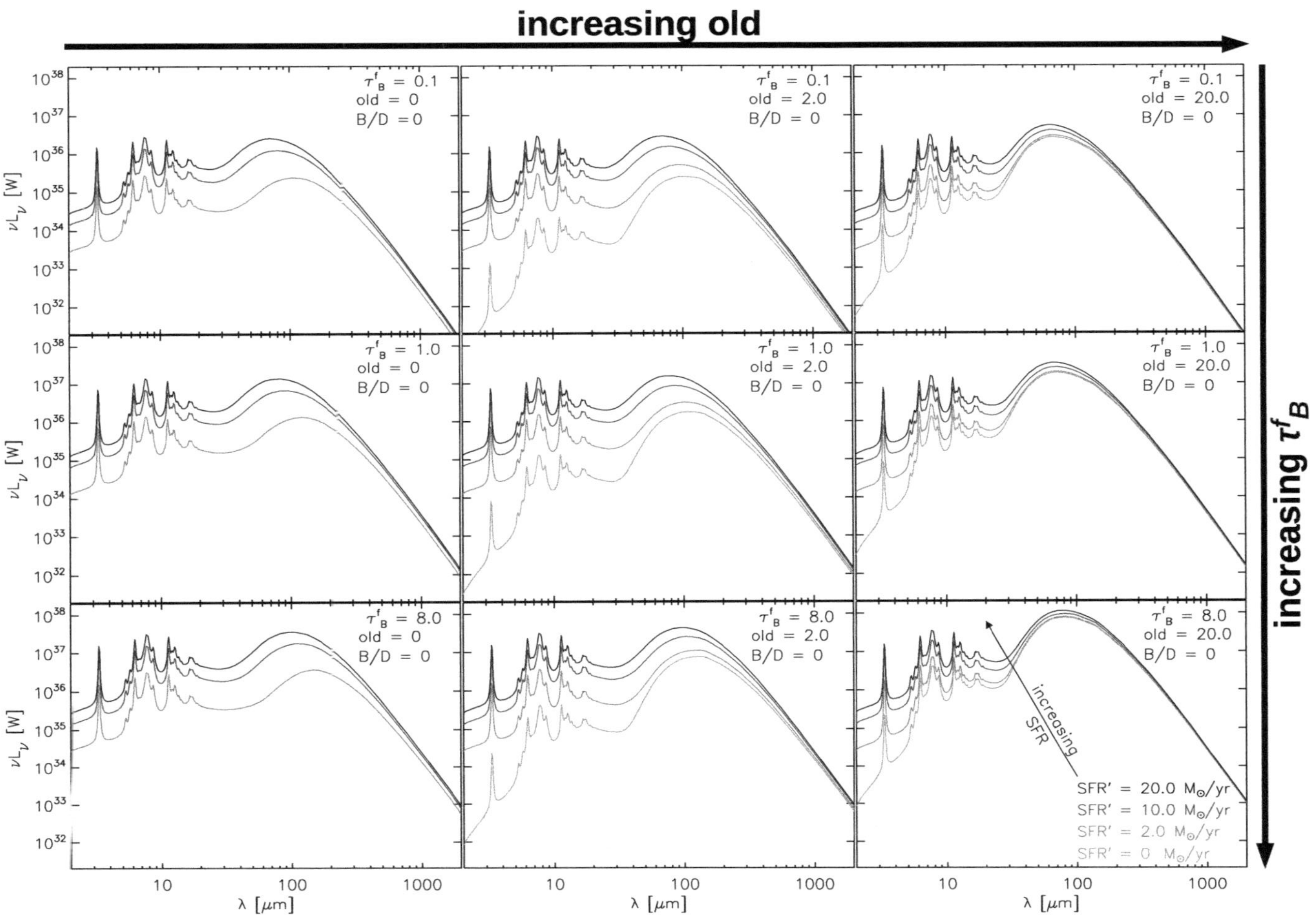

Figure 9.3 Examples of model dust and PAH SEDs. Credit: Popescu et al. (2011), reproduced with permission ©ESO. Colour version available online.

Krüegel and Siebenmorgen (1994), Silva et al. (1998), Efstathiou et al. (2000), Dopita et al. (2005), Siebenmorgen and Krügel (2007), and Groves et al. (2008). Models for starburst dwarfs were developed by Galliano et al. (2003). Other models were developed for AGN torus, but here we restrict this chapter to the discussion of galaxies that are not dominated by an AGN component.

The next step in developing libraries of model SEDs was to generalise radiative-transfer models for which detailed imaging observations were taken into account. This was achieved in Tuffs et al. (2004) for the attenuation of stellar light in the UV/optical, and in Popescu et al. (2011) for both dust attenuation and dust and PAH emission SEDs in the IR/submm (see Fig. 9.3). Unlike libraries of model SED produced with empirical or phenomenological models, radiative-transfer models require knowledge of the size of the object to be fitted (in terms of e.g. the scale-length of the optical-emitting disk). Otherwise there is a degeneracy between the size of the disk and the total luminosities involved in the model. Simulated model images have been also analysed to provide libraries of corrections for the effect of dust on the determination of photometric parameters obtained from UV/optical images of spiral galaxies (Pastrav et al., 2013a,b; Moellenhof et al., 2006). Radiative-transfer model libraries have been used in statistical studies of nearby galaxies and they were able to successfully account for various statistical relations, including relations to star-formation rate measurements (e.g. Driver et al., 2007, 2008, 2012; Graham and Worley, 2008; Masters et al., 2010; Gunawardhana et al., 2011; Grootes et al., 2013; Davies et al., 2016; Grootes et al., 2017).

Libraries of models SEDs created by applying radiative-transfer calculations to numerical simulations of galaxies do exist, but, as mentioned at the beginning of this chapter, they are not discussed here.

9.7 Comparison between Radiative-Transfer Models and Phenomenological Models

A direct comparison between a radiative transfer and a phenomenological model, when applied to an individual object, makes little sense, for a couple of reasons. Firstly, there is not an a priori knowledge of the intrinsic distributions of stellar emissivities and dust in any individual galaxy that could be taken as a standard for comparison, since it is the goal of these models to obtain such knowledge. Secondly, even if such knowledge existed, the phenomenological models would not explicitly take this information into account, therefore the comparison could only be done with respect to the goodness of the fit rather than to the reality of the intrinsic parameters derived from the fit.

Bearing these aspects in mind, the only way to apply a meaningful comparison between models is to look at how well they perform when looking at statistical behaviours of known observational relations. Such a study has been performed by Davies et al. (2016) who looked at the well-known $SFR - M_{\rm star}$ relation using a well-defined local sample of morphologically selected spiral galaxies from the *GAMA*[11] survey (Driver et al., 2011; Liske et al., 2015). Davies et al. (2016) compared different methods for extracting SFRs and found that the most consistent slopes and normalisations of the $SFR - M_{\rm star}$ relations are obtained when using a radiative-transfer method.

9.8 Conclusion

Despite its critical importance, modelling the SEDs of galaxies and deriving SFRs using radiative-transfer codes has made slow progress until now, with only few adepts trying to tackle problems. This is, in part, due to the preconception of large computational costs involved in the calculations. However, many valid approximations and intelligent algorithms have been developed to make calculations tractable, even before the advent of supercomputing facilities. Recently, with the ever better computing resources, this field is now getting more recognition. The main challenge to come is perhaps not so much related to the technical improvement in the radiative-transfer codes, but in the identification of all physical mechanisms that are at play in these complex systems – the star-forming galaxies.

References

Baes, M., Fritz, J., Gadotti, D. A. et al. 2010. Herschel-ATLAS: The Dust Energy Balance in the Edge-On Spiral Galaxy UGC 4754. *Astronomy and Astrophysics*, **518**(July), L39.

Baes, M., Davies, J. I., Dejonghe, H. et al. 2003. Radiative Transfer in Disc Galaxies – III. The Observed Kinematics of Dusty Disc Galaxies. *Monthly Notices of the Royal Astronomical Society*, **343**(4), 1081–1094.

Baes, M., Stamatellos, D., Davies, J. I. et al. 2005. Radiative Equilibrium in Monte Carlo Radiative Transfer Using Frequency Distribution Adjustment. *New Astronomy*, **10**(6), 523–533.

Baes, M., Verstappen, J., De Looze, I. et al. 2011. Efficient Three-Dimensional NLTE Dust Radiative Transfer with SKIRT. *Astrophysical Journal Supplement Series*, **196**(2), 22.

Bahcall, J. N., and Soneira, R. M. 1980. The Universe at Faint Magnitudes. I. Models for the Galaxy and the Predicted Star Counts. *Astrophysical Journal Supplement Series*, **44**(Sep), 73–110.

Bahcall, J. N., and Soneira, R. M. 1984. Comparisons of a Standard Galaxy Model with Stellar Observations in Five Fields. *Astrophysical Journal Supplement Series*, **55**(May), 67–99.

Bianchi, S. 2007. The Dust Distribution in Edge-On Galaxies. Radiative Transfer Fits of V and K'-Band Images. *Astronomy and Astrophysics*, **471**(3), 765–773.

Bianchi, S. 2008. Dust Extinction and Emission in a Clumpy Galactic Disk. An Application of the Radiative Transfer Code TRADING. *Astronomy and Astrophysics*, **490**(1), 461–475.

[11] Galaxy and Mass Assembly; www.gama-survey.org/

Bianchi, S., and Xilouris, E. M. 2011. The Extent of Dust in NGC 891 from Herschel/SPIRE Images. *Astronomy and Astrophysics*, **531**(July), L11.

Camps, P., and Baes, M. 2015. SKIRT: An Advanced Dust Radiative Transfer Code with a User-Friendly Architecture. *Astronomy and Computing*, **9**(Mar.), 20–33.

Camps, P., Misselt, K., Bianchi, S. et al. 2015. Benchmarking the Calculation of Stochastic Heating and Emissivity of Dust Grains in the Context of Radiative Transfer Simulations. *Astronomy and Astrophysics*, **580**(Aug.), A87.

Cohen, M. 1994. Powerful Model for the Point Source Sky: Far-Ultraviolet and Enhanced Midinfrared Performance. *Astronomical Journal*, **107**(Feb.), 582.

da Cunha, E., and Charlot, S. 2011 (June). *MAGPHYS: Multi-Wavelength Analysis of Galaxy Physical Properties*.

da Cunha, E., Charlot, S., and Elbaz, D. 2008. A Simple Model to Interpret the Ultraviolet, Optical and Infrared Emission from Galaxies. *Monthly Notices of the Royal Astronomical Society*, **388**(4), 1595–1617.

Dasyra, K. M., Xilouris, E. M., Misiriotis, A. et al. 2005. Is the Galactic Submillimeter Dust Emissivity Underestimated? *Astronomy and Astrophysics*, **437**(2), 447–456.

Davies, L. J. M., Driver, S. P., Robotham, A. S. G. et al. 2016. GAMA/H-ATLAS: A Meta-Analysis of SFR Indicators – Comprehensive Measures of the SFR-M_* Relation and Cosmic Star Formation History at $z < 0.4$. *Monthly Notices of the Royal Astronomical Society*, **461**(1), 458–485.

De Geyter, G., Baes, M., Camps, P. et al. 2014. The Distribution of Interstellar Dust in CALIFA Edge-On Galaxies via Oligochromatic Radiative Transfer Fitting. *Monthly Notices of the Royal Astronomical Society*, **441**(1), 869–885.

De Geyter, G., Baes, M., De Looze, I. et al. 2015. Dust Energy Balance Study of Two Edge-On Spiral Galaxies in the Herschel-ATLAS Survey. *Monthly Notices of the Royal Astronomical Society*, **451**(2), 1728–1739.

De Geyter, G., Baes, M., Fritz, J. et al. 2013. FitSKIRT: Genetic Algorithms to Automatically Fit Dusty Galaxies with a Monte Carlo Radiative Transfer Code. *Astronomy and Astrophysics*, **550**(Feb.), A74.

De Looze, I., Baes, M., Fritz, J. et al. 2012a. Panchromatic Radiative Transfer Modelling of Stars and Dust in the Sombrero Galaxy. *Monthly Notices of the Royal Astronomical Society*, **419**(1), 895–903.

De Looze, I., Baes, M., Bendo, G. J. et al. 2012b. The Dust Energy Balance in the Edge-On Spiral Galaxy NGC 4565. *Monthly Notices of the Royal Astronomical Society*, **427**(4), 2797–2811.

De Looze, I., Fritz, J., Baes, M. et al. 2014. High-Resolution, 3D Radiative Transfer Modeling. I. The Grand-Design Spiral Galaxy M 51. *Astronomy and Astrophysics*, **571**(Nov), A69.

Dopita, M. A., Groves, B. A., Fischera, J. et al. 2005. Modeling the Pan-Spectral Energy Distribution of Starburst Galaxies. I. The Role of ISM Pressure and the Molecular Cloud Dissipation Timescale. *Astrophysical Journal*, **619**(2), 755–778.

Driver, S. P., Hill, D. T., Kelvin, L. S. et al. 2011. Galaxy and Mass Assembly (GAMA): Survey Diagnostics and Core Data Release. *Monthly Notices of the Royal Astronomical Society*, **413**(2), 971–995.

Driver, S. P., Robotham, A. S. G., Kelvin, L. et al. 2012. Galaxy and Mass Assembly (GAMA): The $0.013 < z < 0.1$ Cosmic Spectral Energy Distribution from 0.1 μm to 1 mm. *Monthly Notices of the Royal Astronomical Society*, **427**(4), 3244–3264.

Driver, S. P., Popescu, C. C., Tuffs, R. J. et al. 2007. The Millennium Galaxy Catalogue: The B-Band Attenuation of Bulge and Disc Light and the Implied Cosmic Dust and Stellar Mass Densities. *Monthly Notices of the Royal Astronomical Society*, **379**(3), 1022–1036.

Driver, S. P., Popescu, C. C., Tuffs, R. J. et al. 2008. The Energy Output of the Universe from 0.1 to 1000 μm. *Astrophysical Journal*, **678**(2), L101.

Efstathiou, A., Rowan-Robinson, M., and Siebenmorgen, R. 2000. Massive Star Formation in Galaxies: Radiative Transfer Models of the UV to Millimetre Emission of Starburst Galaxies. *Monthly Notices of the Royal Astronomical Society*, **313**(4), 734–744.

Galliano, F., Madden, S. C., Jones, A. P. et al. 2003. ISM Properties in Low-Metallicity Environments. II. The Dust Spectral Energy Distribution of NGC 1569. *Astronomy and Astrophysics*, **407**(Aug.), 159–176.

Graham, A. W., and Worley, C. C. 2008. Inclination- and Dust-Corrected Galaxy Parameters: Bulge-to-Disc Ratios and Size-Luminosity Relations. *Monthly Notices of the Royal Astronomical Society*, **388**(4), 1708–1728.

Grootes, M. W., Tuffs, R. J., Popescu, C. C. et al. 2013. GAMA/H-ATLAS: The Dust Opacity-Stellar Mass Surface Density Relation for Spiral Galaxies. *Astrophysical Journal*, **766**(1), 59.

Grootes, M. W., Tuffs, R. J., Popescu, C. C. et al. 2017. Galaxy and Mass Assembly (GAMA): Gas Fueling of Spiral Galaxies in the Local Universe. I. The Effect of the Group Environment on Star Formation in Spiral Galaxies. *Astronomical Journal*, **153**(3), 111.

Groves, B., Dopita, M. A., Sutherland, R. S. et al. 2008. Modeling the Pan-Spectral Energy Distribution of Starburst Galaxies. IV. The Controlling Parameters of the Starburst SED. *Astrophysical Journal Supplement Series*, **176**(2), 438–456.

Gunawardhana, M. L. P., Hopkins, A. M., Sharp, R. G. et al. 2011. Galaxy and Mass Assembly (GAMA): The Star Formation Rate Dependence of the Stellar Initial Mass Function. *Monthly Notices of the Royal Astronomical Society*, **415**(2), 1647–1662.

Krüegel, E., and Siebenmorgen, R. 1994. Dust in Protostellar Cores and Stellar Disks. *Astronomy and Astrophysics*, **288**(Aug.), 929–941.

Kylafis, N. D., and Bahcall, J. N. 1987. Dust Distribution in Spiral Galaxies. *Astrophysical Journal*, **317**(Jun), 637.

Kylafis, N. D., and Xilouris, E. M. 2005 (Apr). Radiation Transfer Models in Galaxies. Pages 3–16 of: Popescu, Cristina C., and Tuffs, Richard J. (eds), *The Spectral Energy Distributions of Gas-Rich Galaxies: Confronting Models with Data*. American Institute of Physics Conference Series, vol. 761.

Liske, J., Baldry, I. K., Driver, S. P. et al. 2015. Galaxy and Mass Assembly (GAMA): End of Survey Report and Data Release 2. *Monthly Notices of the Royal Astronomical Society*, **452**(2), 2087–2126.

MacLachlan, J. M., Matthews, L. D., Wood, K. et al. 2011. The Stability of Low Surface Brightness Disks Based on Multi-Wavelength Modeling. *Astrophysical Journal*, **741**(1), 6.

Masters, K. L., Nichol, R., Bamford, S. et al. 2010. Galaxy Zoo: Dust in Spiral Galaxies. *Monthly Notices of the Royal Astronomical Society*, **404**(2), 792–810.

Misiriotis, A., Kylafis, N. D., Papamastorakis, J. et al. 2000. Is the Exponential Distribution a Good Approximation of Dusty Galactic Disks? *Astronomy and Astrophysics*, **353**(Jan.), 117–123.

Misiriotis, A., Popescu, C. C., Tuffs, R. et al. 2001. Modeling the Spectral Energy Distribution of Galaxies. II. Disk Opacity and Star Formation in 5 Edge-On Spirals. *Astronomy and Astrophysics*, **372**(June), 775–783.

Misiriotis, A., Xilouris, E. M., Papamastorakis, J. et al. 2006. The Distribution of the ISM in the Milky Way. A Three-Dimensional Large-Scale Model. *Astronomy and Astrophysics*, **459**(1), 113–123.

Möllenhoff, C., Popescu, C. C., and Tuffs, R. J. 2006. Modelling the Spectral Energy Distribution of Galaxies. IV. Correcting Apparent Disk Scalelengths and Central Surface Brightnesses for the Effect of Dust at Optical and Near-Infrared Wavelengths. *Astronomy and Astrophysics*, **456**(3), 941–952.

Mosenkov, A. V., Allaert, F., Baes, M. et al. 2016. HERschel Observations of Edge-On Spirals (HEROES). III. Dust Energy Balance Study of IC 2531. *Astronomy and Astrophysics*, **592**(July), A71.

Mosenkov, A. V., Allaert, F., Baes, M. et al. 2018. HERschel Observations of Edge-On Spirals (HEROES). IV. Dust Energy Balance Problem. *Astronomy and Astrophysics*, **616**(Aug), A120.

Natale, G., Popescu, C. C., Tuffs, R. J. et al. 2014. DART-RAY: A 3D Ray-Tracing Radiative Transfer Code for Calculating the Propagation of Light in Dusty Galaxies. *Monthly Notices of the Royal Astronomical Society*, **438**(4), 3137–3162.

Natale, G., Popescu, C. C., Tuffs, R. J. et al. 2015. Predicting the Stellar and Non-Equilibrium Dust Emission Spectra of High-Resolution Simulated Galaxies with DART-RAY. *Monthly Notices of the Royal Astronomical Society*, **449**(1), 243–267.

Natale, G., Popescu, C. C., Tuffs, R. J. et al. 2017. Ray-Tracing 3D Dust Radiative Transfer with DART-Ray: Code Upgrade and Public Release. *Astronomy and Astrophysics*, **607**(Nov.), A125.

Pastrav, B. A., Popescu, C. C., Tuffs, R. J. et al. 2013a. The Effects of Dust on the Derived Photometric Parameters of Disks and Bulges in Spiral Galaxies. *Astronomy and Astrophysics*, **553**(May), A80.

Pastrav, B. A., Popescu, C. C., Tuffs, R. J. et al. 2013b. The Effects of Dust on the Photometric Parameters of Decomposed Disks and Bulges. *Astronomy and Astrophysics*, **557**(Sep.), A137.

Popescu, C. C., and Tuffs, R. J. 2010 (June). The Evolution of Galaxies: An Infrared Perspective. Pages 35–46 of: Debattista, V. P., and Popescu, C. C. (eds), *American Institute of Physics Conference Series*. American Institute of Physics Conference Series, vol. 1240.

Popescu, C. C., Misiriotis, A., Kylafis, N. D. et al. 2000. Modelling the Spectral Energy Distribution of Galaxies. I. Radiation Fields and Grain Heating in the Edge-On Spiral NGC 891. *Astronomy and Astrophysics*, **362**(Oct.), 138–150.

Popescu, C. C., Tuffs, R. J., Dopita, M. A. et al. 2011. Modelling the Spectral Energy Distribution of Galaxies. V. The Dust and PAH Emission SEDs of Disk Galaxies. *Astronomy and Astrophysics*, **527**(Mar), A109.

Popescu, C. C., Yang, R., Tuffs, R. J. et al. 2017. A Radiation Transfer Model for the Milky Way: I. Radiation Fields and Application to High-Energy Astrophysics. *Monthly Notices of the Royal Astronomical Society*, **470**(3), 2539–2558.

Robitaille, T. P. 2011. HYPERION: An Open-Source Parallelized Three-Dimensional Dust Continuum Radiative Transfer Code. *Astronomy and Astrophysics*, **536**(Dec.), A79.

Robitaille, T. P., Churchwell, E., Benjamin, R. A. et al. 2012. A Self-Consistent Model of Galactic Stellar and Dust Infrared Emission and the Abundance of Polycyclic Aromatic Hydrocarbons. *Astronomy and Astrophysics*, **545**(Sep.), A39.

Rowan-Robinson, M., and Efstathiou, A. 1993. Multigrain Dust Cloud Models of Starburst and Seyfert Galaxies. *Monthly Notices of the Royal Astronomical Society*, **263**(Aug.), 675–680.

Saftly, W., Baes, M., De Geyter, G. et al. 2015. Large and Small-Scale Structures and the Dust Energy Balance Problem in Spiral Galaxies. *Astronomy and Astrophysics*, **576**(Apr.), A31.

Sauvage, M., Tuffs, R. J., and Popescu, C. C. 2005. Normal Nearby Galaxies. *Space Science Reviews*, **119**(1–4), 313–353.

Schechtman-Rook, A., Bershady, M. A., and Wood, K. 2012. The Three-Dimensional Distribution of Dust in NGC 891. *Astrophysical Journal*, **746**(1), 70.

Siebenmorgen, R., and Krügel, E. 2007. Dust in Starburst Nuclei and ULIRGs. SED Models for Observers. *Astronomy and Astrophysics*, **461**(2), 445–453.

Siebenmorgen, R., Kruegel, E., and Mathis, J. S. 1992. Radiative Transfer for Transiently Heated Particles. *Astronomy and Astrophysics*, **266**(1), 501–512.

Silva, L., Granato, G. L., Bressan, A. et al. 1998. Modeling the Effects of Dust on Galactic Spectral Energy Distributions from the Ultraviolet to the Millimeter Band. *Astrophysical Journal*, **509**(1), 103–117.

Steinacker, J., Baes, M., and Gordon, K. D. 2013. Three-Dimensional Dust Radiative Transfer*. *Annual Review of Astronomy and Astrophysics*, **51**(1), 63–104.

Thirlwall, J. J., Popescu, C. C., Tuffs, R. J. et al. 2020. A Radiative Transfer Model for the Spiral Galaxy M33. *Monthly Notices of the Royal Astronomical Society*, **495**(1), 835–863.

Tuffs, R. J., and Popescu, C. C. 2005 (Apr). Normal Gas-Rich Galaxies in the Far-Infrared: The Legacy of ISOPHOT. Pages 344–363 of: Popescu, C. C., and Tuffs, R. J. (eds), *The Spectral Energy Distributions of Gas-Rich Galaxies: Confronting Models with Data*. American Institute of Physics Conference Series, vol. 761.

Tuffs, R. J., Popescu, C. C., Völk, H. J. et al. 2004. Modelling the Spectral Energy Distribution of Galaxies. III. Attenuation of Stellar Light in Spiral Galaxies. *Astronomy and Astrophysics*, **419**(June), 821–835.

Viaene, S., Baes, M., Tamm, A. et al. 2017. The Herschel Exploitation of Local Galaxy Andromeda (HELGA). VII. A SKIRT Radiative Transfer Model and Insights on Dust Heating. *Astronomy and Astrophysics*, **599**(Mar.), A64.

Wainscoat, R. J., Cohen, M., Volk, K. et al. 1992. A Model of the 8–25 Micron Point Source Infrared Sky. *Astrophysical Journal Supplement Series*, **83**(Nov.), 111.

Weingartner, J. C., and Draine, B. T. 2001. Dust Grain-Size Distributions and Extinction in the Milky Way, Large Magellanic Cloud, and Small Magellanic Cloud. *Astrophysical Journal*, **548**(1), 296–309.

Williams, T. G., Baes, M., De Looze, I. et al. 2019. High-Resolution Radiative Transfer Modelling of M33. *Monthly Notices of the Royal Astronomical Society*, **487**(2), 2753–2770.

Witt, A. N. 1977. Multiple Scattering in Reflection Nebulae. I. A Monte Carlo Approach. *Astrophysical Journal Supplement Series*, **35**(Sep), 1–6.

Witt, A. N., and Johnson, M. W. 1973. The Interstellar Radiation Density between 1250 and 4250 Λ. *Astrophysical Journal*, **181**(Apr), 363–368.

Wood, K. 1997. Scattering and Dichroic Extinction: Polarimetric Signatures of Galaxies. *Astrophysical Journal*, **477**(1), L25–L28.

Xilouris, E. M. 2005 (Apr). Calculating the Submillimeter Emissivity of Dust Grains in Spiral Galaxies. Pages 141–144 of: Popescu, Cristina C., and Tuffs, Richard J. (eds),

The Spectral Energy Distributions of Gas-Rich Galaxies: Confronting Models with Data. American Institute of Physics Conference Series, vol. 761.

Xilouris, E. M., Alton, P. B., Davies, J. I. et al. 1998. Optical and NIR Modelling of NGC 891. *Astronomy and Astrophysics*, **331**(Mar.), 894–900.

Xilouris, E. M., Byun, Y. I., Kylafis, N. D. et al. 1999. Are Spiral Galaxies Optically Thin or Thick? *Astronomy and Astrophysics*, **344**(Apr.), 868–878.

Xilouris, E. M., Kylafis, N. D., Papamastorakis, J. et al. 1997. The Distribution of Stars and Dust in Spiral Galaxies: The Edge-On Spiral UGC 2048. *Astronomy and Astrophysics*, **325**(Sep.), 135–143.

Yusef-Zadeh, F., Morris, M., and White, R. L. 1984. Bipolar Reflection Nebulae: Monte Carlo Simulations. *Astrophysical Journal*, **278**(Mar.), 186–194.

Zubko, V., Dwek, E., and Arendt, R. G. 2004. Interstellar Dust Models Consistent with Extinction, Emission, and Abundance Constraints. *Astrophysical Journal Supplement Series*, **152**(2), 211–249.

10

Measuring the Star-Formation Rate in Active Galactic Nuclei

BRENT GROVES

10.1 Introduction

Active galactic nuclei (AGN) are thought to play major role in the evolution of galaxies, impacting both star formation and gas accretion onto galaxies. The supermassive black holes (SMBH) at the centres of galaxies powering the AGN show a similar evolution in growth as their host galaxy stellar masses (Madau and Dickinson, 2014). The correlation of the mass of the SMBH with their host galaxies' central velocity dispersion (the $M_{\bullet} - \sigma$ relation, Magorrian et al., 1998) and bulge stellar mass (the $M - M_{\rm Bulge}$ relation, Kormendy and Ho, 2013) observed in the local universe are also demonstrative that there exists a co-evolution of black hole growth and stellar mass growth. Conversely, in simulations AGN play a central role in the quenching of star formation in galaxies and maintaining their quiescence (e.g. Springel et al., 2005; Croton et al., 2006). Clearly, there is a need to determine the star-formation rates (SFRs) in AGN host galaxies to understand and disentangle the growth of SMBH and their hosts.

The issue in measuring the SFRs in AGN is that AGN can be bright (bolometric luminosities $> 10^{46}$ erg s^{-1}; 10^{14} L$_{\odot}$, Barger et al., 2005), overwhelming the host galaxy emission, and emit strongly at all wavelengths from the X-ray to radio (see e.g. Elvis et al., 1994). Therefore, all of the SFR tracers discussed in the other chapters can be impacted by non-stellar emission if an AGN is present. The key to measuring the SFR in AGN is thus to subtract all of the AGN emission from the SFR tracer.

The first problem is to identify whether a galaxy hosts an AGN. While for some objects (e.g. Quasars) such identification is easy as the AGN emission overwhelms the emission at most wavelengths, for other galaxies (e.g. Seyfert IIs) the presence of an AGN may only be detectable at certain wavelengths. AGN identification is necessary, for example, in determining the evolution of the SFR density across redshift, where AGN emission may bias the results (see e.g. Karim et al., 2011), or for determining the host properties of AGN (Kauffmann et al., 2003). Once an AGN is identified, it can either be separated from a pure star-forming sample as in the previous cases, or the contribution of the AGN to the SFR tracer being used can be estimated and subtracted.

Subtraction of AGN emission from the SFR tracers is difficult as it requires knowing the luminosity of the AGN at the particular wavelength used as a tracer. The bolometric luminosity of an AGN is purely a function of the accretion rate and the efficiency of this accretion (or rather the fraction of the gravitational energy released in the accretion process). However, the distribution of this energy across wavelength is dependent upon many parameters, including the mass of the SMBH, the relative accretion rate, and the geometry of the gas and galaxy around the black hole (Antonucci, 1993; Urry and Padovani, 1995). While detection of an AGN at certain wavelengths can assist in determining the contribution at other wavelengths (e.g. for the correlation of the X-ray luminosity and [OIII] emission-line luminosity (Heckman et al., 2005)), such correlations are empirical and suffer the uncertainties associated with their determination. As AGN are, by definition, central objects, they can suffer higher extinction due to dust and gas as compared to star formation, which can be distributed throughout the galaxy. This differential extinction can complicate the determination and subtraction of the contribution of AGN to a SFR tracer.

As AGN are also extremely compact objects (the emitting accretion disk around the SMBH is <0.01 pc), the intrinsic emission (i.e. not reprocessed) from AGN can vary on the scales of days, with these variations arising from small changes in the accretion rate (Ulrich et al., 1997). This variation in the emission further complicates the subtraction of the AGN contribution to several SFR tracers, especially in cases where diagnostics using multiple wavelengths observed over multiple epochs are used.

We discuss in this chapter the methods used to identify whether a galaxy hosts an AGN and to determine the fractional contribution of the AGN to various SFR tracers. We split these methods into the various wavelengths associated with AGN identification and the SFR tracers discussed in this book. While we acknowledge this will be an incomplete review of the literature associated with this topic, this chapter will give a broad overview and understanding of the methods and issues involved in determining the SFR in AGN host galaxies.

10.2 The Physics of an AGN and Its Emission

AGNs are powered by the accretion of material onto SMBHs at the centres of galaxies. This accretion is thought to take the form of a thin accretion disk (thought to be $10^{-3}-10^{-2}$ pc in radius) surrounded by a hot ($>10^6$ K) corona. The accretion disk converts the fraction of the gravitational energy released in the accretion process into thermal emission, approximately as a black body peaking at ultraviolet wavelengths. Some of this ultraviolet (UV) emission is upscattered by the surrounding corona by inverse Compton scattering into hard ($>$few keV) X-rays. The UV emission is also reprocessed by the surrounding gas into emission lines (possibly velocity broadened if near to the SMBH) or by the surrounding dust into mid-infrared (MIR) and far-infrared emission (FIR). Some fraction of the accretion energy can be emitted non-radiatively as a wind or, in the presence of magnetic fields, a jet. The jet emits strongly in radio wavelengths due to synchrotron emission. For a comprehensive review of the physics of AGN see Osterbrock and Ferland (2006), or for a more recent review see Alexander and Hickox (2012).

The emission from the AGN can be further modified through the absorption and scattering by gas and dust in the host galaxy. Type-I AGN present bright UV and optical emission and broad emission lines directly from the accretion disk and surrounds, while in Type-II AGN this emission is either faint or absent (Seyfert, 1943). Unifying these two types of AGN is the idea that dust is associated with the accretion disk, either directly, in the form of dusty clouds or winds, or in the form of a co-planar dusty torus. The galactic disk can obscure some solid angle fraction of the AGN, and it is purely the viewing angle that differentiates these two AGN types (Antonucci, 1993). The dust nearest to the accretion disk is thought to be heated up to sublimation temperatures ($T_{dust} \sim 2000$ K) and destroyed, and emit strongly in the MIR. The destruction of dust in the inner regions of the AGN return metals from the dust back to the gas, such that heavily depleted metals, such as iron and magnesium, can emit strongly in the broad-line regions.

Taken together, an AGN emits across the full electromagnetic spectrum, as demonstrated by Fig. 10.1. This figure shows the mean spectral energy distributions of three samples

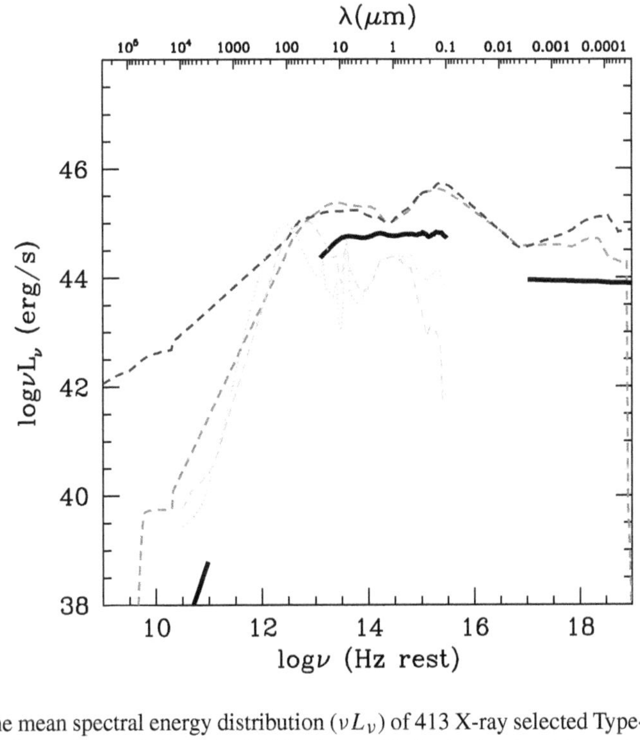

Figure 10.1 The mean spectral energy distribution (νL_ν) of 413 X-ray selected Type-I AGN from the COSMOS field (black lines: Elvis et al., 2012) along with the Radio Loud (green) and Radio Quiet (red) Quasars from Elvis et al. (1994). Note the almost flat appearance across all wavelengths. As a comparison, galaxy templates from Polletta et al. (2007) are shown (blue). Original figure modified from Elvis et al. (2012). © AAS. Reproduced with permission. Colour version available online.

of AGN that have minimal dust obscuration along the line of sight: Type-I X-ray AGN, Radio Quiet Quasars and Radio Loud Quasars. Figure 10.1 demonstrates that AGN can emit strongly at all the wavelengths typically used as SFR tracers.

10.3 X-ray Identification

Hard ($L_{2-10\,\mathrm{keV}} > L_{0.5-2.0\,\mathrm{keV}}$) and strong ($L_\mathrm{X} > 10^{42}\,\mathrm{erg\,s^{-1}}$) X-ray emission from a galaxy is a clear indicator for an AGN. In star-forming galaxies X-ray emission arises from thermal emission from hot, $\sim 10^6$ K gas from stellar winds and supernovae shocks, as well as from high-mass X-ray binaries (material from a massive star accreting onto compact object). However, the inverse-Compton emission from an AGN can easily reach high luminosities ($L_\mathrm{X} > 10^{44}\,\mathrm{erg\,s^{-1}}$) and overwhelm the weaker emission from these other sources. AGNs have a much higher X-ray-to-optical flux ratio relative to star forming galaxies (see e.g. Fig. 10.1 and Maccacaro et al., 1988), and thus X-rays provide a strong contrast against stars for easier detection.

The Compton process that produces the X-ray emission in AGN also appears to be universal and directly associated with the accretion rate onto the SMBH (see e.g. Gibson et al., 2008). Thus, the presence of strong and hard X-rays is sufficient to identify the presence of an AGN and has been used in many surveys where other diagnostics are difficult to obtain (see Brandt and Alexander, 2015, for an overview of both the X-ray physics of AGN and large X-ray surveys).

However, weak X-ray emission ($L_\mathrm{X} < 10^{41}\,\mathrm{erg\,s^{-1}}$) does not necessarily indicate the absence of an AGN. If there is a significant column of gas between the AGN and the observer ($N_\mathrm{H} > 10^{24}\,\mathrm{cm^{-2}}$) then the transmission becomes Compton thick ($N_\mathrm{H} > 1/\sigma_\mathrm{T}$). At such high column densities AGN X-ray emission becomes indistinguishable from that arising from star formation.

As X-ray emission in star-forming galaxies arises from high-mass X-ray binaries and hot gas associated with massive stars and supernovae, it is possible to use X-ray luminosity as a SFR tracer (see e.g. Aird et al., 2017). However, there are still large uncertainties associated with using such a tracer, and given the AGN dominance at these wavelengths, subtracting the AGN contribution to a galaxy in X-rays is difficult. Thus, X-ray emission remains one of the better identifiers of AGN in galaxies.

10.4 Ultraviolet and Optical Continuum

The UV and optical continuum traces the AGN in a similar manner to the UV tracing the massive stars when used as a SFR tracer (as discussed in particular in Chapter 6 and Chapter 7). The appearance of the AGN is also similar to that of massive stars: a power-law continuum rising towards the UV, peaking near the Lyman-limit (see Fig. 10.1). This similarity in appearance means that identifying an AGN and separating its contribution to the UV continuum in the integrated spectrum of a galaxy is difficult, unless the galaxy has already been identified as an AGN using other tracers.

Furthermore, as the UV continuum of AGN arises from the relatively small accretion disk, the UV emission can vary on the scales of days. Therefore, the contribution of the AGN will vary from epoch to epoch, further complicating the subtraction of any AGN emission from the UV.

However, as for using UV alone as a SFR tracer, the UV is generally a poor identifier of AGN as it can be readily extinguished by dust. Given that AGNs lie at the centres of galaxies, their UV and optical light typically experience higher levels of extinction than massive stars. If a galaxy is known to host an AGN, but tracers of the nuclear emission are not seen (i.e. Type-II AGN without broad emission lines, see Section 10.5), it is assumed that it suffers high levels of nuclear extinction. In these cases, the UV emission from the host galaxy can be used as a direct star formation tracer, with negligible contribution from the obscured AGN (see e.g. Zakamska et al., 2006).

In cases where the nuclear continuum is not obscured (i.e. Type-I AGN) the UV and continuum emission from the AGN can still be subtracted from the host galaxy if the galaxy can be spatially resolved. As the UV and optical emission directly arises from the AGN accretion disk (excluding scattered light) it can be considered as being emitted by a central point source. Given this, the host galaxy can be detected by subtracting a scaled point spread function from the image of the AGN (see Jahnke and Wisotzki, 2003). This scaling and subtraction can be performed on multiband images or even spectra (see Jahnke et al., 2004, 2007), such that the UV slope or optical continuum can be detected, and therefore the SFR can be determined as in Chapter 7).

10.5 Emission Lines

As AGNs emit a hard and strong extreme UV continuum ($\lambda < 91.2$ nm) they are able to ionise the surrounding interstellar medium (ISM). This leads to AGN emission-line nebulae, in the same way massive stars create HII regions. This similarity means it is necessary to account for the AGN contribution to emission-line SFR tracers such as Hα. The AGN emission-line nebulae are generally separated into two regions based upon their emission-line widths, broad-line regions ($\sigma > 1000\,\mathrm{km\,s^{-1}}$) and narrow-line regions ($\sigma < 1000\,\mathrm{km\,s^{-1}}$).

The broad-line region (BLR) arises close to the central SMBH (<1 pc), where the lines are Doppler broadened likely due to orbital motions around the SMBH, but also possibly outflows. The gas density in this region is high ($n_e > 10^9$ cm^{-3}), therefore only emission lines from permitted (electric dipole) transitions are observed (i.e. CIV 154.9nm, Hβ, etc. Osterbrock, 1978). The large line widths of the broad lines make them clear diagnostics for the presence of an AGN. As the BLR is close to the SMBH, the broad lines are obscured in a similar manner to the UV and optical continuum, and are commonly used diagnostics to differentiate Type-I and Type-II AGN (see Section 10.2). As the BLR is so close to the central SMBH, the emission from this region can vary on the scales of months to years (this variation can be used to estimate the SMBH mass, e.g. Peterson, 1993; Peterson et al., 2004). While the BLR emission is broader than HII regions, subtracting the BLR emission is necessary to correctly determine the line luminosity in recombination line SFR tracers like Hα.

The narrow-line region (NLR) arises from gas distributed further out from the SMBH, up to ~kpc in distance. Its velocity is dominated by both the galaxy rotation, internal gas dispersion and outflows (see e.g. Cecil et al., 2002). The NLR gas is of lower density than the BLR, though higher than typical HII regions ($n_e \sim 10^3$ cm^{-3}). At these densities both permitted and forbidden (magnetic dipole) transitions are observed (e.g. [OIII] λ500.7nm, [NII] λ658.3nm). As the NLR emission extends out into the galactic disk where star formation occurs, appears similar in width as HII regions, and emits similar lines, subtracting the NLR contribution to the SFR tracers is difficult.

As the AGN emits an ionising continuum that is harder ($h\nu > 54$ eV), it is able to ionise species to higher levels than possible in HII regions. Highly ionised, narrow emission lines like [NeV] λ342.6 nm (excitation potential 97.12 eV) are clear indicators for the presence of an AGN. Even in highly obscured AGN, the presence of near- and mid-IR lines such as [SiVII] soace λ2.48 μm (excitation potential 205.3 eV) or [NeV] λ14.3 μm can still clearly signify the presence of an AGN.

The hard extreme-ultraviolet and soft X-ray photons also create warm partially ionised regions in the NLR that lead to relatively stronger low-ionisation lines than found in HII region such as [NII] λ658.3nm (see e.g. Groves et al., 2004). This relative difference is the basis for line-ratio diagnostic diagrams such as the 'BPT' diagram of [NII]/Hα versus [OIII]/Hβ (Baldwin et al., 1981; Veilleux and Osterbrock, 1987). These line-ratio diagnostic diagrams clearly distinguish AGN-dominated versus star formation dominated galaxies when using integrated spectra (see e.g. Kewley et al., 2001; Kauffmann et al., 2003; and Fig. 10.2, left) and provide another diagnostic for the presence of an AGN when the direct emission is obscured (i.e. Type-II AGN).

Furthermore, integrated and nuclear galaxy spectra show a continuous distribution on these diagnostic diagrams (see Fig. 10.2, left), suggesting that the fractional contribution of the AGN to the emission lines can be determined. By assuming that the AGN branch of the BPT diagnostic diagram arises from a linear mixture of AGN excited emission and HII region emission, the line ratios can be used to determine the fractional contribution of the AGN to the Hα emission line. Once the AGN fraction is known, the contribution can be subtracted and the SFR can be determined from the remaining Hα luminosity.

This is demonstrated in Fig. 10.2, right, which shows the distribution on the BPT diagram of individual, 1-arcsec in size regions in the nearby Seyfert II galaxy NGC 5728 measured with an integral field spectrograph (Davies et al., 2016). Choosing two basis spectra representing a maximal AGN spectrum and maximal HII region spectrum based on

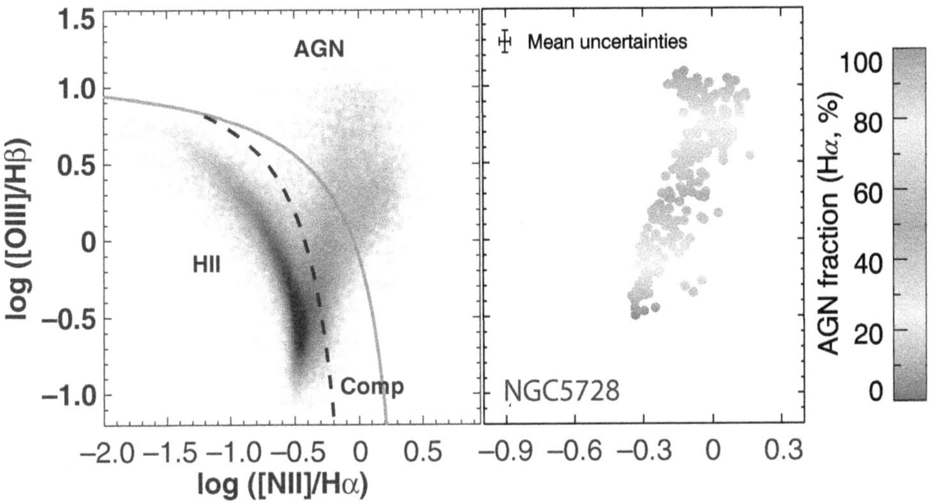

Figure 10.2 The BPT line ratio diagnostic diagram showing:
Left: the distribution of SDSS galaxies revealing the clear mixing of star forming galaxies and Type-II AGN. The lower dashed curve shows the diagnostic line of Kauffmann et al. (2003) separating AGN and star-forming galaxies, while the upper solid curve shows the 'maximal-starburst' line of Kewley et al. (2001), above which AGN must contribute. Between these curves are composite galaxies that likely have both exciting sources contributing to their spectra. From Kewley et al. (2006), by permission of Oxford University Press on behalf of the Royal Astronomical Society.
Right: The distribution of spectra from individual 1-arcsec regions from integral field spectrographic observations of the Seyfert-II galaxy NGC 5728, overlaid with the same two diagnostic curves. The colour indicates the fractional contribution of the AGN to the Hα emission, determined by assuming that all pixel spectra can be constructed from a linear combination of a maximal AGN and maximal HII region From Davies et al. (2016), by permission of Oxford University Press on behalf of the Royal Astronomical Society. Colour version available online.

the continuous trend visible in the BPT diagram, Davies et al. (2016) were able to determine the fractional contribution of the AGN to the Hα line in every pixel. From this they could create maps of the NLR-Hα and SFR maps (star formation associated Hα) for this galaxy.

This methodology can be expanded to any integral-field spectrographic observations of galaxies hosting AGN, or even to integrated galaxy spectra based on the SDSS BPT diagrams (see Fig. 10.2, left). The main issues with this method are: the choice of the basis spectra, differential extinction (between AGN and star formation), and secondary parameters that create dispersion (such as the gas-phase metallicity). The basis spectra can be chosen from either the extrema, as in Davies et al. (2016), but also using principal component analysis, or using theoretical model grids (such as Groves et al., 2004). While extinction should not greatly impact the mixing sequence seen on the BPT diagram (see Veilleux and Osterbrock, 1987), it will still impact the determination of the

SFRs and AGN Hα extent. Assuming uniform extinction (i.e. Hα/Hβ traces extinction) somewhat ameliorates, but does not fix, this problem. Secondary contributions to the dispersion in the BPT diagram can only be corrected for through the assumption of further basis spectra or additional principal components, but the problem can be reduced by minimising the contribution of these effects (such as concentrating on the nuclei of massive galaxies).

10.6 Mid-Infrared Emission

The MIR ($\sim$3–30 μm) emission from galaxies is dominated by the declining emission from stars (Rayleigh-Jeans emission), the emission from hot (~ 1000 K) and warm (~ 100 K) dust, and broad emission features thought to arise from polycyclic aromatic hydrocarbons (PAHs; see e.g. Leger and D'Hendecourt, 1985). The MIR is one of the more important diagnostics for the presence of AGN. As stellar emission is declining in the MIR, the dust emission, which is a strong indicator for AGN, is rising. In addition, the MIR is only weakly attenuated by dust. Nevertheless, in galaxies with high dust columns even the MIR can be attenuated, and a strong silicate-absorption feature at 9.7 μm can be seen (see e.g. Alonso-Herrero et al., 2001; Levenson et al., 2007). The MIR spectral range also includes several high-ionisation emission lines ([OIV] 25.9μm and [NeV] 14.3μm) that are also good indicators for the presence of AGN, again due to the relatively weak attenuation at these wavelengths.

AGNs are strong MIR emitters because of the dust in or near to the accretion disk at the centre of the AGN: the dusty 'torus'. Dust this near to the accretion disk is heated by the intense radiation field up to sublimation temperatures ($T_{dust} \sim 2000$ K), above which dust is destroyed. This hot dust emits strongly at MIR wavelengths, peaking at $\sim 3\mu$m (Pier and Krolik, 1992; Nenkova et al., 2002; Hönig et al., 2006; Schartmann et al., 2008). Dust further away, or deeper into the dusty torus, will see a progressively diluted and/or absorbed emission from the accretion disk, and therefore be heated to cooler temperatures and progressively emit at longer wavelengths. The sum of the emission from the progressively cooler dust leads to a power-law like continuum rising with wavelength in the MIR for AGN (Sturm et al., 2000; Laurent et al., 2000; Netzer et al., 2007; Sajina et al., 2012). This dust continuum then appears to turn over at $\sim$20–40 μm and is declining by 100 μm, though the exact turnover point in the infrared spectrum is dependent on the luminosity of the AGN and surrounding dust geometry (Netzer et al., 2007; Mullaney et al., 2011).

This strong and red MIR continuum (as compared to the blue Rayleigh-Jeans emission from stars) is one of more robust indicators for the presence of AGN. The MIR has been used as an identifier of AGN both spectroscopically (Laurent et al., 2000) and photometrically (Haas et al., 2004). MIR colour cuts look for the very red continuum of AGN and even allow for AGN to be detected at multiple redshifts (Stern et al., 2005), and in particular are useful in large area surveys covering galaxies at multiple redshifts (e.g. Assef et al., 2013; Stefanon et al., 2017).

The dust from the torus is sufficiently hot that broad absorption features at 9.7μm and 18μm from silicate-dominated dust (Draine and Lee, 1984) can actually be seen in emission in the most luminous AGNs and Type-I AGNs (Hao et al., 2005; Siebenmorgen et al., 2005). In weaker Type-I or in Type-II AGNs, these silicate features are either not seen or seen in absorption. As AGNs are bright and lie at the core of galaxies they can experience much higher levels of attenuation than star formation which is typically more distributed. A bright AGN beneath a high column of dust can lead to extremely deep silicate absorption features, which can also be a possible indicator for the presence of an AGN (Spoon et al., 2007). Ultra-luminous infrared galaxies show such strong absorption features (Hao et al., 2007), however, these features could also arise from centrally concentrated starbursts rather than AGN.

PAHs are large molecules of $\sim$10–100s of carbon atoms joined in aromatic bonds, and are often considered to be small dust grains that may even be the building blocks of larger carbonaceous dust (see e.g. Tielens, 2010, for more details about these molecules). In star-forming galaxies, the MIR spectrum is dominated by the broad-emission features arising from vibrational transitions in PAHs (Smith et al., 2007). These features can be used as SFR tracers as these molecules are mainly excited by UV photons (see Calzetti et al., 2007, and this volume, Chapters 6 and 7). However, there is clear evidence that these features and the PAHs that cause them are affected by AGN. There is a strong anti-correlation between the equivalent width of the PAH lines (such as 7.7μm or 11.8μm features) and the presence of high ionisation MIR emission lines like [Nev] 14.3μm (Genzel et al., 1998), indicating that the PAH equivalent width can be used as an AGN indicator (e.g. Genzel et al., 1998; Spoon et al., 2007).

In nearby resolved HII regions it is clear that hydrogen-ionising continuum destroys PAHs. Yet, there are also suggestions that the anti-correlation of the PAH equivalent widths with the presence of an AGN could be predominantly due to the dilution of the PAH emission by the underlying AGN MIR continuum, and not by the destruction of PAHs by the hard-ionising continuum of the AGN (Alonso-Herrero et al., 2014). If this is the case, it suggests that the PAH emission-line flux can still be used as a SFR tracer if it can be separated from the underlying AGN continuum (see e.g. Alonso-Herrero et al., 2012).

While the MIR is one of the more robust wavelength ranges for the identification of AGN, using the same wavelengths for the determination of SFRs for the host galaxies is difficult. One issue is that, as a wavelength range that can be dominated by AGN emission, the subtraction of the AGN emission to measure the relatively weaker emission from star formation is dominated by the uncertainty in what the intrinsic shape of the AGN continuum is. The blue part of the MIR range can be dominated by attenuation by dust in the surrounding torus and host galaxy, with a range of shapes observed, and some correlation with AGN type (i.e. more attenuation in Type-II as compared to Type-I, Alonso-Herrero et al. (2012)). At the red end of the MIR it is uncertain at what point does the torus emission turn-over and the host galaxy emission dominate (see Section 10.7, or models by Hönig et al., 2006).

In measuring the SFR in AGN host galaxies using the MIR there are two possible methods. The first is to use a pure SFR tracer such as the PAH emission-line fluxes or other line fluxes like [NeII] 12.4 μm, making sure to remove any underlying AGN continuum (e.g. Lutz et al., 2008). The second is to use a model or template AGN MIR spectrum (e.g. Polletta et al., 2007) to fit to the observed MIR spectrum and then determine the SFR using the residual rest-frame 24 μm, as discussed in the previous chapters (see work by Alonso-Herrero et al., 2001, and subsequent papers). In both cases, it is possible to use another tracer to determine the contribution of the AGN to the MIR, such as the X-ray emission (see e.g. Lutz et al., 2004), however, they are both limited in our understanding of the intrinsic shape of the AGN MIR continuum.

10.7 Far-Infrared Emission

The FIR ($\sim$30–1000 μm) emission in galaxies arises from dust heated by diffuse radiation, mostly UV due to the steep wavelength dependence of dust opacity. The FIR is used as a SFR tracer in a bolometric approach, that is it assumes that the dust absorbs a significant fraction of the UV emitted in galaxies which arises predominantly from massive stars (see e.g. Kennicutt and Evans, 2012, and Chapters 4, 6, and 7). The main issue with using the FIR as a SFR tracer in the host galaxies of AGN is the unknown contribution of the AGN emission to the diffuse dust heating.

Most works take the FIR in AGN hosts to still be a proxy for the SFR (see e.g. Rosario et al., 2012; Bernhard et al., 2016). This is based on the assumption that the AGN infrared emission is dominated by the torus, meaning that the AGN emission peaks in the MIR and falls off past $\sim$ 30μm (see e.g. Hönig et al., 2006; Alonso-Herrero et al., 2012, and Section 10.6). There is strong evidence for this in QSOs and AGN that dominate their host galaxy at all wavelengths (see Fig. 10.1 and e.g. Netzer et al., 2007; Hatziminaoglou et al., 2010; Mullaney et al., 2011). While AGNs show bluer MIR colours and warmer MIR-FIR colours than inactive starforming galaxies, there is a consistency in the sub-mm colours of AGNs and inactive starburst host galaxies (Rodriguez Espinosa et al., 1987; Hatziminaoglou et al., 2010; Rosario et al., 2012), a consistency in the infrared spectrum of QSOs once a reasonable host galaxy spectrum has been subtracted (Netzer et al., 2007), and even examples of rare QSOs with pure-AGN dominated infrared spectra (Hony et al., 2011).

However, it is also clear that dusty clouds exposed to a more diluted or absorbed form of the AGN radiation field will also emit at the same FIR wavelengths. The narrow-line region clouds (NLR; Section 10.5) must have dust (the lack of strong iron and silicate lines indicates metals must be depleted), and therefore must emit in the infrared (Groves et al., 2006). What prevents the NLR of AGN being a dominant source of FIR emission is that the NLR only covers a small fraction of the AGN ($\sim$5–10% in most AGN), and therefore cannot be bolometrically dominant (see arguments in Netzer et al., 2007). Therefore, our current understanding, and typical assumption, is that the dusty torus covers a large fraction of the AGN (30–50%), the NLR a smaller fraction ($\sim$5–10%) with the remainder escaping directly as UV emission (ignoring the BLR) from the host galaxy (Mor et al., 2009). This then leads to the AGN infrared emission being dominated by the torus with a smaller contribution from the NLR.

In any case, the FIR in AGN-host galaxies can still be contaminated by AGN emission, especially in the 30–100μm regime in luminous AGN. This means some contribution of the AGN to the FIR must be subtracted using AGN templates (see e.g. Netzer et al., 2007; Sajina et al., 2008; Mullaney et al., 2011; Symeonidis et al., 2016), which can either be fit to observed MIR spectra and photometry or scaled by the observed correlation of the AGN infrared with X-rays (see e.g. Lutz et al., 2004; Ichikawa et al., 2017). The main issue is the choice of AGN infrared template, there are suggestions that the FIR spectrum of AGN is much cooler than simple torus models suggest (Symeonidis et al., 2016), or may be luminosity dependent (Mullaney et al., 2011). Yet most current models and results appear to indicate that for most AGN-host galaxies, the FIR can be considered a robust tracer of the SFR, and only with the more luminous AGN (where such contamination is obvious in either the MIR or X-rays) should the FIR be considered only as an upper limit or even poor measure for the SFR of the AGN host.

10.8 Radio Continuum

The radio emission in galaxies arises from the combination of synchrotron radiation and thermal or Brehmsstrahlung radiation. Both emission mechanisms are associated with star formation: the thermal emission arises from HII regions and diffuse ionised gas, while the synchrotron arises from supernovae (Condon, 1992). There is a strong correlation of the radio luminosity with the FIR luminosity, with little scatter over several orders of magnitude (Helou et al., 1985; Bell, 2003). This correlation has been used as both further evidence and calibration for the radio luminosity as a SFR tracer. The radio is considered to be an unbiased tracer of the SFR as it is unaffected by extinction (such as optical-UV measures) or limited by a bolometric assumption (such as the FIR).

For AGNs there is a dichotomy with the radio emission, with two distinct populations appearing (see Fig. 10.1): radio-loud AGNs (RLAGNs) and radio-quiet AGNs (RQAGNs). In RLAGNs the radio emission arises from synchrotron emission from jets of relativistic plasma accelerated and collimated through magnetic fields (Urry and Padovani, 1995). In RQAGNs, the mechanism which creates the emission is still uncertain. There is currently still some debate whether RLAGNs and RQAGNs are actually two distinct populations. There is a large body of evidence which supports the idea of two separate populations (e.g. Kimball et al., 2011; Padovani et al., 2015), but some have suggested that the radio luminosity function of AGNs is more continuous (e.g. Bonchi et al., 2013)

In RLAGNs, the radio emission is far in excess to what would be expected from the other wavelengths for inactive galaxies, and this excess is a clear and robust indicator of such AGN. RLAGNs can be selected and classified in comparison with optical photometry (e.g. Best et al., 2005), relative to the FIR emission (e.g. Morić et al., 2010), or even against X-ray emission for RLAGN/RQAGN classification (e.g. Tasse et al., 2011).

There is a further separation within the RLAGNs: high-excitation RLAGN, which show X-ray, optical and infrared indications for AGN (such as Quasars) and possibly star-forming galaxy hosts; and low-excitation RLAGNs, that appear to be quiescent early-type galaxies at all other wavelengths apart from the radio (see e.g. Best and Heckman, 2012). High-excitation RLAGNs also appear to be much brighter in radio luminosity ($L_{1.4\,\mathrm{GHz}} > 10^{26}\,\mathrm{W\,Hz^{-1}}$) than low-excitation RLAGNs, achieving radio luminosities that would require unphysical SFRs. In both excitation cases, the use of the radio emission in RLAGNs to trace the SFR is fraught with difficulty due to the overwhelming brightness of the AGN relative to any star formation.

In RQAGNs the origin of the radio emission has been associated with both star formation in the host galaxy and non-thermal processes due to the AGN (see e.g. Condon et al., 2013; Zakamska et al., 2016, and references therein). If the radio emission from RQAGN is dominated by star formation, then the radio emission can still be used as a SFR tracer as discussed in Chapters 7 and 5, and has been used as such in several surveys (Kimball et al., 2011; Bonzini et al., 2015). In nearby Seyfert galaxies, radio jets can be seen at the centres of galaxies (see e.g. Rampadarath et al., 2015) and VLBI studies have found that even in some radio-quiet QSOs such jets can contribute significantly (>50%) to the radio emission (Herrera Ruiz et al., 2016). The radio luminosity has also appeared to correlate with narrow-line region kinematics suggesting the radio emission is associated with the non-thermal processes (winds or jets) in the AGN (Zakamska and Greene, 2014). Conversely, in many RQAGNs a correlation is observed between the radio and FIR emission, with radio/FIR ratios similar to that observed in normal star-forming galaxies, suggesting that the radio emission is dominated by star formation (Morić et al., 2010; Bonzini et al., 2015).

Therefore, in radio-loud AGNs, the radio emission is a clear diagnostic of the presence of an AGN in a galaxy, yet also prevents the use of the radio emission as a SFR tracer. In radio-quiet AGNs (identified through the X-ray or optical), the radio emission can be possibly used as a SFR tracer, especially in weaker AGN. However, as with the FIR (previous section), caution must be used as there is a still uncertain contribution of the AGN to the radio emission in these objects.

10.9 Summary

AGNs emit strongly across all wavelengths and, therefore, impact all SFR tracers. However, as nuclear sources their emission is centrally concentrated, and results in a different multi-wavelength spectrum from that of star formation, so AGN emission can be spatially and spectrally distinguished from star formation under the correct circumstances.

As the AGN emission can vary due to both internal parameters (such as the accretion rate and black hole mass) and external parameters (such as the inclination and obscuring gas and dust column), no simple formulae exist for subtracting the AGN to determine the SFR of the AGN-host galaxy. However, at each wavelength methods exist to minimise or subtract the AGN contribution to that wavelength so that it can be used as a SFR tracer, which we have discussed in this chapter. In general, the FIR emission and radio (in radio quiet AGN) seem to be least affected by AGN emission and thus the most direct tracers of the SFR in AGN-host galaxies. However, caution is still advised as the actual contribution of the AGN to these wavelengths is still currently debated and uncertain.

References

Aird, J., Coil, A. L., and Georgakakis, A. 2017. X-Rays across the Galaxy Population – I. Tracing the Main Sequence of Star Formation. *Monthly Notices of the Royal Astronomical Society*, **465**(3), 3390–3415.

Alexander, D. M., and Hickox, R. C. 2012. What Drives the Growth of Black Holes? *New Astronomy Reviews*, **56**(4), 93–121.

Alonso-Herrero, A., Ramos Almeida, C., Esquej, P. et al. 2014. Nuclear 11.3 μm PAH Emission in Local Active Galactic Nuclei. *Monthly Notices of the Royal Astronomical Society*, **443**(3), 2766–2782.

Alonso-Herrero, A., Pereira-Santaella, M., Rieke, G. H. et al. 2012. Local Luminous Infrared Galaxies. II. Active Galactic Nucleus Activity from Spitzer/Infrared Spectrograph Spectra. *Astrophysical Journal*, **744**(1), 2.

Alonso-Herrero, A., Quillen, A. C., Simpson, C. et al. 2001. The Nonstellar Infrared Continuum of Seyfert Galaxies. *Astronomical Journal*, **121**(3), 1369–1384.

Antonucci, R. 1993. Unified Models for Active Galactic Nuclei and Quasars. *Annual Review of Astronomy and Astrophysics*, **31**(Jan.), 473–521.

Assef, R. J., Stern, D., Kochanek, C. S. et al. 2013. Mid-Infrared Selection of Active Galactic Nuclei with the Wide-Field Infrared Survey Explorer. II. Properties of WISE-Selected Active Galactic Nuclei in the NDWFS Boötes Field. *Astrophysical Journal*, **772**(1), 26.

Baldwin, J. A., Phillips, M. M., and Terlevich, R. 1981. Classification Parameters for the Emission-Line Spectra of Extragalactic Objects. *Publications of the Astronomical Society of the Pacific*, **93**(Feb.), 5–19.

Barger, A. J., Cowie, L. L., Mushotzky, R. F. et al. 2005. The Cosmic Evolution of Hard X-ray-Selected Active Galactic Nuclei. *Astronomical Journal*, **129**(2), 578–609.

Bell, Eric F. 2003. Estimating Star Formation Rates from Infrared and Radio Luminosities: The Origin of the Radio-Infrared Correlation. *Astrophysical Journal*, **586**(2), 794–813.

Bernhard, E., Mullaney, J. R., Daddi, E. et al. 2016. An Enhanced Fraction of Starbursting Galaxies among High Eddington Ratio AGNs. *Monthly Notices of the Royal Astronomical Society*, **460**(1), 902–916.

Best, P. N., and Heckman, T. M. 2012. On the Fundamental Dichotomy in the Local Radio-AGN Population: Accretion, Evolution and Host Galaxy Properties. *Monthly Notices of the Royal Astronomical Society*, **421**(2), 1569–1582.

Best, P. N., Kauffmann, G., Heckman, T. M. et al. 2005. A Sample of Radio-Loud Active Galactic Nuclei in the Sloan Digital Sky Survey. *Monthly Notices of the Royal Astronomical Society*, **362**(1), 9–24.

Bonchi, A., La Franca, F., Melini, G. et al. 2013. On the Radio Luminosity Distribution of Active Galactic Nuclei and the Black Hole Fundamental Plane. *Monthly Notices of the Royal Astronomical Society*, **429**(3), 1970–1980.

Bonzini, M., Mainieri, V., Padovani, P. et al. 2015. Star Formation Properties of Sub-mJy Radio Sources. *Monthly Notices of the Royal Astronomical Society*, **453**(1), 1079–1094.

Brandt, W. N., and Alexander, D. M. 2015. Cosmic X-ray Surveys of Distant Active Galaxies. The Demographics, Physics, and Ecology of Growing Supermassive Black Holes. *Astronomy and Astrophysics Reviews*, **23**(Jan.), 1.

Calzetti, D., Kennicutt, R. C., Engelbracht, C. W. et al. 2007. The Calibration of Mid-Infrared Star Formation Rate Indicators. *Astrophysical Journal*, **666**(2), 870–895.

Cecil, G., Dopita, M. A., Groves, B. et al. 2002. Spatial Resolution of High-Velocity Filaments in the Narrow-Line Region of NGC 1068: Associated Absorbers Caught in Emission? *Astrophysical Journal*, **568**(2), 627–638.

Condon, J. J. 1992. Radio Emission from Normal Galaxies. *Annual Review of Astronomy and Astrophysics*, **30**(Jan), 575–611.

Condon, J. J., Kellermann, K. I., Kimball, A. E. et al. 2013. Active Galactic Nucleus and Starburst Radio Emission from Optically Selected Quasi-Stellar Objects. *Astrophysical Journal*, **768**(1), 37.

Croton, D. J., Springel, V., White, S. D. M. et al. 2006. The Many Lives of Active Galactic Nuclei: Cooling Flows, Black Holes and the Luminosities and Colours of Galaxies. *Monthly Notices of the Royal Astronomical Society*, **365**(1), 11–28.

Davies, R. L., Groves, B., Kewley, L. J. et al. 2016. Dissecting Galaxies: Spatial and Spectral Separation of Emission Excited by Star Formation and AGN Activity. *Monthly Notices of the Royal Astronomical Society*, **462**(2), 1616–1629.

Draine, B. T., and Lee, H. M. 1984. Optical Properties of Interstellar Graphite and Silicate Grains. *Astrophysical Journal*, **285**(Oct.), 89.

Elvis, M., Hao, H., Civano, F. et al. 2012. Spectral Energy Distributions of Type 1 Active Galactic Nuclei in the COSMOS Survey. I. The XMM-COSMOS Sample. *Astrophysical Journal*, **759**(1), 6.

Elvis, M., Wilkes, B. J., McDowell, J. C. et al. 1994. Atlas of Quasar Energy Distributions. *Astrophysical Journal Supplement Series*, **95**(Nov.), 1.

Genzel, R., Lutz, D., Sturm, E. et al. 1998. What Powers Ultraluminous IRAS Galaxies? *Astrophysical Journal*, **498**(2), 579–605.

Gibson, R. R., Brandt, W. N., and Schneider, D. P. 2008. Are Optically Selected Quasars Universally X-ray Luminous? X-ray-UV Relations in Sloan Digital Sky Survey Quasars. *Astrophysical Journal*, **685**(2), 773–786.

Groves, B., Dopita, M., and Sutherland, R. 2006. The Infrared Emission from the Narrow Line Region. *Astronomy and Astrophysics*, **458**(2), 405–416.

Groves, B. A., Dopita, M. A., and Sutherland, R. S. 2004. Dusty, Radiation Pressure-Dominated Photoionization. I. Model Description, Structure, and Grids. *Astrophysical Journal Supplement Series*, **153**(1), 9–73.

Haas, M., Siebenmorgen, R., Leipski, C. et al. 2004. Mid-Infrared Selection of AGN. *Astronomy and Astrophysics*, **419**(May), L49–L53.

Hao, L., Spoon, H. W. W., Sloan, G. C. et al. 2005. The Detection of Silicate Emission from Quasars at 10 and 18 Microns. *Astrophysical Journal*, **625**(2), L75–L78.

Hao, L., Weedman, D. W., Spoon, H. W. W. et al. 2007. The Distribution of Silicate Strength in Spitzer Spectra of AGNs and ULIRGs. *Astrophysical Journal*, **655**(2), L77–L80.

Hatziminaoglou, E., Omont, A., Stevens, J. A. et al. 2010. HerMES: Far Infrared Properties of Known AGN in the HerMES Fields. *Astronomy and Astrophysics*, **518**(Jul), L33.

Heckman, T. M., Ptak, A., Hornschemeier, A. et al. 2005. The Relationship of Hard X-ray and Optical Line Emission in Low-Redshift Active Galactic Nuclei. *Astrophysical Journal*, **634**(1), 161–168.

Helou, G., Soifer, B. T., and Rowan-Robinson, M. 1985. Thermal Infrared and Non-thermal Radio: Remarkable Correlation in Disks of Galaxies. *Astrophysical Journal*, **298**(Nov), L7–L11.

Herrera Ruiz, N., Middelberg, E., Norris, R. P. et al. 2016. Unveiling the Origin of the Radio Emission in Radio-Quiet Quasars. *Astronomy and Astrophysics*, **589**(May), L2.

Hönig, S. F., Beckert, T., Ohnaka, K. et al. 2006. Radiative Transfer Modeling of Three-Dimensional Clumpy AGN Tori and Its Application to NGC 1068. *Astronomy and Astrophysics*, **452**(2), 459–471.

Hony, S., Kemper, F., Woods, P. M. et al. 2011. The Spitzer Discovery of a Galaxy with Infrared Emission Solely Due to AGN Activity. *Astronomy and Astrophysics*, **531**(July), A137.

Ichikawa, K., Ricci, C., Ueda, Y. et al. 2017. The Complete Infrared View of Active Galactic Nuclei from the 70 Month Swift/BAT Catalog. *Astrophysical Journal*, **835**(1), 74.

Jahnke, K., and Wisotzki, L. 2003. The B-Band Luminosities of Quasar Host Galaxies. *Monthly Notices of the Royal Astronomical Society*, **346**(1), 304–318.

Jahnke, K., Kuhlbrodt, B., and Wisotzki, L. 2004. Quasar Host Galaxy Star Formation Activity from Multicolour Data. *Monthly Notices of the Royal Astronomical Society*, **352**(2), 399–415.

Jahnke, K., Wisotzki, L., Courbin, F. et al. 2007. Spatial Decomposition of On-Nucleus Spectra of Quasar Host Galaxies. *Monthly Notices of the Royal Astronomical Society*, **378**(1), 23–40.

Karim, A., Schinnerer, E., Martínez-Sansigre, A. et al. 2011. The Star Formation History of Mass-Selected Galaxies in the COSMOS Field. *Astrophysical Journal*, **730**(2), 61.

Kauffmann, G., Heckman, T. M., Tremonti, C. et al. 2003. The Host Galaxies of Active Galactic Nuclei. *Monthly Notices of the Royal Astronomical Society*, **346**(4), 1055–1077.

Kennicutt, R. C., and Evans, N. J. 2012. Star Formation in the Milky Way and Nearby Galaxies. *Annual Review of Astronomy and Astrophysics*, **50**(Sep), 531–608.

Kewley, L. J., Dopita, M. A., Sutherland, R. S. et al. 2001. Theoretical Modeling of Starburst Galaxies. *Astrophysical Journal*, **556**(1), 121–140.

Kewley, L. J., Groves, B., Kauffmann, G. et al. 2006. The Host Galaxies and Classification of Active Galactic Nuclei. *Monthly Notices of the Royal Astronomical Society*, **372**(3), 961–976.

Kimball, A. E., Kellermann, K. I., Condon, J. J. et al. 2011. The Two-Component Radio Luminosity Function of Quasi-Stellar Objects: Star Formation and Active Galactic Nucleus. *Astrophysical Journal*, **739**(1), L29.

Kormendy, J., and Ho, L. C. 2013. Coevolution (Or Not) of Supermassive Black Holes and Host Galaxies. *Annual Review of Astronomy and Astrophysics*, **51**(1), 511–653.

Laurent, O., Mirabel, I. F., Charmandaris, V. et al. 2000. Mid-Infrared Diagnostics to Distinguish AGNs from Starbursts. *Astronomy and Astrophysics*, **359**(July), 887–899.

Leger, A., and D'Hendecourt, L. 1985. Are Polycyclic Aromatic Hydrocarbons the Carriers of the Diffuse Interstellar Bands in the Visible? *Astronomy and Astrophysics*, **146**(1), 81–85.

Levenson, N. A., Sirocky, M. M., Hao, L. et al. 2007. Deep Mid-Infrared Silicate Absorption as a Diagnostic of Obscuring Geometry toward Galactic Nuclei. *Astrophysical Journal*, **654**(1), L45–L48.

Lutz, D., Maiolino, R., Spoon, H. W. W. et al. 2004. The Relation between AGN Hard X-ray Emission and Mid-Infrared Continuum from ISO Spectra: Scatter and Unification Aspects. *Astronomy and Astrophysics*, **418**(May), 465–473.

Lutz, D., Sturm, E., Tacconi, L. J. et al. 2008. Star Formation in the Hosts of High-z QSOs: Evidence from Spitzer PAH Detections. *Astrophysical Journal*, **684**(2), 853–861.

Maccacaro, T., Gioia, I. M., Wolter, A. et al. 1988. The X-ray Spectra of the Extragalactic Sources in the Einstein Extended Medium-Sensitivity Survey. *Astrophysical Journal*, **326**(Mar.), 680.

Madau, P., and Dickinson, M. 2014. Cosmic Star-Formation History. *Annual Review of Astronomy and Astrophysics*, **52**(Aug.), 415–486.

Magorrian, J., Tremaine, S., Richstone, D. 1998. The Demography of Massive Dark Objects in Galaxy Centers. *Astronomical Journal*, **115**(6), 2285–2305.

Mor, R., Netzer, H., and Elitzur, M. 2009. Dusty Structure Around Type-I Active Galactic Nuclei: Clumpy Torus Narrow-Line Region and Near-Nucleus Hot Dust. *Astrophysical Journal*, **705**(1), 298–313.

Morić, I., Smolčić, V., Kimball, A. et al. 2010. A Closer View of the Radio-FIR Correlation: Disentangling the Contributions of Star Formation and Active Galactic Nucleus Activity. *Astrophysical Journal*, **724**(1), 779–790.

Mullaney, J. R., Alexander, D. M., Goulding, A. D. et al. 2011. Defining the Intrinsic AGN Infrared Spectral Energy Distribution and Measuring Its Contribution to the Infrared Output of Composite Galaxies. *Monthly Notices of the Royal Astronomical Society*, **414**(2), 1082–1110.

Nenkova, M., Ivezić, Ž., and Elitzur, M. 2002. Dust Emission from Active Galactic Nuclei. *Astrophysical Journal*, **570**(1), L9–L12.

Netzer, H., Lutz, D., Schweitzer, M. et al. 2007. Spitzer Quasar and ULIRG Evolution Study (QUEST). II. The Spectral Energy Distributions of Palomar-Green Quasars. *Astrophysical Journal*, **666**(2), 806–816.

Osterbrock, D. E. 1978. Physical State of the Emission-Line Region. *Physica Scripta*, **17**(Mar), 285–292.

Osterbrock, D. E., and Ferland, G. J. 2006. *Astrophysics of Gaseous Nebulae and Active Galactic Nuclei.*

Padovani, P., Bonzini, M., Kellermann, K. I. et al. 2015. Radio-Faint AGN: A Tale of Two Populations. *Monthly Notices of the Royal Astronomical Society*, **452**(2), 1263–1279.

Peterson, B. M., Ferrarese, L., Gilbert, K. M. et al. 2004. Central Masses and Broad-Line Region Sizes of Active Galactic Nuclei. II. A Homogeneous Analysis of a Large Reverberation-Mapping Database. *Astrophysical Journal*, **613**(2), 682–699.

Peterson, B. M. 1993. Reverberation Mapping of Active Galactic Nuclei. *Publications of the Astronomical Society of the Pacific*, **105**(Mar.), 247.

Pier, E. A., and Krolik, J. H. 1992. Infrared Spectra of Obscuring Dust Tori around Active Galactic Nuclei. I. Calculational Method and Basic Trends. *Astrophysical Journal*, **401**(Dec.), 99.

Polletta, M., Tajer, M., Maraschi, L. et al. 2007. Spectral Energy Distributions of Hard X-ray Selected Active Galactic Nuclei in the XMM-Newton Medium Deep Survey. *Astrophysical Journal*, **663**(1), 81–102.

Rampadarath, H., Morgan, J. S., Soria, R. et al. 2015. A High-Resolution Wide-Field Radio Survey of M51. *Monthly Notices of the Royal Astronomical Society*, **452**(1), 32–53.

Rodriguez E., J. M., Rudy, R. J., and Jones, B. 1987. Star Formation in Seyfert Galaxies. *Astrophysical Journal*, **312**(Jan.), 555.

Rosario, D. J., Santini, P., Lutz, D. et al. 2012. The Mean Star Formation Rate of X-ray Selected Active Galaxies and Its Evolution from z ~2.5: Results from PEP-Herschel. *Astronomy and Astrophysics*, **545**(Sep.), A45.

Sajina, A., Yan, L., Fadda, D. et al. 2012. Spitzer- and Herschel-Based Spectral Energy Distributions of 24 μm Bright z ~0.3-3.0 Starbursts and Obscured Quasars. *Astrophysical Journal*, **757**(1), 13.

Sajina, A., Yan, L., Lutz, D. et al. 2008. Spitzer Mid-Infrared Spectroscopy of Infrared Luminous Galaxies at z ~2. III. Far-IR to Radio Properties and Optical Spectral Diagnostics. *Astrophysical Journal*, **683**(2), 659–682.

Schartmann, M., Meisenheimer, K., Camenzind, M. et al. 2008. Three-Dimensional Radiative Transfer Models of Clumpy Tori in Seyfert Galaxies. *Astronomy and Astrophysics*, **482**(1), 67–80.

Seyfert, C. K. 1943. Nuclear Emission in Spiral Nebulae. *Astrophysical Journal*, **97**(Jan.), 28.

Siebenmorgen, R., Haas, M., Kruegel, E. et al. 2005. Discovery of 10 Mu.m Silicate Emission in Quasars: - Evidence of the AGN Unification Scheme. *Astronomische Nachrichten*, **326**(Aug), 556–556.

Smith, J. D. T., Draine, B. T., Dale, D. A. et al. 2007. The Mid-Infrared Spectrum of Star-Forming Galaxies: Global Properties of Polycyclic Aromatic Hydrocarbon Emission. *Astrophysical Journal*, **656**(2), 770–791.

Spoon, H. W. W., Marshall, J. A., Houck, J. R. et al. 2007. Mid-Infrared Galaxy Classification Based on Silicate Obscuration and PAH Equivalent Width. *Astrophysical Journal*, **654**(1), L49–L52.

Springel, V., Di Matteo, T., and Hernquist, L. 2005. Modelling Feedback from Stars and Black Holes in Galaxy Mergers. *Monthly Notices of the Royal Astronomical Society*, **361**(3), 776–794.

Stefanon, M., Yan, H., Mobasher, B. et al. 2017. CANDELS Multi-Wavelength Catalogs: Source Identification and Photometry in the CANDELS Extended Groth Strip. *Astrophysical Journal Supplement Series*, **229**(2), 32.

Stern, D., Eisenhardt, P., Gorjian, V. et al. 2005. Mid-Infrared Selection of Active Galaxies. *Astrophysical Journal*, **631**(1), 163–168.

Sturm, E., Lutz, D., Tran, D. et al. 2000. ISO-SWS Spectra of Galaxies: Continuum and Features. *Astronomy and Astrophysics*, **358**(Jun), 481–493.

Symeonidis, M., Giblin, B. M., Page, M. J. et al. 2016. AGN Are Cooler Than You Think: The Intrinsic Far-IR Emission from QSOs. *Monthly Notices of the Royal Astronomical Society*, **459**(1), 257–276.

Tasse, C., Röttgering, H., and Best, P. N. 2011. The Dependence of X-ray AGN Activity on Host Galaxy Properties and Environment. *Astronomy and Astrophysics*, **525**(Jan.), A127.

Tielens, A. G. G. M. 2010. *The Physics and Chemistry of the Interstellar Medium.*

Ulrich, M.-H., Maraschi, L., and Urry, C. M. 1997. Variability of Active Galactic Nuclei. *Annual Review of Astronomy and Astrophysics*, **35**(Jan.), 445–502.

Urry, C. M., and Padovani, P. 1995. Unified Schemes for Radio-Loud Active Galactic Nuclei. *Publications of the Astronomical Society of the Pacific*, **107**(Sep), 803.

Veilleux, S., and Osterbrock, D. E. 1987. Spectral Classification of Emission-Line Galaxies. *Astrophysical Journal Supplement Series*, **63**(Feb.), 295.

Zakamska, N. L., and Greene, J. E. 2014. Quasar Feedback and the Origin of Radio Emission in Radio-Quiet Quasars. *Monthly Notices of the Royal Astronomical Society*, **442**(1), 784–804.

Zakamska, N. L., Lampayan, K., Petric, A. et al. 2016. Star Formation in Quasar Hosts and the Origin of Radio Emission in Radio-Quiet Quasars. *Monthly Notices of the Royal Astronomical Society*, **455**(4), 4191–4211.

Zakamska, N. L., Strauss, M. A., Krolik, J. H. et al. 2006. Type II Quasars from the Sloan Digital Sky Survey. V. Imaging Host Galaxies with the Hubble Space Telescope. *Astronomical Journal*, **132**(4), 1496–1516.

11

High-Energy Star-Formation Rate Indicators

ANDREAS ZEZAS

11.1 Introduction

All star-formation rate (SFR) indicators discussed in the previous chapters are based on the measurement of the direct or reprocessed stellar emission. However, an alternative way to probe stellar populations is though stellar death: the stellar remnants (compact object populations, supernova remnants; SNRs) bear the imprint of the star-formation activity responsible for the formation of their parent stellar populations. These objects are primarily traced by their high-energy emission produced from X-ray binaries, long and short γ-ray bursts, SNRs, and their effect on their surrounding interstellar medium (ISM). More recently the detection of gravitational waves from merging compact object systems opened a new window in our exploration of compact object populations and hence the stellar populations that are responsible for their formation. The formation timescales of these objects range from a few Myr (e.g. core-collapse supernovae and their SNRs, long γ-ray bursts, high-mass X-ray binaries), to several Gyr (e.g. type Ia supernovae and their SNRs, gravitational-wave sources, low-mass X-ray binaries). Therefore, these new probes of stellar populations can be used as probes of star-formation activity in different timescales.

11.2 X-ray Emission from Galaxies

The high-energy emission of galaxies arises from two components: discrete sources and diffuse emission. The discrete sources are associated with X-ray binaries, SNRs, and in the case of our nearest galaxies, stars. The diffuse emission consists of a component associated with the hot ($\sim 10^6$–10^7K) ISM, and unresolved emission from fainter discrete sources. It is clear therefore, that the X-ray emission from galaxies traces their stellar populations. This has been recognised from the first studies of the integrated galactic X-ray emission with the *Einstein Observatory*[1] which showed strong correlations between their X-ray luminosity and B-band or K-band luminosity tracing young and old stellar populations respectively (Fabbiano, 1989). Subsequent studies with *ROSAT*[2] showed strong

[1] *Einstein Observatory (HEAO2)*; https://heasarc.gsfc.nasa.gov/docs/einstein/heao2_about.html
[2] Röntgensatellit; www.mpe.mpg.de/ROSAT

correlation of the diffuse, thermal, X-ray luminosity of star-forming galaxies with their B-band luminosity (Stevens and Strickland, 1998), also indicating a direct link with young stellar populations.

However, it was only with the launch of the *Chandra X-ray Observatory*[3] and *XMM-Newton*[4] in 1999 that it became possible to study in detail the connection between the X-ray emission of galaxies and their stellar content. This is because of the superb spatial resolution of *Chandra* (0.5″) which allowed us to resolve the discrete X-ray sources (typically X-ray binaries, but also SNRs, active galactic nuclei, or stars) and study their populations in our local universe, or detect the X-ray emission from normal galaxies at cosmological distances. *Chandra*'s spatial resolution, matching that of optical, ultraviolet (UV), and near-infrared observations, also enables the identification of multi-wavelength counterparts to the X-ray sources, a key parameter for determining their nature. On the other hand, the ∼3 times larger effective area of *XMM-Newton* allowed the study of the integrated emission of large numbers of galaxies, and often the spectral decomposition of the hot gaseous component (associated with optically thin thermal plasma) and the X-ray binary component (associated with a power-law spectral-model component).

These investigations produced scaling relations between: the soft (typically 0.5–2.0 keV) or hard (typically 2.0–8.0 keV) band luminosity of star-forming galaxies and their SFR and/or stellar mass. For the nearest galaxies it has also been possible to study the correlation between the number of X-ray binaries and SFR in star-forming galaxies; or the number of X-ray binaries and stellar mass in galaxy bulges or elliptical galaxies (Gilfanov, 2004; Mineo et al., 2014). These scaling relations have been extended beyond our local universe, through the detection of individual galaxies at redshifts of $z \sim 0.5$–1.0 (typically, and in some cases up to $z \sim 3.0$; Lehmer et al., 2016; Aird et al., 2017), or stacking analysis at higher redshifts ($z \sim 2$–3; Basu-Zych et al., 2013; Fornasini et al., 2018).

The main effort of the community in the past decade has been focused on calibrating the galactic X-ray emission and source populations against SFR and stellar mass. A significant aspect of this effort is related to the theoretical study of X-ray binaries either as individual systems (especially in the case of very luminous systems) or as populations through population synthesis modelling with the goal of understanding the parameters that affect the evolution of X-ray binaries and their populations.

11.2.1 Discrete Sources

Even from the early days of X-ray astronomy it became clear that our Galaxy hosts a large population of discrete X-ray sources associated with stellar remnants (black holes, neutron stars, white dwarfs) which are accreting material from a companion star. These X-ray binaries exhibit significant variability in the form of luminosity and/or spectral variations, the most extreme manifestation of which is transient behaviour. During outbursts X-ray

[3] *Chandra X-ray Observatory (Chandra);* https://chandra.harvard.edu
[4] X-ray Multi Mirror Mission (XMM-Newton); https://sci.esa.int/web/xmm-newton

binaries have typical luminosities of $>10^{37}$ erg s^{-1}(although there are also sub-luminous outbursts with luminosities $\sim 10^{36}$ erg s^{-1}). The duration of these outbursts varies from a few days to several years, depending on the characteristics of each system. The same holds for their duty cycle: some sources show recurrent outbursts, while others have only exhibited only one or two outbursts during the ~50 year long history of X-ray astronomy (e.g. Done et al., 2007; Tetarenko et al., 2016). In addition, there is a population of persistent sources with typical luminosities $\sim 10^{36}$–10^{38} erg s^{-1}, which however, also show intensity and spectral variations.

X-ray binaries can be characterised on the basis of their donor stars into two broad categories: those accreting material from an early-type star (OB star, or a supergiant), and those accreting from a later-type star (typically of M,K spectral type). These are referred-to as High-Mass X-ray binaries (HMXBs) and Low-Mass X-ray binaries (LMXBs) respectively since their donor stars have masses typically above $\sim 8M_\odot$ and below ~ 1–$2M_\odot$; Intermediate-Mass X-ray binaries (with donor star mass between 2–$8M_\odot$) are more rare due to evolutionary reasons.

X-ray binaries are the result of complex processes involving the evolution of the individual stars in the binary system (on their nuclear timescale) and the parallel evolution of the system's orbital parameters. The latter is driven by the loss and/or exchange of mass and angular momentum between the two stars (e.g. Bhattacharya and van den Heuvel, 1991; Eggleton, 2006; Tauris and van den Heuvel, 2006) and effects such as tidal evolution, magnetic braking (Ivanova and Taam, 2003), common envelope evolution (Ivanova et al., 2013), and supernova kicks (Podsiadlowski et al., 2004). These play a dramatic role on the fate of a binary system and they determine whether it will merge, disassemble, and whether and when, the mass-transfer rate onto the compact object is high enough to produce observable X-ray emission. The mass transfer can take place either via stellar winds (in the case of early-type donor stars), or through Roche-lobe overflow (RLOF; in the case of supergiant donors, or late-type stars).

HMXBs are short-lived since their lifetime is driven by the relatively short lifetime of their donor stars (~ 10–100 Myr). Typically the onset of the X-ray emitting phase in HMXBs takes place soon after the formation of the compact object (at the end of the life of the more massive of the two stars). This is because of the short nuclear timescale of the donor stars resulting in fast evolution to a supergiant phase, or their strong stellar winds, initiating mass transfer immediately after the formation of the compact object.

On the other hand, the longer lifetimes of the LMXB donor stars allow them to survive for longer periods and they provide the necessary time for orbital decay to bring the star and the compact object close enough to initiate mass transfer through RLOF. The latter is required since the stellar winds of low-mass stars are too weak to result in any appreciable mass transfer. As a result the formation timescales of LMXBs are much longer than that of HMXBs.

Because the HMXBs form in timescales of ~10–100 Myr, while LMXBs form in Gyr timescales (Fragos et al., 2013) the relative number of the two sub-populations depends on the star-formation history of the galaxy. In our Galaxy we have discovered ~82 and

HMXBs and ~108 LMXBs respectively (Krivonos et al., 2012), the majority of which are transients with duty cycles ranging from months to years (or more for those that have exhibited only one outburst). In addition to the luminous X-ray binaries, binary systems of massive stars (Wolf-Rayet or O-type) may also produce X-ray emission due to the interaction of their stellar winds and the formation of shocks in their interface (Rauw and Nazé, 2016, colliding wind binaries;, but they are very rare, and they have relatively low luminosity ($\lesssim 10^{36}$ erg s^{-1})). Normal stars also emit X-rays as a result of the magnetic activity on their surface; however, their luminosity typicaly does not exceed $\sim 10^{32}$ erg s^{-1} and therefore are not significant contributors to the integrated X-ray emission of a galaxy.

Other types of discrete X-ray sources are isolated pulsars and SNRs. Isolated rotation-powered pulsars produce sychrotron emission through the acceleration of particles in their magnetosphere. Their non-thermal luminosities range from $\sim 10^{36}$ erg s^{-1}for the younger systems (e.g. Crab pulsar) to $\sim 10^{32}$ erg s^{-1}or less for the older ones ($> 10^5$ yr). In addition to the non-thermal emission there is a thermal component from the surface of the neutron star with luminosity $\sim 10^{30} - 10^{32}$ erg s^{-1}. SNRs are classified into two main categories: those dominated by the thermal emission from the reverse shock (thermal SNRs; e.g. Reynolds, 2008; Vink, 2012) and those dominated by the emission of the central pulsar and its synchrotron nebula (plerions; Gaensler and Slane, 2006). The X-ray luminosity of both types of systems can be up to $\sim 10^{36}$ erg s^{-1}, but the small number of X-ray emitting SNRs (32 detected so far in our Galaxy;[5] Ferrand and Safi-Harb, 2012) make them a small contributor to the overall X-ray luminosity of our Galaxy.

11.2.2 Diffuse Emission

Imaging observations of nearby galaxies (first with the *Einstein Observatory*, later with *ROSAT*, *ASCA*,[6] and currently with the *Chandra X-ray Observatory* and *XMM-Newton*) showed that in addition to the discrete X-ray source component, there is also diffuse emission which in fact can dominate the X-ray emission of star-forming galaxies in the soft (0.5–2.0 keV) band (e.g. Fabbiano, 2006).

The diffuse, soft X-ray, emission is characterised by an optically thin thermal plasma spectrum with typical temperature in the ~0.5–2.0 keV range (Grimes et al., 2005; Mineo et al., 2012b; Smith et al., 2018). This hot gaseous component results from the cumulative effects of supernovae and stellar winds which deposit mechanical energy to the interstellar medium (ISM) and heat it to multi-million degree temperatures (Grimes et al., 2005; Strickland and Heckman, 2009). The luminosity of this component depends on the SFR of the galaxy, the degree of thermalisation of the mechanical energy injected in the galaxy's ISM, and the retention of the resulting hot ISM. This component can take the form of an outflow along the galaxy's minor axis (a superwind; Strickland et al., 2004) which in

[5] www.physics.umanitoba.ca/snr/SNRcat/
[6] Advanced Satellite for Cosmology and Astrophysics; https://heasarc.gsfc.nasa.gov/docs/asca/

some cases can dominate the soft X-ray emission of the galaxy and it may have important implications for its evolution (e.g. Veilleux et al., 2005).

In addition to the diffuse thermal emission there is also a diffuse non-thermal component which dominates at energies above $\sim$2 keV. This component results from the unresolved emission of fainter X-ray binaries. Its spectrum is generally described by a power law with typical photon index of $\Gamma \sim$1.0–2.5. Since the unresolved X-ray binaries can be comprised of HMXBs, as well as, LMXBs, the luminosity of this component scales with the SFR and the stellar mass of a galaxy.

11.2.3 X-ray Emission above 10 keV

X-ray binaries are strong X-ray sources at energies up to $\sim$100 keV. Their spectrum and luminosity depends on their accretion state (which does change) and the nature of the compact object (e.g. Remillard and McClintock, 2006; Done et al., 2007). With the launch of the *NuSTAR*[7] hard X-ray telescope it has been possible to image for the first time nearby galaxies at energies above 10 keV (see e.g. Hornschemeier et al., 2016, for an overview). These observations showed that their hard X-ray emission consists of bright X-ray sources and a diffuse unresolved component. Characterisation of the discrete X-ray sources on the basis of their X-ray spectral shape and hardness ratios (or X-ray colours) indicates that at luminosities above $\sim 5 \times 10^{39}$ erg s^{-1}(in the 4–25 keV band) they are dominated by black-hole X-ray binaries while the fraction of neutron-star X-ray binaries increases at lower luminosities (Vulic et al., 2018). Furthermore, a study of the spatial distribution of the different X-ray binary populations in M31, showed that accreting pulsars are preferentially associated with the spiral arms as expected for a population of young high-mass X-ray binaries. On the other hand the bulge region of the galaxy hosts predominantly neutron-star X-ray binaries, while black-hole systems are present both in the disk and the bulge (e.g. Vulic et al., 2018).

The diffuse component is expected to consist of a population of unresolved X-ray binaries and diffuse emission produced by inverse-Compton scattering of optical/infrared photons by relativistic electrons associated with SNRs (Lacki and Thompson, 2013). Analysis of the existing data in the case of NGC 253 show that all the diffuse emission can be accounted for by faint X-ray binaries and there is no evidence for an inverse-Compton component, at least at the sensitivity of the existing observations (Wik et al., 2014).

11.3 Scaling Relations between X-ray Emission and Stellar Populations

Since the discovery of X-ray emission from galaxies, significant efforts have been made to explore the correlation of their X-ray luminosity with different metrics of their stellar content (e.g. SFR, stellar mass; Fabbiano, 1989, 2006). The most recent such studies are

[7] Nuclear Spectroscopic Telescope Array; www.nustar.caltech.edu

based on *Chandra Observatory* and *XMM-Newton* data which have observed large numbers of nearby galaxies. These correlations are reported as scaling relations between either the X-ray luminosity or the number of X-ray binaries in a galaxy, and its SFR and/or stellar mass. A summary of these scaling relations is presented in Table 11.1.

Grimm et al. (2003) first reported a linear correlation between the number of X-ray binaries in star-forming galaxies (or equivalently their integrated X-ray luminosity), and their SFR. In a follow-up work Gilfanov (2004) reported a similar correlation between the number of LMXBs in early-type galaxies, or their integrated X-ray luminosity, and their stellar mass. The correlation of X-ray luminosity and SFR was refined by Mineo et al. (2012a) who expanded the galaxy sample, and took care to focus on galaxies with high specific SFR ($\mathrm{sSFR} \equiv \mathrm{SFR}/\mathrm{M_{star}} > 10^{-10}\,\mathrm{M_\odot\,yr^{-1}\,M_\odot^{-1}}$). The latter ensures that the sample is dominated by HMXBs which are expected to correlate with SFR, instead of LMXBs which correlate with older stellar populations and are better traced by the stellar mass. Subsequent studies (e.g. Lehmer et al., 2008, 2010; Mineo et al., 2014) extended these $\mathrm{L_X}$–SFR scaling relations to SFR as high as $\sim\!10^3\mathrm{M_\odot\,yr^{-1}}$, by including (ultra) luminous infrared galaxies ((U)LIRGs) and objects detected in deep X-ray surveys, and explored their redshift evolution up to $z \sim 2$ (Basu-Zych et al., 2013; Lehmer et al., 2016; Aird et al., 2017). These studies also introduced joint relations between X-ray luminosity, SFR, and stellar mass, which account for both the HMXB *and* the LMXB components. The SFRs in these studies are generally based on infrared (IR) SFR indicators, although in some cases IR-UV hybrid (see Chapter 6) indicators, or even SED-fitting (see Chapter 8) have been used (e.g. Basu-Zych et al., 2013; Aird et al., 2017).

More recently, extension of these scaling relations to energies above 10.0 keV based on *NuSTAR* observations, showed that while the hard X-ray luminosity of normal galaxies follows the general scaling relations with SFR and sSFR (albeit with different scaling factors than in the 0.5–2.0 keV band), dwarf galaxies with large populations of black holes or ultra-luminous X-ray sources (ULXs[8]) are more efficient producers of X-ray emission (Vulic et al., 2018).

One important trait of the $\mathrm{L_X}$–SFR–$\mathrm{M_{star}}$ relations is that while there is a rather tight correlation at high luminosities and SFRs, there is significant scatter in the low SFR, or high specific SFR regimes. This can be the result of stochastic effects, source variability, and/or variations in the star-formation history from galaxy to galaxy. The X-ray emission of galaxies with integrated X-ray luminosities even in the 10^{39}–$10^{40}\,\mathrm{erg\,s^{-1}}$ range can be dominated by a few luminous sources (and quite often a single source). These sources are in the high-luminosity end of the X-ray luminosity function of X-ray binaries (see Section 11.4) and therefore their number is dominated by stochastic sampling of the X-ray luminosity function (XLF) (c.f. Chapters 2 and 5; Gilfanov et al., 2004; Mineo et al., 2012a; Anastasopoulou et al., 2019). Furthermore, the significant variability of X-ray binaries will further increase the scatter of the X-ray luminosity in these galaxies

[8] ULXs are generally defined as sources with X-ray luminosities above $10^{39}\,\mathrm{erg\,s^{-1}}$, i.e. the Eddington luminosity for a $\sim\!5\mathrm{M_\odot}$ black hole (see Kaaret et al., 2017, for an extensive review).

Table 11.1. *Scaling relations between X-ray emission, SFR, and stellar mass*

Scaling Relation	Scatter (dex)	β	α	Band	Notes
$\log \frac{L_X}{10^{40} \mathrm{erg\,s^{-1}}} = \beta + \log \frac{L_{FIR}}{10^{43} \mathrm{erg\,s^{-1}}}$	0.27	-3.68		0.5–2.0 keV	Ranalli et al. (2003) [a]
$\log \frac{L_X}{10^{40} \mathrm{erg\,s^{-1}}} = \beta + \log \frac{L_{1.4\,GHz}}{10^{29} \mathrm{erg\,s^{-1}Hz^{-1}}}$	0.24	11.08		0.5–2.0 keV	Ranalli et al. (2003) [b]
$\log \frac{L_X}{10^{40} \mathrm{erg\,s^{-1}}} = \beta + \log \frac{L_{FIR}}{10^{43} \mathrm{erg\,s^{-1}}}$	0.29	-3.62		2.0–10.0 keV	Ranalli et al. (2003) [a]
$\log \frac{L_X}{10^{40} \mathrm{erg\,s^{-1}}} = \beta + \log \frac{L_{1.4\,GHz}}{10^{29} \mathrm{erg\,s^{-1}Hz^{-1}}}$	0.29	11.13		2.0–10.0 keV	Ranalli et al. (2003) [b]
$\frac{L_X}{\mathrm{erg\,s^{-1}}} = \beta\, SFR$	0.4	2.61×10^{39}		0.5–8.0 keV	Mineo et al. (2012a)
$\frac{L_X}{\mathrm{erg\,s^{-1}}} = \beta\, SFR$	0.4	$(4.0 \pm 0.4) \times 10^{39}$		0.5–8.0 keV	Mineo et al. (2014)
$\frac{L_X^{gas}}{\mathrm{erg\,s^{-1}}} = \beta\, SFR$	0.34	$(5.2 \pm 0.2) \times 10^{38}$		0.5–2.0 keV	Mineo et al. (2012b) [c]
$\frac{L_X}{\mathrm{erg\,s^{-1}}} = \alpha(1+z)^{\gamma} M_{star}$	0.16	$\log\beta = 39.38 \pm 0.03$	$\log\alpha = 29.04 \pm 0.17$	0.5–2.0 keV	Lehmer et al. (2016) [d]
$+\beta(1+z)^{\delta} SFR$		$\delta = 0.99 \pm 0.26$	$\gamma = 3.78 \pm 0.82$		
$\frac{L_X}{erg\,s^{-1}} = \alpha(1+z)^{\gamma} M_{star}$	0.16	$\log\beta = 39.28 \pm 0.05$	$\log\alpha = 29.37 \pm 0.15$	2.0–10.0 keV	Lehmer et al. (2016) [d]
$+\beta(1+z)^{\delta} SFR$		$\delta = 1.31 \pm 0.13$	$\gamma = 2.03 \pm 0.60$		
$\frac{L_X}{\mathrm{erg\,s^{-1}}} = \beta\, SFR + \alpha\, M_{star}$	0.34	$(1.62 \pm 0.22) \times 10^{39}$	$(9.05 \pm 0.37) \times 10^{28}$	2.0–10.0 keV	Lehmer et al. (2010) [e]

(Continued)

Table 11.1. *Continued*

Scaling Relation	Scatter (dex)	β	α	Band	Notes
$\frac{L_X}{erg\,s^{-1}} = \alpha(1+z)^{\gamma} M_{star}$		$\log\beta = 39.50 \pm 0.06$	$\log\alpha = 28.81 \pm 0.08$	2.0–10.0 keV	Aird et al. (2017) [f]
$+\beta(1+z)^{\delta} SFR^{\theta}$		$\delta = 0.67 \pm 0.31$	$\gamma = 3.90 \pm 0.36$		
		$\theta = 0.86 \pm 0.05$			
$\log \frac{L_X}{\mathrm{erg\,s^{-1}}} = \alpha + \beta \log SFR$		(0.83 ± 0.05)	(39.48 ± 0.05)	2.0–10.0 keV	Aird et al. (2017) [g]
$+C \log(1+z)$		$C = 1.34 \pm 0.20$			
$\log \frac{L_X}{\mathrm{erg\,s^{-1}}} = \beta\, SFR + \alpha\, M_{star}$		$39.73^{+0.15}_{-0.10}$	$29.15^{+0.07}_{-0.05}$	0.5–8.0 keV	Lehmer et al. (2019) [h]
$\frac{L_X}{\mathrm{erg\,s^{-1}}} = \beta\, SFR + \alpha\, M_{star}$		$(1.90 \pm 0.84) \times 10^{39}$	$(3.56 \pm 1.16) \times 10^{28}$	4.0–25.0 keV	Vulic et al. (2018) [i]
$\frac{L_X}{\mathrm{erg\,s^{-1}}} = \beta\, SFR + \alpha\, M_{star}$		$(0.51 \pm 0.19) \times 10^{39}$	$(0.34 \pm 0.22) \times 10^{28}$	12.0–25.0 keV	Vulic et al. (2018) [i]

The SFR and stellar mass (M_{star}) are measured in $M_{\odot}\,yr^{-1}$ and $M_{\odot}$ respectively.

[a] Based on ASCA and BeppoSAX data. The far-IR (FIR) luminosity is based on the combination of the 60μ and 100μ IRAS fluxes.

[b] Based on ASCA and BeppoSAX data.

[c] Relation for the thermal gas component based on spectral fitting of the X-ray data.

[d] Sample includes local galaxies and *Chandra* Deep Field-South detections.

[e] Sample includes local galaxies and (U)LIRGS.

[f] Sample includes galaxies in $0.1 < z < 2$ from the Ultra-VISTA (http://ultravista.org/) and CANDELS (Cosmic Assembly Near-infrared Deep Extragalactic Legacy Survey; http://arcoiris.ucolick.org/candels/) samples.

[g] Sample includes galaxies from the full X-ray Ultra-VISTA and CANDELS samples.

[h] 'Cleaned' sample; Based on integration of the measured X-ray binary X-ray luminosity function.

[i] Based on *NuSTAR* data.

Table 11.1. *Scaling relations between X-ray emission, SFR, and stellar mass*

Scaling Relation	Scatter (dex)	β	α	Band	Notes
$\log \frac{L_X}{10^{40} \mathrm{erg\,s^{-1}}} = \beta + \log \frac{L_{FIR}}{10^{43} \mathrm{erg\,s^{-1}}}$	0.27	-3.68		0.5–2.0 keV	Ranalli et al. (2003) [a]
$\log \frac{L_X}{10^{40} \mathrm{erg\,s^{-1}}} = \beta + \log \frac{L_{1.4\,\mathrm{GHz}}}{10^{29} \mathrm{erg\,s^{-1}\,Hz^{-1}}}$	0.24	11.08		0.5–2.0 keV	Ranalli et al. (2003) [b]
$\log \frac{L_X}{10^{40} \mathrm{erg\,s^{-1}}} = \beta + \log \frac{L_{FIR}}{10^{43} \mathrm{erg\,s^{-1}}}$	0.29	-3.62		2.0–10.0 keV	Ranalli et al. (2003) [a]
$\log \frac{L_X}{10^{40} \mathrm{erg\,s^{-1}}} = \beta + \log \frac{L_{1.4\,\mathrm{GHz}}}{10^{29} \mathrm{erg\,s^{-1}\,Hz^{-1}}}$	0.29	11.13		2.0–10.0 keV	Ranalli et al. (2003) [b]
$\frac{L_X}{\mathrm{erg\,s^{-1}}} = \beta\, SFR$	0.4	2.61×10^{39}		0.5–8.0 keV	Mineo et al. (2012a)
$\frac{L_X}{\mathrm{erg\,s^{-1}}} = \beta\, SFR$	0.4	$(4.0 \pm 0.4) \times 10^{39}$		0.5–8.0 keV	Mineo et al. (2014)
$\frac{L_X^{gas}}{\mathrm{erg\,s^{-1}}} = \beta\, SFR$	0.34	$(5.2 \pm 0.2) \times 10^{38}$		0.5–2.0 keV	Mineo et al. (2012b) [c]
$\frac{L_X}{\mathrm{erg\,s^{-1}}} = \alpha(1+z)^{\gamma} M_{star}$	0.16	$\log\beta = 39.38 \pm 0.03$	$\log\alpha = 29.04 \pm 0.17$	0.5–2.0 keV	Lehmer et al. (2016) [d]
$+\beta(1+z)^{\delta} SFR$		$\delta = 0.99 \pm 0.26$	$\gamma = 3.78 \pm 0.82$		
$\frac{L_X}{erg\,s^{-1}} = \alpha(1+z)^{\gamma} M_{star}$	0.16	$\log\beta = 39.28 \pm 0.05$	$\log\alpha = 29.37 \pm 0.15$	2.0–10.0 keV	Lehmer et al. (2016) [d]
$+\beta(1+z)^{\delta} SFR$		$\delta = 1.31 \pm 0.13$	$\gamma = 2.03 \pm 0.60$		
$\frac{L_X}{\mathrm{erg\,s^{-1}}} = \beta\, SFR + \alpha\, M_{star}$	0.34	$(1.62 \pm 0.22) \times 10^{39}$	$(9.05 \pm 0.37) \times 10^{28}$	2.0–10.0 keV	Lehmer et al. (2010) [e]

(Continued)

Table 11.1. *Continued*

Scaling Relation	Scatter (dex)	β	α	Band	Notes
$\frac{L_X}{erg\,s^{-1}} = \alpha(1+z)^{\gamma} M_{star}$		$\log\beta = 39.50 \pm 0.06$	$\log\alpha = 28.81 \pm 0.08$	2.0–10.0 keV	Aird et al. (2017) [f]
$+\beta(1+z)^{\delta} SFR^{\theta}$		$\delta = 0.67 \pm 0.31$	$\gamma = 3.90 \pm 0.36$		
		$\theta = 0.86 \pm 0.05$			
$\log \frac{L_X}{\mathrm{erg\,s^{-1}}} = \alpha + \beta \log SFR$		(0.83 ± 0.05)	(39.48 ± 0.05)	2.0–10.0 keV	Aird et al. (2017) [g]
$+C \log(1+z)$		$C = 1.34 \pm 0.20$			
$\log \frac{L_X}{\mathrm{erg\,s^{-1}}} = \beta\, SFR + \alpha\, M_{star}$		$39.73^{+0.15}_{-0.10}$	$29.15^{+0.07}_{-0.05}$	0.5–8.0 keV	Lehmer et al. (2019) [h]
$\frac{L_X}{\mathrm{erg\,s^{-1}}} = \beta\, SFR + \alpha\, M_{star}$		$(1.90 \pm 0.84) \times 10^{39}$	$(3.56 \pm 1.16) \times 10^{28}$	4.0–25.0 keV	Vulic et al. (2018) [i]
$\frac{L_X}{\mathrm{erg\,s^{-1}}} = \beta\, SFR + \alpha\, M_{star}$		$(0.51 \pm 0.19) \times 10^{39}$	$(0.34 \pm 0.22) \times 10^{28}$	12.0–25.0 keV	Vulic et al. (2018) [i]

The SFR and stellar mass (M_{star}) are measured in $M_{\odot}\,yr^{-1}$ and $M_{\odot}$ respectively.

[a] Based on ASCA and BeppoSAX data. The far-IR (FIR) luminosity is based on the combination of the 60μ and 100μ IRAS fluxes.

[b] Based on ASCA and BeppoSAX data.

[c] Relation for the thermal gas component based on spectral fitting of the X-ray data.

[d] Sample includes local galaxies and *Chandra* Deep Field-South detections.

[e] Sample includes local galaxies and (U)LIRGS.

[f] Sample includes galaxies in $0.1 < z < 2$ from the Ultra-VISTA (http://ultravista.org/) and CANDELS (Cosmic Assembly Near-infrared Deep Extragalactic Legacy Survey; http://arcoiris.ucolick.org/candels/) samples.

[g] Sample includes galaxies from the full X-ray Ultra-VISTA and CANDELS samples.

[h] 'Cleaned' sample; Based on integration of the measured X-ray binary X-ray luminosity function.

[i] Based on *NuSTAR* data.

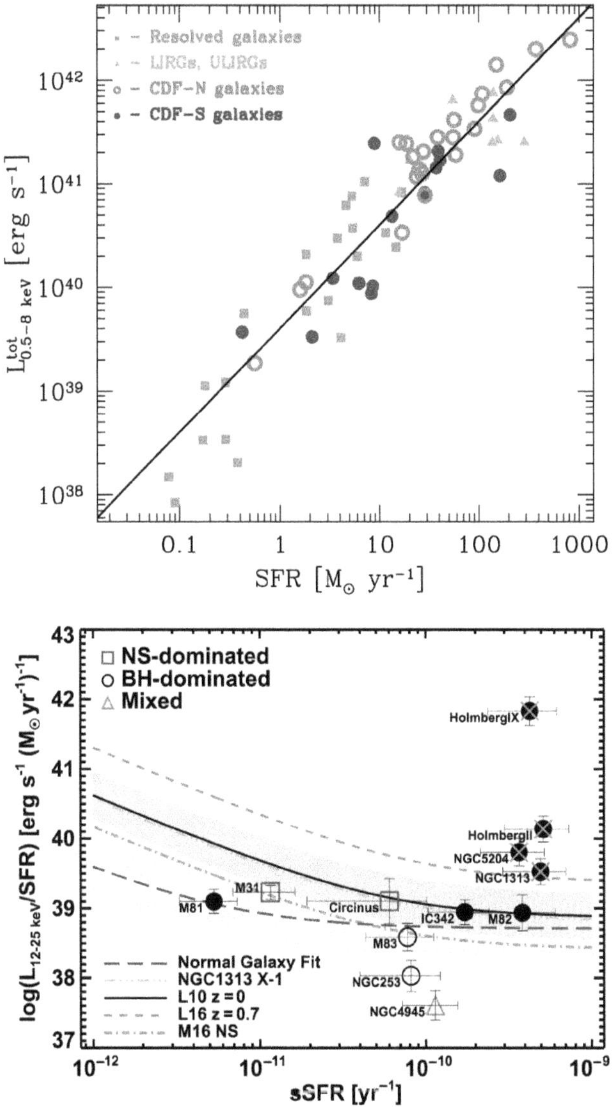

Figure 11.1 *Top panel*: Scaling relation between the 0.5–8.0 keV X-ray luminosity and SFR. The sample includes nearby galaxies, luminous IR galaxies, and higher-redshift galaxies detected in deep X-ray surveys. From Mineo et al. (2014), by permission of Oxford University Press on behalf of the Royal Astronomical Society. Colour version available online.
Bottom panel: Scaling relation between the hard X-ray luminosity (12–25 keV) per SFR against the specific SFR (defined as SFR per stellar mass). The dashed line and the yellow shaded band indicate the best fit relation and the corresponding uncertainty. Neutron-star and black-hole dominated galaxies are shown with squares and circles respectively. The fit excludes dwarf galaxies indicated with X. The other lines show local and high-z scaling relations from the works of Lehmer et al. (2010) and Lehmer et al. (2016). Adopted from Vulic et al. (2018). © AAS. Reproduced with permission. Colour version available online.

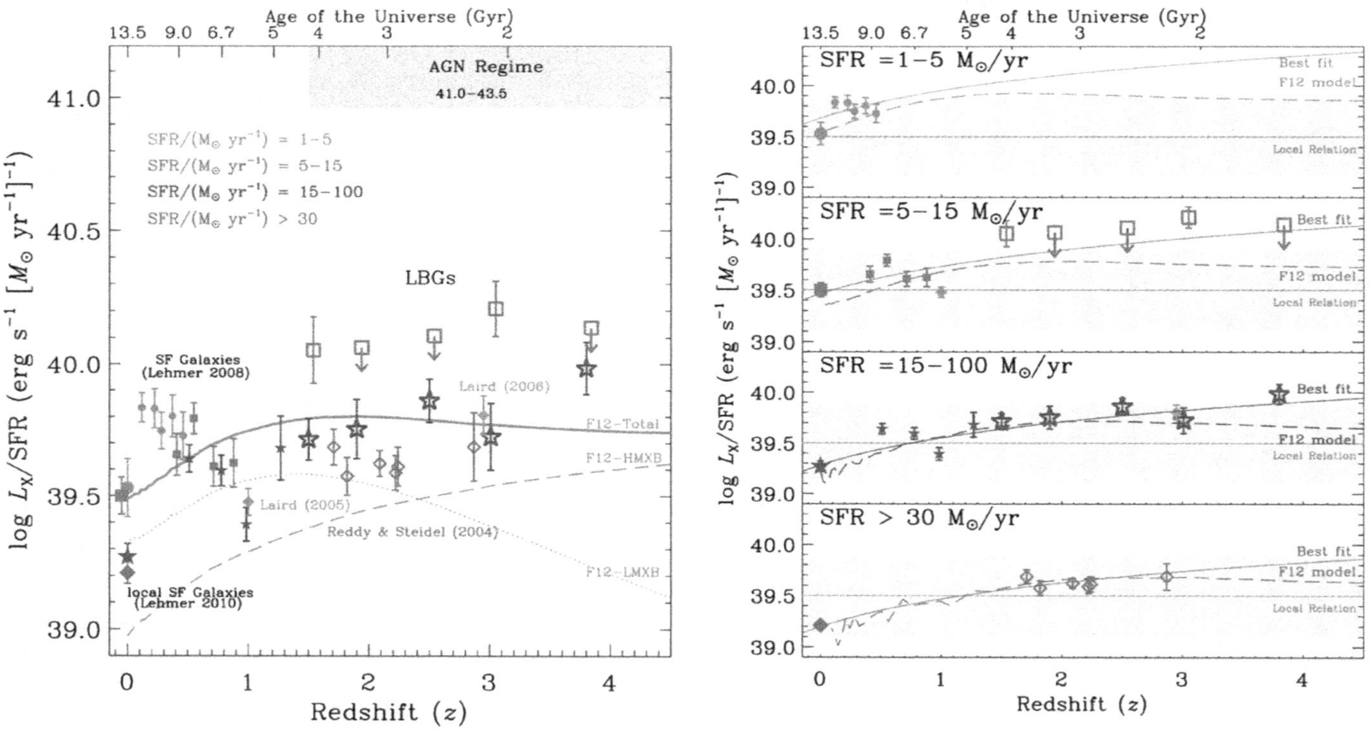

Figure 11.2 Redshift evolution of the X-ray luminosity – SFR scaling relation. Different symbols indicate samples probing different SFR regimes. The lines show the expected evolution of the X-ray emission from X-ray binaries and their HMXB and LMXB subpopulations from the models of Fragos et al. (2013). Adopted from (Basu-Zych et al., 2013) (see this paper for further details). © AAS. Reproduced with permission. Colour version available online.

(e.g. Gilfanov et al., 2004). For example, the integrated X-ray luminosity of the prototypical starburst galaxy M82 is $\sim 10^{41}\,\mathrm{erg\,s^{-1}}$, with the exact value varying depending on the luminosity of the two variable ULXs which dominate its X-ray emission (e.g. Matsumoto and Tsuru, 1999; Ptak and Griffiths, 1999; Brightman et al., 2016).

An additional source of scatter in these scaling relations is the variations of the X-ray binary populations as a function of their age and metallicity as discussed in Section 11.5. Since dwarf star-forming galaxies have experienced recent star-formation in the past $\lesssim 100\,\mathrm{Myr}$ (in addition to star-forming activity extending over several Gyr; e.g.i McQuinn et al., 2010; Weisz et al., 2011; Cignoni et al., 2019), the bulk of their X-ray emission will be associated with HMXBs. The formation rate of HMXBs, and hence their total X-ray luminosity, strongly depends on their age and metallicity. Therefore, differences in the recent star-formation history and/or metallicity of galaxies can result in significant variations of their integrated X-ray luminosity. This effect is more pronounced in dwarf galaxies which are dominated by individual star-formation events; on the other hand, more massive galaxies tend to have more secular evolution, which tends to reduce the effect of age variations on the X-ray output of a galaxy.

As discussed in Kouroumpatzakis et al. (2020) the sensitivity of the different SFR indicators on the age of the stellar populations (c.f. Chapter 7) may lead to a mismatch between the age of the XRB populations and the age of the stellar populations contributing to the considered SFR indicator. In this respect the Hα emission is the most appropriate indicator for measuring the scaling of HMXBs with young stellar populations since they probe stellar populations of similar timescales.

In fact, recent investigations of the galactic X-ray emission in sub-galactic scales (e.g. Fig. 11.3) showed that while the scaling relations of the X-ray luminosity with SFR and/or stellar mass are, on average, consistent with those inferred in galaxy-wide scales, there are local variations (expressed as scatter) which could be the result of age, and possibly metallicity, differences (e.g. Anastasopoulou et al., 2019; Kouroumpatzakis et al., 2020). In addition, in the extremely low SFRs probed in the smallest spatial scales ($\sim$1–2 kpc) there is a flattening in the X-ray luminosity – SFR relation resulting from the contribution of the underlying LMXB population (Kouroumpatzakis et al., 2020).

Elliptical galaxies are dominated by LMXBs resulting from past star-forming activity over timescales of several Gyr. As such they provide excellent laboratories for studying the dependence of the LMXB populations on the total mass and age of old stellar populations. Several studies show that there is a linear scaling between LMXB populations and the optical luminosity or stellar mass of their host galaxies (Humphrey and Buote, 2008; Boroson et al., 2011; Zhang et al., 2012). These relations are in agreement with the scaling of X-ray luminosity and stellar mass in spiral galaxies (Table 11.1). One complication is that LMXBs can be produced through two different channels: formation of binary systems in the field, and dynamical formation in dense stellar systems (e.g. globular clusters or dense galaxy bulges, e.g. Fabian et al., 1975; Voss and Gilfanov, 2007). The latter are then either ejected through three (or more) body interactions, or released in the field as the globular clusters slowly dissolve under the effect of the tidal field of the galaxy. While scaling relations between LMXBs, stellar luminosity (or stellar mass), and globular cluster

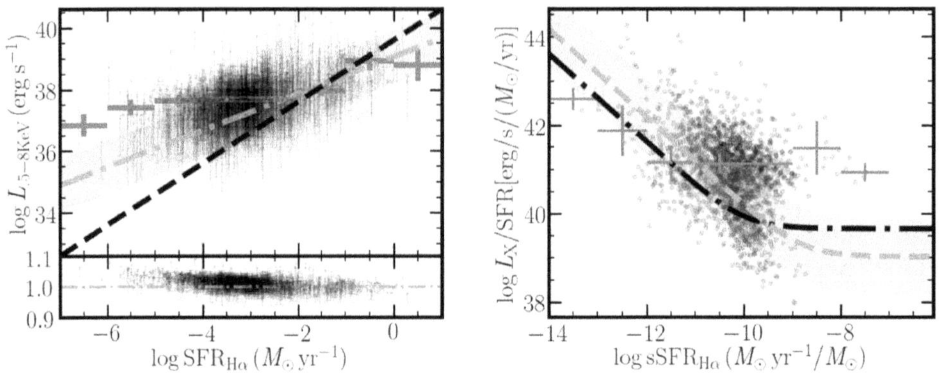

Figure 11.3 *Left panel:* The L_X–SFR scaling relation at $1 \times 1\,\mathrm{kpc}^2$ scales using Hα-based SFR. The thick points indicate the median X-ray luminosity in 0.5 dex-wide SFR bins. The dashed line shows the relation of Mineo et al. (2014) which is clearly steeper than the trend shown in the data. The flattening observed at very low SFR is interpreted as the result of the underlying, unresolved, LMXB population. The bottom panel of this figure shows the residuals with respect to the best-fit linear relation (dash-dot line), clearly exhibiting an excess at low SFR. Colour version available online.
Right panel: The L_X/SFR–sSFR scaling relation at $1 \times 1\,\mathrm{kpc}^2$ scales using Hα-based SFR. The black line shows the best-fit relation and corresponding scatter (gray band). For reference the scaling relation of (Lehmer et al., 2016) is also shown with the dash-dot line (see Kouroumpatzakis et al., 2020, for more details). Colour version available online.

specific frequency[9] show that the LMXB populations also depend on the globular cluster populations (e.g. Irwin, 2005; Boroson et al., 2011; Zhang et al., 2012), the relative importance of globular cluster vs field formation is currently unclear.

11.4 X-ray Binary Luminosity Functions

X-ray luminosity functions (XLFs) have become a very important tool for parameterising the X-ray binary populations. This has been possible thanks to the high spatial resolution of the *Chandra X-ray Observatory* which has enabled for the first time the systematic study of X-ray binary populations in different galaxies. These works led to the derivation of the X-ray luminosity functions (XLFs) of High-mass and Low-mass X-ray binaries in star-forming and early-type galaxies respectively (see Fabbiano, 2006, for an overview). The first result that became clear from these works is that the XLFs of HMXBs can be characterised in general by a power-law of the form $\frac{dN}{dL} = \mathrm{A}L^{-\alpha}$, with an index of α in the range ∼1.5 – 1.8 (e.g. Grimm et al., 2003; Zezas et al., 2007; Mineo et al., 2012a; Wang et al., 2016) and a normalisation constant A giving the number of sources at a reference luminosity. It was further found that the X-ray luminosity functions of LMXBs in early-type galaxies and the bulges of spiral galaxies are characterised by a broken power-law

[9] The globular cluster specific frequency is defined as the number of globular clusters per V-band luminosity of their host galaxy.

$$\frac{dN}{dL} = \mathrm{A}\left\{\begin{array}{ll} L^{-\alpha_1} & L > L_{\mathrm{b}} \\ L_{\mathrm{b}}^{\alpha_2-\alpha_1} L^{-\alpha_2} & L > L_{\mathrm{b}} \end{array}\right\} \tag{11.1}$$

with typical indices of $\alpha_1 \sim 1.5 - 2.5$ above a break-point at $L_{\mathrm{b}} \sim 3 \times 10^{38}\,\mathrm{erg\,s^{-1}}$ and $\alpha_2 \sim 2.0 - 3.0$ at lower luminosities (e.g. Gilfanov, 2004; Kim et al., 2006; Kim and Fabbiano, 2010; Zhang et al., 2012; Lehmer et al., 2014; Lin et al., 2015; Peacock and Zepf, 2016). The normalisation of the XLF provides an accurate census of the populations of the X-ray binaries since it accounts for incompleteness and sampling effects. As expected, the above studies of the XLFs in different galaxies found that the number of HMXBs scales with SFR while the number of LMXBs scales with stellar mass.

Spatially resolved studies of spiral galaxies showed that the XLFs of binaries in different regions of galaxy disks (e.g. spiral arms, inter-arm regions) become steeper with increasing distance from the spiral arms (Soria and Wu, 2003; Swartz et al., 2003), which is interpreted as a signature of the 'aging' of the X-ray binary populations (see also Lehmer et al., 2017, who measured the evolution of the XLF slope as function of age). This is supported by a systematic study of nearby galaxies (Lehmer et al., 2019) showing that the overall XLF becomes steeper with decreasing specific SFR, as the result of the decreasing contribution of HMXBs with decreasing specific SFR. Similar behaviour is observed in the XLFs of sources characterised as LMXBs or HMXBs based on their optical counterparts (Zezas, 2008; Binder et al., 2012).

One point of concern in investigations of the XLF is the significant temporal and spectral variability of X-ray binaries. However, monitoring studies of a few nearby galaxies, showed that the variability of individual sources does not influence the X-ray luminosity function of the overall population (Zezas et al., 2007; Zezas, 2008; Sell et al., 2011). This suggests that individual observations of a galaxy provide a representative picture of the X-ray luminosity function of the X-ray binaries it hosts.

11.5 Age and Metallicity Dependence of X-ray Binary Formation Efficiency and Luminosity Functions

The onset of the X-ray emitting phase during the evolution of a binary stellar system depends on the combination of several parameters including the initial mass, metallicity and orbital separation of the two stars, the magnitude of the supernova kick, and the outcome of the common envelope phase, if the system goes through one (see e.g. Tauris and van den Heuvel, 2006; Belczynski et al., 2008; Bhattacharya and van den Heuvel, 1991, for overviews). The mass and metallicity of the two stars will determine their nuclear timescale and mass-loss rate. The exchange of mass and angular momentum between the two objects is driven by their initial separation in addition to their mass and metallicity. The survival of the system after the supernova is driven by the mass lost during the explosion and the kick imparted onto the resulting compact object. Similarly, the onset and the outcome of a common envelope phase, which has dramatic effect on the fate of a binary system, depends on the orbital separation and the mass of the star. All these key factors will determine whether and when at any point in the evolution of a binary stellar system the compact object will accrete material at an appreciable rate to become a visible X-ray binary.

Since the initial binary systems are forming with a distribution of stellar masses and initial orbital separations, each one of them will reach the X-ray binary phase at a different time. Therefore, their number is a strong function of the time ellapsed since the formation of their parent stellar population. In addition the parameters of the binary system at the onset of the X-ray emitting phase (mass of the donor star and compact object, orbital separation, eccentricity) will determine its duration, whether it will be a transient or persistent system and to some degree its (average) X-ray luminosity. These factors result in strong age dependence of the X-ray emission from X-ray binary populations (see Fig. 11.4; e.g. Fragos et al., 2013; Wiktorowicz et al., 2017).

The first efforts to systematically measure the formation rate of HMXBs as a function of their age has been in the case of the Magellanic Clouds. Initial studies found that HMXBs in the Small Magellanic Clouds (SMC) are located preferentially in regions with young stellar populations with ages of ~30–60 Myr (Shtykovskiy and Gilfanov, 2007; Antoniou et al.,

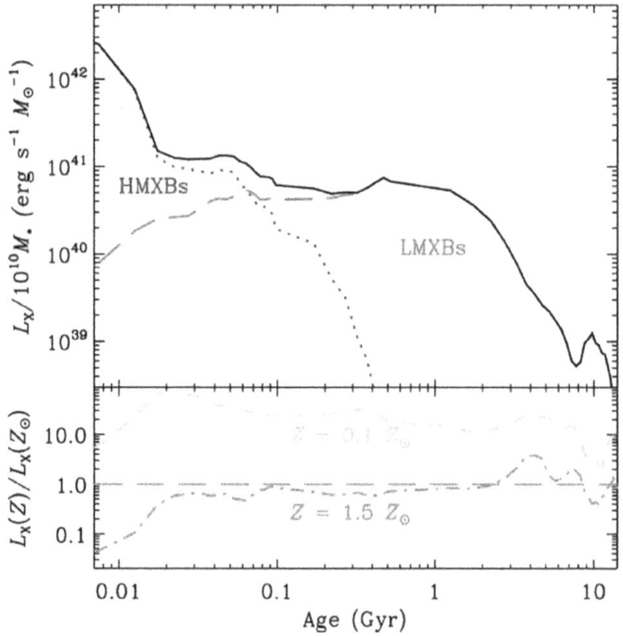

Figure 11.4 *Top panel:* Evolution of the X-ray luminosity of a population of X-ray binaries formed in an instantaneous burst of star-formation as a function of time since their formation. The solid black line shows the bolometric X-ray luminosity per unit (parent) stellar mass based on X-ray binary population synthesis models. The dotted (blue) and dashed (red) lines show the contribution of the HMXB and LMXB subpopulations.
Bottom panel: The ratio of the X-ray luminosity evolution for stellar populations of sub-solar ($0.1Z_\odot$) and super-solar ($1.5Z_\odot$) metallicity with respect to solar metallicity (shown in the upper pannel). From Fragos et al. (2013). © AAS. Reproduced with permission. Colour version available online.

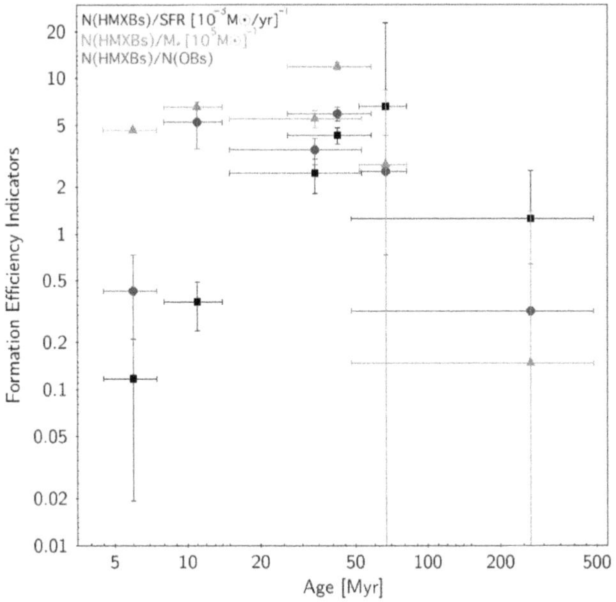

Figure 11.5 The evolution of the formation rate of HMXBs in the Small Magellanic Cloud as function of time. The figure shows three different metrics of the HMXB formation rate: number of HMXBs per SFR of their parent star-formation episode (black squares), number of HMXBs per total stellar mass formed in their parent star-formation episode (red triangles), and number of HMXBs per present-day OB stars (in units of 10^{-3} OB stars; blue circles). From Antoniou et al. (2019), © AAS. Reproduced with permission. Colour version available online.

2010), while in the case of the Large Magellanic Cloud (LMC) they are associated with an even younger stellar population at ~ 10 Myr (Shtykovskiy and Gilfanov, 2005; Antoniou and Zezas, 2016). This difference in the age of the HMXBs in the two galaxies is attributed to their different star-formation histories: the LMC is dominated by a more recent burst of star formation leading to a stronger population of HMXBs with supergiant donors (present at ages $\leq\sim$ 10 Myr), in contrast to the Be-HMXBs which dominate at ages ~30–50 Myr (e.g. Dray, 2006; Antoniou et al., 2010; Linden et al., 2010).

More recently Antoniou et al. (2019) measured the formation efficiency of the overall population of active HMXBs (down to luminosities of $\sim 5 \times 10^{32}$ erg s^{-1}) in regions of the SMC dominated by stellar populations of different ages. This is based on the spatially resolved star-formation history of the SMC (Harris and Zaritsky, 2004), and therefore it provides a direct measurement of the formation rate of HMXBs present today against their *parent* stellar populations. This work showed that the formation rate of HMXBs as a function of the SFR or total stellar mass of their parent stellar populations rises at ~10 Myr, shows a peak at ~30 – 50 Myr and a steep decline at timescales of ≲70 Myr (see Fig. 11.5; Antoniou et al., 2019). This is a more direct measurement of the formation efficiency of X-ray binaries than those based on the scaling relations discussed in Section 11.3, the

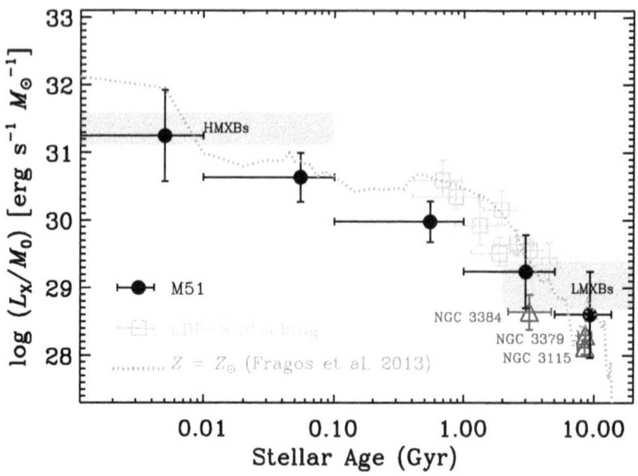

Figure 11.6 The evolution of the X-ray luminosity per unit (parent) stellar mass in M51 (black points). The dotted line shows the prediction from X-ray binary population synthesis models for solar metallicity (see Fig. 11.4; Fragos et al., 2013). The open squares and triangles show measurements from higher-redshift galaxies in *Chandra* deep surveys, and elliptical galaxies respectively. Adapted from Lehmer et al. (2017). Colour version available online.

reason being that the latter are typically based on broad-band or monochromatic SFR indicators and hence they probe a wide range of stellar population ages (e.g. Chapters 3, 5, 6, and 7; Kennicutt and Evans, 2012; Kouroumpatzakis et al., 2020).

Subsequent studies in other spiral galaxies (NGC300, NGC2403, M33, M31) based on the combination of *Chandra* and *Hubble Space Telescope*[10] data support the enhanced production of X-ray binaries at timescales of ∼20–40 Myr (Williams et al., 2013, 2018), and the presence of a formation channel for younger X-ray binaries at ≤10 Myr (in M33; Garofali et al., 2018) like in the case of the LMC. Similarly, a study of the connection between the X-ray binary populations in M51 and the age of their local stellar populations based on SED fits (see Fig. 11.6; Lehmer et al., 2017; Eufrasio et al., 2017) agrees with these results, while it extends our measurements of the X-ray binary formation efficiency to timescales beyond ∼ 100 Myr, which are hard to probe using spatially resolved stellar population studies. A follow-up study by Lehmer et al. (2019) demonstrated that XLFs (including sources with $L_X > 5 \times 10^{36}$ erg s^{-1}) of aging populations become steeper with time and with decreasing normalisation (i.e. number of X-ray binaries per stellar mass of the populations associated with their formation). The combination of these results with the detailed studies of X-ray binaries in Local Group galaxies, which probe lower-luminosity populations and provide higher resolution star-formation histories in the <50 Myr timescales that are relevant for the HMXB populations, give a comprehensive picture of the X-ray binary formation from Myr to Gyr timescales.

[10] *Hubble Space Telescope (HST)*; www.stsci.edu/hst

Another factor that affects the populations of X-ray binaries and their X-ray emission is stellar metallicity. The weaker stellar winds in lower metallicity stars affect the evolution of X-ray binaries in two ways: they lead to the formation of more massive compact objects, and the lower mass and angular momentum loss from the system results in tighter orbits. The net result of these effects is that lower metallicity systems may have higher X-ray luminosities since they are more likely to accrete through Roche-lobe overflow (see Fig. 11.4; Dray, 2006; Linden et al., 2010; Fragos et al., 2013). This effect becomes particularly important in the case of the ULXs: a series of recent studies has demonstrated that ULXs are indeed more common in lower metallicity systems (e.g. Mapelli et al., 2010; Prestwich et al., 2013; Brorby et al., 2016; Kaaret et al., 2017; Kovlakas et al., 2020), or in other words low-metallicity galaxies tend to have flatter XLFs (e.g. Lehmer et al., 2019). This has important implications for the X-ray output of galaxies at cosmological distances. Basu-Zych et al. (2013, 2016) find that high-redshift Lyman-break galaxies, as well as, local Lyman-break analogs have higher integrated X-ray luminosity per SFR (L_X/SFR) ratios than local galaxies (see Fig. 11.2). The same work shows that the evolution of the L_X/SFR ratio can be well reproduced with X-ray binary population synthesis models when accounting for the redshift evolution of metallicity.

11.6 X-ray Binary Population Synthesis Models

Similar to stellar population synthesis (c.f. Chapters 3 and 5), X-ray binary population synthesis models are a key tool for studying the formation and evolution of individual X-ray binaries, as well as, their populations. The general approach followed in most X-ray binary population synthesis codes is to follow the evolution of individual binary systems with initial parameters (e.g. mass, orbital separation, rotation) sampled from appropriate distributions (e.g. Belczynski et al., 2008). Because only a small fraction of the binary systems are going to become X-ray binaries at any given time, this approach requires the simulation of a very large population of initial systems. In order to predict the X-ray binary populations of a galaxy these models are then convolved with its star-formation history (e.g. Belczynski et al., 2004, 2008; Tzanavaris et al., 2013; Olejak et al., 2020). Selection from these modelled populations of those that match a set of observable parameters (e.g. X-ray luminosity, orbital periods, compact object types, donor stars, etc.) allows for the theoretical study of the X-ray binary populations, their evolutionary paths, and the indirect determination of their types.

Pilot studies comparing predictions from X-ray binary population synthesis models for different assumptions of initial parameters (e.g. initial mass ratio distribution) or binary evolution parameters (supernova kick velocity distribution, common envelope ejection efficiency, etc.) with the XLFs of X-ray binaries in nearby galaxies (or their integrated luminosity in the case of more distant galaxies), showed that even with this minimal set of information it is possible to constrain X-ray binary evolution parameters (Basu-Zych et al., 2013; Tzanavaris et al., 2013; Lehmer et al., 2016). One point that became clear from these studies, is that generally models for similar parameter values are consistent with the observed XLFs even for different galaxies. In addition, while the population synthesis models depend on several parameters (e.g. Tzanavaris et al., 2013; Breivik et al., 2019), the comparison with the observed XLFs of the overall X-ray binary populations can constrain only a few of them (most notably the common envelope ejection efficiency and distribution of binary initial separation).

Since the first general purpose binary population synthesis models (e.g. `BSE`;[11] Hurley et al., 2002) there has been a systematic effort to develop new, more efficient, models with improved treatment of binary evolution and/or mass transfer. These include for example: treatment of complex phases in binary evolution (e.g. common envelope), the distribution of supernova kick velocities, improvements in accretion physics. This effort has been driven mainly by the need to model compact object mergers (gravitational-wave sources and short γ-ray bursts), type Ia supernovae, or the effect of binaries in the photometric properties of stellar populations. Examples of this new generation of codes include `StarTrack` (Belczynski et al., 2008), `binary.c` (Izzard et al., 2018), `COMPASS` (Barrett et al., 2018), `BPASS` (Eldridge et al., 2017, see also Chapter 3), and `Cosmic` (Breivik et al., 2019).

Very recently there has also been an effort to develop interfaces for some of these codes based on state-of-the-art statistics methods: e.g. `dart_board` (Andrews et al., 2018); `STROOPWAFFLE` (Broekgaarden et al., 2019). These methods allow for the inference of binary evolution parameters and the more efficient study of individual systems, or particular classes of systems. In addition, they allow the inclusion of spatially resolved star-formation histories, making it possible to study the spatial distribution of X-ray binaries and set additional constraints on the supernova-kick distributions based on the displacement of the X-ray binaries from the star-forming regions (Andrews et al., 2018).

11.7 X-ray Emission as SFR Indicator: Promise and Complications

The results from the systematic study of the resolved X-ray binary populations and the integrated X-ray emission of nearby, as well as, more distant galaxies show very tight correlations with SFR, demonstrating that they can be used as probes of star-forming activity. However, the more recent observational and theoretical investigations have also revealed some of the limitations of these calibrations.

The most important complication stems from the strong age dependence of the HMXB populations. As discussed in Section 11.5, theoretical models (see Fig. 11.4), as well as, observational results (see Figs. 11.5 and 11.6), show that the formation rate of HMXB populations changes significantly in timescales as short as $\sim$10 Myr after a star-formation episode, while it drops dramatically at timescales $>$50 Myr. While this makes them excellent indicators of the recent (almost instantaneous) SFR, it also complicates their calibration as SFR indicators since they should be compared against other indicators probing similar timescales. Given the evolutionary timescales of X-ray binaries (see Fig. 11.4), this requires star-formation histories with resolution $\frac{\Delta t}{t}$ (i.e. $\Delta t \sim 1$ Myr at timescales $<$10 Myr, and $\Delta t \sim 10$ Myr at timescales of $\sim$10–100 Myr). Therefore,

[11] Binary-Star Evolution; http://astronomy.swin.edu.au/~jhurley/

the most appropriate calibration is against star-formation histories inferred from resolved stellar populations (see Section 11.5; Shtykovskiy and Gilfanov, 2007; Antoniou and Zezas, 2016; Antoniou et al., 2019). However, this is generally impractical since currently it can only be achieved with deep *HST* observations which are not available for large samples of galaxies. This inadvertently leads to the use of broad-band or monochromatic SFR indicators for the empirical calibrations of the X-ray–SFR relations. The most commonly available SFR indicators (IR and/or UV) provide biased results since they probe stellar populations with lifetimes up to $\sim 100\,\mathrm{Myr}$ or longer (Chapter 7; Kennicutt and Evans, 2012). These are much longer than the lifetimes of HMXBs, resulting in contamination by star-forming activity (or UV emission) unrelated to the HMXB populations (Kouroumpatzakis et al., 2020). This effect is exacerbated by the complex star-formation histories of star-forming galaxies, and since it depends on the star-formation history of each galaxy it is not easy to correct for. On the other hand, while recent studies have started to explore SED-fitting based SFRs and star-formation histories (e.g. Eufrasio et al., 2017; Lehmer et al., 2017), these methods either generally rely on pre-defined prescriptions for the star-formation history (see Chapters 7 and 8), or those that provide non-parametric star-formation histories they do not have the ~few Myr resolution required to trace the evolution of HMXBs. The only SFR indicator that probes stellar populations with ages similar to the formation timescales of HMXBs is the Hα emission or any other tracer of the ionised ISM (Kouroumpatzakis et al., 2020).

When we consider the effect of metallicity the above picture becomes more complicated. Observational (Mapelli et al., 2010; Prestwich et al., 2013; Brorby et al., 2016) as well as theoretical studies (e.g. Fig. 11.4; Linden et al., 2010; Dray, 2006; Fragos et al., 2013) show that there is strong dependence of the X-ray luminosity of a binary population on metallicity, with lower-metallicity galaxies having higher X-ray luminosities. However, since the X-ray luminosity–SFR scaling relations are generally based on large samples of galaxies the effect of metallicity contributes to the scatter of these scaling relations.

Another complication in the empirical derivation of X-ray–SFR scaling relations is the underlying population of LMXBs. These are produced by star-forming activity that took place over timescales much longer than the lifetime of HMXBs (up to several Gyr; Fig. 11.4). If the accumulated stellar mass from past star-forming activity is large enough compared to the young (<50 Myr) stellar populations, the LMXB populations may dominate over the HMXBs. Therefore, the contamination becomes more significant for galaxies with lower specific SFR (e.g. Lehmer et al., 2019), with the correction depending on the star-formation history of each galaxy *and* the SFR indicator used (Kouroumpatzakis et al., 2020). Based on the HMXB and LMXB scaling relations against SFR and stellar mass respectively (Table 11.1) one can estimate that HMXBs dominate at sSFR $\gtrsim 5 \times 10^{-11} \mathrm{M_\odot\,yr^{-1}\,M_\odot^{-1}}$. This is consistent with the turn-up point in the $\mathrm{L_X/SFR}$–$\mathrm{M_*}$ relation (see Fig. 11.3; see also Lehmer et al., 2010). Therefore, the X-ray emission of galaxies can be used as a proxy for recent star-forming activity but only in the case of starburst galaxies where the contribution of the LMXB component is negligible. For the lower sSFR galaxies one can, in principle, correct for the LMXB contribution by using the $\mathrm{L_X}$–$\mathrm{M_{star}}$ relation if the stellar mass is known.

Theoretical calibration of the X-ray binary populations (their numbers or their integrated X-ray luminosity) as a SFR indicator can also be calculated with X-ray binary population synthesis models in the same way as stellar population synthesis models are used to calibrate broad-band or monochromatic SFR indicators. Such calibrations have been calculated (e.g. Fig. 11.4; Fragos et al., 2013) and they are in good agreement with observational results (e.g. Fig. 11.6; Lehmer et al., 2017). However, as it is clear from Fig. 11.4 they are a strong function of both age and metallicity. Therefore, any such calibration should provide a time-average scaling accounting for different star-formation scenarios over the lifetime of HMXBs (Kouroumpatzakis et al., 2020).

As discussed in Section 11.3 variability of individual X-ray binaries can influence the integrated X-ray luminosity of a galaxy. Monitoring observations of a few nearby galaxies showed that the shape of the X-ray luminosity function of their X-ray binaries is not affected by the variability of the individual sources (Zezas et al., 2007; Sell et al., 2011). Therefore, the integrated X-ray emission of galaxies with high SFR (i.e. more than a few $M_\odot\,yr^{-1}$) that host large populations of X-ray binaries is less influenced by effects such as stochastic sampling of their X-ray luminosity function and/or variability of individual bright sources. However, the X-ray output of galaxies with low SFR can be dominated by a few individual sources (such examples are most ULX-hosting dwarf star-forming galaxies; M82, Holmberg-II, Holmberg-IX etc.). In this case their X-ray luminosity is not a reliable indicator of their SFR since it can be easily affected by the variability of the individual sources (e.g. as in the case of M82, Brightman et al., 2016). However, it can be a reliable indicator for samples of dwarf galaxies where one measures their total X-ray luminosity since the scaling relations hold *on average* even at very low SFR regardless of stochastic effects (c.f. Anastasopoulou et al., 2019; Lehmer et al., 2019; Kouroumpatzakis et al., 2020).

Finally, because LMXBs have formation timescales ranging from a few Myr to several Gyr, their integrated emission cannot be used as probe of the SFR at these timescales. The very broad distribution of their formation timescales (e.g. Figs. 11.4 and 11.6) will smear any information on the star-formation history of their host galaxy. Instead they are good proxies of its stellar mass.

11.8 Supernovae and Supernova Remnants

Core-collapse supernovae and their remnants (SNRs) being the end-points of massive stars ($M > 8M_\odot$) are direct probes of very recent star-forming activity. While the rarity of the individual supernovae does not allow their use as SFR indicators in individual galaxies, they can be used to infer the volume density of star-forming activity by studying their populations in a given redshift bin (e.g. Madau and Dickinson, 2014).

Instead, since SNRs are a record of supernovae exploded in the past $\sim$50–100 kyr they can provide a better census of recently exploded stars. For example, in our Galaxy we have identified 294 SNRs (Green, 2019), in contrast to 5 supernovae recorded in the past 10^3 years; e.g. Dragicevich et al. (1999). Given the very short lifetimes of their progenitors

SNRs are an excellent proxy for the 'instantaneous' SFR. However, the detection of SNRs is subject to several selection effects:

1. The luminosity and spectral-energy distribution of an SNR is a strong function of its age. Generally, young SNRs are stronger X-ray sources, while older SNRs emit in the optical part of the spectrum. Therefore in order to obtain a complete picture of the SNR population one has to combine populations detected in several wavebands (e.g. Leonidaki et al., 2013; Long, 2017) which however have different detection thresholds and selection biases.
2. Older SNRs are more extended and characterised by lower surface brightness, making their detection against their surrounding or background ISM more challenging.
3. Identification of SNRs is generally based on their optical forbidden-line emission (typically the $S_{II}/H\alpha$ ratio; e.g. Mathewson and Clarke, 1973), which in turn is a strong function of their shock velocity and metallicity. This inadvertently leads to a bias against lower-metallicity and/or SNRs with slower shocks (Kopsacheili et al., 2020).
4. Only core-collapse supernovae are related to recent star-forming activity. Disentangling the type of a supernova based on the properties of its remnant is a difficult task. Traditionally this is done based on the abundance of different metals in the remnant: core collapse supernovae show enhanced abundances in α elements, while Type-Ia supernovae are characterised by higher Fe and Ni abundances. Recently Lopez et al. (2011) pioneered the use of the SNR shape in order to discriminate between core-collapse or Type-Ia progenitors: the latter tend to produce more circular and symmetric remnants than the former. This difference is seen in the continuum (X-ray, IR) or line emission (from X-ray data). However, both methods can only be applied in our nearest galaxies where we can obtain data with the required sensitivity and resolution.

Despite these complications, there have been efforts to compare supernovae or SNR populations against other SFR indicators (e.g. Tammann et al., 1994; Leonidaki et al., 2013) and to measure their connection with stellar populations of different ages (e.g. Maoz and Badenes, 2010). Future observatories (the *James Webb Space Telescope*,[12] 30m-class telescopes, the *ATHENA X-ray Observatory*[13]) will provide the necessary increase in resolution or sensitivity required to overcome many of the above-mentioned limitations, and enable the systematic use of SNRs in nearby galaxies as proxies of their most recent star-forming activity. In addition, systematic transient searches (e.g. *ZTF*;[14] Graham et al., 2019) or the *LSST*,[15] as well as, all sky surveys such as *WFIRST*[16] or the *Euclid* mission[17] in the future, are transforming this field by providing large numbers of supernovae in relatively unbiased surveys. The high cadence offered by these surveys will also allow the characterisation of

[12] James Webb Space Telescope (JWST); www.stsci.edu/jwst/
[13] The Athena Advanced Telescope for High-Energy Astrophysics; www.the-athena-X-ray-observatory.eu
[14] Zwicky Transient Factory; www.ztf.caltech.edu
[15] Large Synoptic Survey Telescope; www.lsst.org
[16] Wide Field Infrared Survey Telescope; https://wfirst.gsfc.nasa.gov
[17] www.euclid-ec.org

the supernova type on the basis of their light curve. Furthermore, large-scale spectroscopic surveys allow the identification and classification of supernovae in different galaxies and the exploration of their rates as function of the stellar populations (age, metallicity) of their host galaxy (e.g. Galbany et al., 2014; Graur et al., 2015).

11.9 γ-ray Emission and Star Formation

Both the point-source and the diffuse components of the galactic γ-ray emission are closely linked with star formation. The discrete sources are SNRs, pulsars, and X-ray binaries (The Fermi-LAT collaboration, 2019), however, they contribute a small fraction of the integrated γ-ray emission in comparison to the diffuse component ($\sim$10–20% in our Galaxy; Strong, 2007). The diffuse γ-ray emission is the result of three main mechanisms:

- The decay of π^0 particles produced from the interaction of cosmic-ray particles (mainly protons and α particles) with the ISM.
- Bremsstrahlung emission from relativistic electrons in the ISM.
- Inverse-Compton scattering of the galaxy radiation field off relativistic electrons.

Since the main source of cosmic rays and relativistic particles in a normal (non-AGN) galaxy are SNRs one would expect to have a linear correlation between the γ-ray luminosity and SFR, under two conditions: there is no leakage of cosmic rays to the intergalactic space and a constant fraction of their energy is converted to γ-ray emission. This regime is usually referred-to as the 'calorimetric limit'. The strong connection between star formation and γ-ray emission is further enforced by the fact that the interaction between cosmic rays and the ISM is most efficient in the dense molecular clouds which are also the sites of star formation. Based on this physical mechanism one can calculate the γ-ray luminosity of a galaxy by making three basic assumptions (e.g. Fields et al., 2010):

- The SFR is proportional to the gas surface density, Σ, following a Schmitt-Kennicutt law with index α: $SFR \propto \Sigma^{\alpha}$ (Kennicutt, 1998).
- The supernova rate is proportional to the SFR.
- The cosmic-ray flux is proportional to the supernova rate, which is proportional to the SFR.

In most galaxies only a fraction of the cosmic-ray flux remains in the galaxy and interacts with the ISM (e.g. Pavlidou and Fields, 2001; Fields et al., 2010). In this case, where leakage dominate cosmic-ray losses (the so-called non-calorimetric limit), the γ-ray luminosity is proportional to the cosmic-ray flux and the gas mass. Combining the above relations we find that in this limit the γ-ray luminosity of a galaxy at redshift z scales with the SFR as $L_\gamma \propto (1+z)^{-\beta} SFR^{w+1}$ where $\beta = 2(1 - 1/\alpha)$, and $w = 1/\alpha$ (e.g. Fields et al., 2010; Komis et al., 2019). In the case of starburst galaxies, however, cosmic-ray losses though inelastic collisions can dominate over escape losses, and the above formulation will overestimate the expected γ-ray luminosity. In this case (calorimetric limit) the γ-ray luminosity is proportional to the cosmic-ray flux and it also depends on the index of the proton cosmic-ray energy spectrum (Wang and Fields, 2018).

Based on the 8 bona-fide star-forming galaxies detected with the *Fermi γ-ray Space Telescope*[18] it has been possible to derive empirical scaling relations between γ-ray emission and SFR (see Fig. 11.7; Komis et al., 2019, for the most recent calibrations). Since the volume density of star-forming galaxies increases towards higher redshifts and greatly

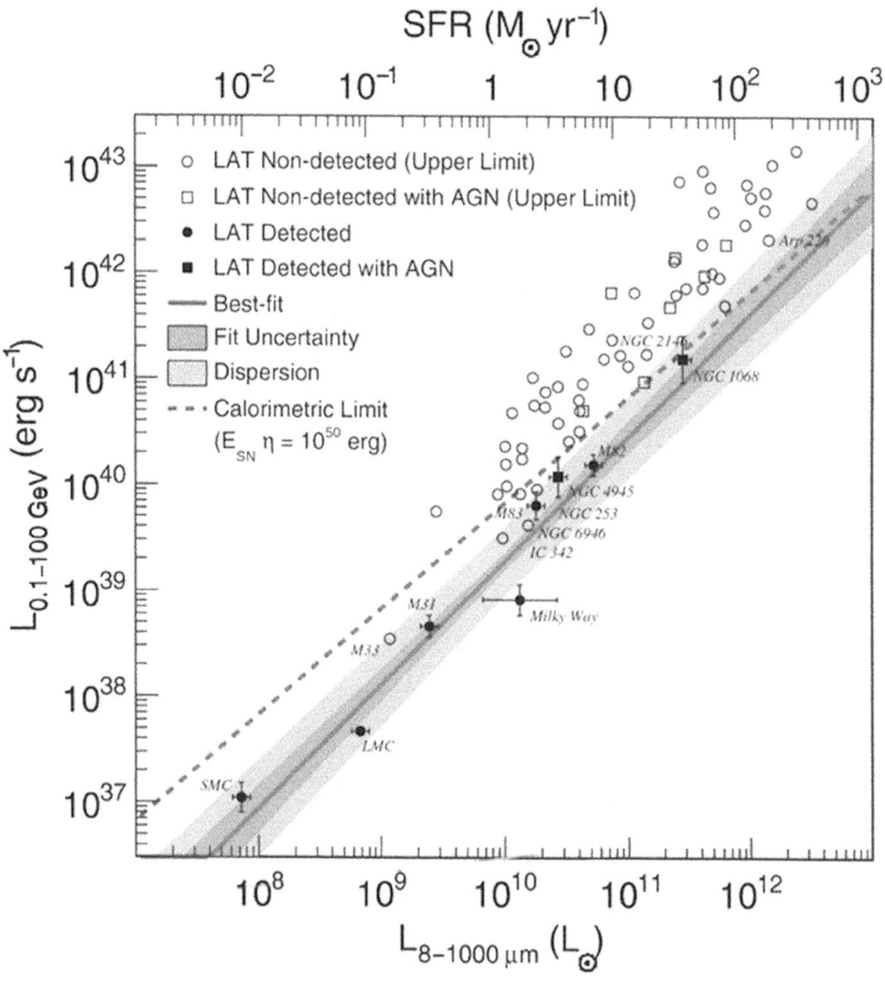

Figure 11.7 Scaling relation between the γ-ray burst and total-IR emission (or SFR) for the sample of nearby galaxies observed with *Fermi*-LAT (circles indicate upper limits). The dashed line shows the calorimetric limit. Colour version available online.

[18] https://fermi.gsfc.nasa.gov

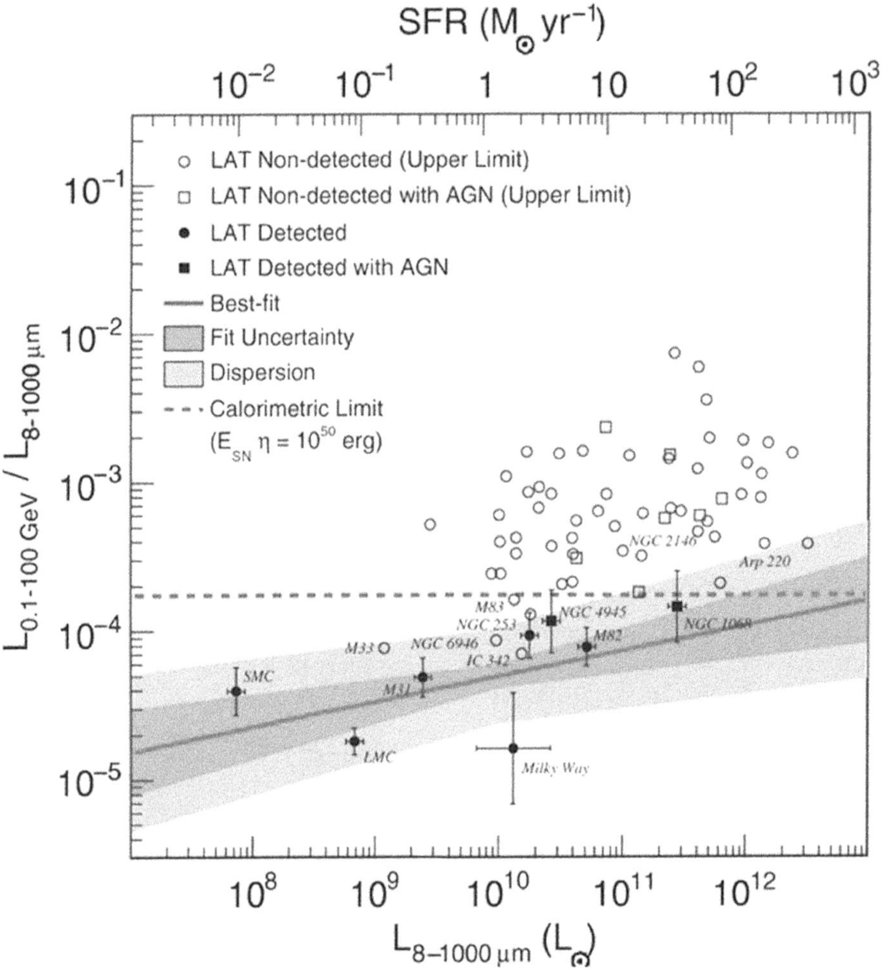

Figure 11.7 (*Cont.*) Residual plot with respect to the best-fit linear model.
From Ackermann et al. (2012) by permission of Oxford University Press on behalf of the Royal Astronomical Society. Colour version available online.

outnumbers that of blazars and other γ-ray-bright AGN, normal galaxies are considered as potentially significant contributors to the γ-ray background. Models of the γ-ray background based on these scaling relations and the cosmological evolution of star formation show that galaxies at $z < 3$–4, can produce up to 100% of the observed γ-ray background, although the exact number is highly uncertain (e.g. Komis et al., 2019). The main source of uncertainty results from the small number of galaxies detected in the γ-rays which does not allow a reliable empirical measurement of the proportionality constant in the L_γ–SFR relation, and the complications introduced by galaxies in the 'starburst' regime where collisional cosmic-ray losses may become important resulting in a different slope of the L_γ–SFR relation (e.g. Wang and Fields, 2018). Obviously future, more sensitive γ-ray observations will provide more accurate L_γ–SFR scaling relations and will allow the investigation of the effect of collisional cosmic-ray losses in galaxies with higher specific SFR, or higher compactness of their star-forming regions (see e.g. Martínez-Galarza et al., 2016; Groves et al., 2008, for a definition of compactness).

11.10 γ-ray Bursts as Star-Formation Rate Probes

γ-ray bursts are powerful outbursts (fluence $\sim 10^{-8} - 10^{-3}$ erg cm^2; Ajello et al., 2019) of γ-ray emission with duration ranging from less than a second up to several hours. Their spectrum extends up to TeV energies (Fishman and Meegan, 1995). The burst is followed by a slowly declining afterglow which can last for several days (see e.g. van Paradijs et al., 2000), which plays a critical role in the localisation of the burst and the identification of its host galaxy. The afterglow is dominated by emission in progressively longer wavelengths: X-ray emission in the first hours, to optical, IR, and radio wavelengths in the following days.

γ-ray bursts are categorised in two types depending on their duration: those that last up to ~2 sec (short γ-ray bursts) and those with longer duration (long γ-ray bursts). The short γ-ray bursts originate from the merger of a neutron star with another neutron star or a black hole (Mészáros, 2002) producing a kilonova (Metzger, 2019, and references therein). This connection was first established after the identification of short γ-ray bursts with early-type galaxies (Gehrels et al., 2005) and confirmed more recently with the association of the short γ-ray burst GRB 170817A with the detection of gravitational-wave event GW170817 resulting from the merger of two neutron stars (Abbott et al., 2017). The long γ-ray bursts instead, are associated with the collapse of massive stars (Woosley and Bloom, 2006). This is supported by the direct association of some long γ-ray bursts with core-collapse supernovae and their positional coincidence with star-forming galaxies. Therefore, long γ-ray bursts have the closest connection with on-going star-forming activity, while short γ-ray bursts, because of the delay until the merger of the two compact objects, probe past star-forming activity over timescales of a few hundred Myr up to several Gyr (Leibler and Berger, 2010; Wanderman and Piran, 2015).

Owing to their very high luminosity we can detect γ-ray bursts out to very high redshifts (the current record is $z \sim 8.1$[19]). This potentially makes them excellent probes of the star-forming activity in the early universe. The volume density of long γ-ray bursts gives a measurement of the SFR density as a function of redshift. These measurements are in general agreement with the star-formation history of the universe as inferred from other more traditional methods (e.g. UV or far-IR observations of galaxies; e.g. Yüksel et al., 2008; Kistler et al., 2009; Trenti et al., 2012). However, there are some differences that can be attributed to the metallicity and/or IMF dependence of γ-ray burst production, and/or selection effects (e.g. Trenti et al., 2013; Chary et al., 2016). For example, long γ-ray bursts

[19] https://swift.gsfc.nasa.gov/archive/grb_table.html/

are preferentially found in low-metallicity environments ($Z \lesssim 0.7\,Z_\odot$). As a result they do not trace well the star-forming activity at low redshifts ($z \lesssim 2$), but they are expected to be better tracers of the star-forming activity in the low metallicity environments found at higher redshifts (e.g. Palmerio et al., 2019).

However, there are some observational limitations that complicate the use of γ-ray bursts as SFR tracers. While, the *Neil Gehrels Swift Observatory*[20] has transformed the study of γ-ray bursts by providing a more sensitive all-sky γ-ray burst-monitor and fast X-ray follow-up, it has not been possible to detect an afterglow for several γ-ray bursts,[21] either because of observational constraints or because of significant obscuration in their host galaxy (e.g. Greiner et al., 2016). Furthermore, the faintness of many γ-ray burst host galaxies hampers the measurement of their redshift.

The brightness of the γ-ray burst afterglows allows them to be used as beacons to probe the insterstellar medium of their host galaxy. For example, UV spectroscopy of γ-ray burst afterglows has beed used to constrain the metallicity of high redshift galaxies, measure the extinction curve along the line-of-sight (e.g. Zafar et al., 2018, and references therein), or even measure the dust-to-gas ratio at redshifts $z > 5$ (e.g. Schady, 2017, and references therein). The shape of the extinction curve at very high z gives information on the formation of dust which is linked to star formation.

While long γ-ray bursts are direct probes of on-going star-forming activity, using short γ-ray bursts to infer the star-formation density requires accounting for the delay function from the formation of the binary stellar system until the coalescence of the two compact objects. This delay can range from $\sim 10\,\mathrm{Myr}$ up to several Gyr, depending on the initial parameters of the system (mass, orbital parameters) and its evolutionary path (e.g. de Mink and Belczynski, 2015; Wanderman and Piran, 2015; Andrews and Zezas, 2019). This obviously complicates the use of short γ-ray bursts as SFR probes since the observed rates include contributions from stellar populations born in different epochs.

11.11 Gravitational Waves as Star-Formation Rate Probes

The detection of gravitational waves from two merging black holes by *LIGO*[22] on 14 August 2015 (Abbott et al., 2016) opened a new avenue to study the populations of stellar remnants through their mergers. The populations of compact objects bear the imprint of the star-forming activity responsible for the formation of their progenitors. As with the short γ-ray bursts, in order to make the connection between gravitational-wave sources and their parent stellar populations one has to model the evolution of the binary

[20] https://swift.gsfc.nasa.gov

[21] X-ray afterglows have been detected for 1,086 out of the 1,342 γ-ray bursts observed up to January 2020; https://swift.gsfc.nasa.gov/archive/grb_table.html/

[22] LASER Interferometer Gravitational-wave Observatory; www.ligo.org

system and account for: (a) the systems that do not survive through the different binary evolution phases (i.e. they are disrupted, or merging) and (b) the delay between the formation of the initial binary and the coalescence of the compact-object binary system. However, in the case of systems that are sources of gravitational-waves, the analysis of the gravitational-wave signal (waveform and intensity) can provide information on the mass of the compact objects, their spin, orbital parameters, and the redshift of the source (Veitch et al., 2015). The information on the binary compact-object system is important for further constraining the initial parameters of the binary stellar system (e.g. initial mass) and inferring the time since the formation of the primordial system. The direct measurement of the redshift also removes the requirement to identify the host of the gravitational-wave source which can be non-trivial given their poor localisation.

Therefore, a significant degree of effort goes into the calculation of the delay-time distribution and the gravitational-wave progenitor selection function (i.e. fraction of the binary population that produces compact-object mergers) using binary evolution models (e.g. Belczynski et al., 2002). Combination of the gravitational-wave progenitor selection function and the delay-time distribution with information on the SFR, stellar mass, and metallicity of galaxies, has been already used to develop prescriptions for the identification of the hosts of gravitational-wave sources (e.g. Artale et al., 2020a,b, 2019). In the case of black-hole binary mergers, in particular, the timescales are fairly short (typically less than $\sim$ 10 Myr, but with a tail extending to longer timescales; e.g. Eldridge and Stanway, 2016). This makes the rate of these mergers almost instantaneous SFR indicators and it simplifies the translation of the gravitational-wave merger rate to a star-formation volume density at a given redshift. SFR measurements based on gravitational wave detections also have the advantage that they are not affected by the presence of dust. Although, only a few gravitational-wave events have been detected so-far, the increased sensitivity of the next generation of gravitational-wave detectors will provide thousands of events up to $z \sim 10$ allowing the accurate determination of the cosmic SFR density evolution independently of observations in the electromagnetic spectrum (e.g Vitale et al., 2019).

11.12 Summary

High-energy (X-ray and γ-ray) and gravitational-wave observations provide a new window for the study of current and past star-forming activity. The first results from the calibration and use of these indicators are very positive, allowing us to probe heavily obscured objects and star-forming activity in the very early universe. The recent advances in stellar and binary evolution models have greatly improved our understanding of the systematics in calibrating these SFR indicators. However, while the connection between X-ray emission and star-formation activity is fairly well understood and calibrated, the γ-ray emission and gravitational waves are still in their infancy, mainly because of the small volume of available data. An important aspect of these indicators is that they provide tests for the formation and evolution of binary stellar systems (with implications for traditional SFR indicators; c.f. Chapter 3), and the properties of the ISM at high redshifts. The promise from on-going and future X-ray observatories (e.g. the *eROSITA Telescope*[23] survey and the *ATHENA X-ray Observatory* respectively), monitoring campaigns (e.g. ZTF, LSST), and the next generation of gravitational wave detectors will allow us to utilise the full potential of the SFR indicators discussed in this Chapter.

References

Abbott, B. P., Abbott, R., Abbott, T. D. et al. 2016. Observation of Gravitational Waves from a Binary Black Hole Merger. *Physical Review Letters*, **116**(6), 061102.

Abbott, B. P., Abbott, R., Abbott, T. D. et al. 2017. Gravitational Waves and Gamma-Rays from a Binary Neutron Star Merger: GW170817 and GRB 170817A. *Astrophysical Journal*, **848**(2), L13.

Ackermann, M., Ajello, M., Allafort, A. et al. 2012. GeV Observations of Star-Forming Galaxies with the Fermi Large Area Telescope. *Astrophysical Journal*, **755**(2), 164.

Aird, J., Coil, A. L., and Georgakakis, A. 2017. X-rays Across the Galaxy Population - I. Tracing the Main Sequence of Star Formation. *Monthly Notices of the Royal Astronomical Society*, **465**(3), 3390–3415.

Ajello, M., Arimoto, M., Axelsson, M. et al. 2019. A Decade of Gamma-Ray Bursts Observed by Fermi-LAT: The Second GRB Catalog. *Astrophysical Journal*, **878**(1), 52.

Anastasopoulou, K., Zezas, A., Gkiokas, V. et al. 2019. Do Sub-Galactic Regions Follow the Galaxy-Wide X-ray Scaling Relations? The Example of NGC 3310 and NGC 2276. *Monthly Notices of the Royal Astronomical Society*, **483**(1), 711–733.

Andrews, J. J., and Zezas, A. 2019. Double Neutron Star Formation: Merger Times, Systemic Velocities, and Travel Distances. *Monthly Notices of the Royal Astronomical Society*, **486**(3), 3213–3227.

Andrews, J. J., Zezas, A., and Fragos, T. 2018. dart_board: Binary Population Synthesis with Markov Chain Monte Carlo. *Astrophysical Journal Supplement Series*, **237**(1), 1.

Antoniou, V., and Zezas, A. 2016. Star Formation History and X-ray Binary Populations: The Case of the Large Magellanic Cloud. *Monthly Notices of the Royal Astronomical Society*, **459**(1), 528–553.

Antoniou, V., Zezas, A., Hatzidimitriou, D. et al. 2010. Star Formation History and X-ray Binary Populations: The Case of the Small Magellanic Cloud. *Astrophysical Journal*, **716**(2), L140–L145.

Antoniou, V., Zezas, A., Drake, J. J. et al. 2019. Deep Chandra Survey of the Small Magellanic Cloud. III. Formation Efficiency of High-Mass X-ray Binaries. *Astrophysical Journal*, **887**(1), 20.

Artale, M. C., Mapelli, M., Giacobbo, N. et al. 2019. Host Galaxies of Merging Compact Objects: Mass, Star Formation Rate, Metallicity, and Colours. *Monthly Notices of the Royal Astronomical Society*, **487**(2), 1675–1688.

Artale, M. C., Bouffanais, Y., Mapelli, M. et al. 2020a. An Astrophysically Motivated Ranking Criterion for Low-Latency Electromagnetic Follow-Up of Gravitational Wave Events. *Monthly Notices of the Royal Astronomical Society*, **495**(2), 1841–1852.

[23] *eROSITA Telescope* (extended ROentgen Survey with an Imaging Telescope Array) on board the *Spektr-RG Space Observatory*; www.mpe.mpg.de/eROSITA

Artale, M. C., Mapelli, M., Bouffanais, Y. et al. 2020b. Mass and Star Formation Rate of the Host Galaxies of Compact Binary Mergers Across Cosmic Time. *Monthly Notices of the Royal Astronomical Society*, **491**(3), 3419–3434.

Barrett, J. W., Gaebel, S. M., Neijssel, C. J. et al. 2018. Accuracy of Inference on the Physics of Binary Evolution from Gravitational-Wave Observations. *Monthly Notices of the Royal Astronomical Society*, **477**(4), 4685–4695.

Basu-Zych, A. R., Lehmer, B., Fragos, T. et al. 2016. Exploring the Overabundance of ULXs in Metal- and Dust-Poor Local Lyman Break Analogs. *Astrophysical Journal*, **818**(2), 140.

Basu-Zych, A. R., Lehmer, B. D., Hornschemeier, A. E. et al. 2013. The X-ray Star Formation Story as Told by Lyman Break Galaxies in the 4 Ms CDF-S. *Astrophysical Journal*, **762**(1), 45.

Belczynski, K., Kalogera, V., and Bulik, T. 2002. A Comprehensive Study of Binary Compact Objects as Gravitational Wave Sources: Evolutionary Channels, Rates, and Physical Properties. *Astrophysical Journal*, **572**(1), 407–431.

Belczynski, K., Kalogera, V., Rasio, F. A. et al. 2008. Compact Object Modeling with the StarTrack Population Synthesis Code. *Astrophysical Journal Supplement Series*, **174**(1), 223–260.

Belczynski, K., Kalogera, V., Zezas, A. et al. 2004. X-ray Binary Populations: The Luminosity Function of NGC 1569. *Astrophysical Journal*, **601**(2), L147–L150.

Bhattacharya, D., and van den Heuvel, E. P. J. 1991. Formation and Evolution of Binary and Millisecond Radio Pulsars. *Physics Reports*, **203**(1–2), 1–124.

Binder, B., Williams, B. F., Eracleous, M. et al. 2012. The Chandra Local Volume Survey: The X-ray Point-Source Catalog of NGC 300. *Astrophysical Journal*, **758**(1), 15.

Boroson, B., Kim, D.-W., and Fabbiano, G. 2011. Revisiting with Chandra the Scaling Relations of the X-ray Emission Components (Binaries, Nuclei, and Hot Gas) of Early-Type Galaxies. *Astrophysical Journal*, **729**(1), 12.

Breivik, K., Coughlin, S. C., Zevin, M. et al. 2019. COSMIC Variance in Binary Population Synthesis. *arXiv e-prints*, Nov, arXiv:1911.00903.

Brightman, M., Harrison, F., Walton, D. J. et al. 2016. Spectral and Temporal Properties of the Ultraluminous X-ray Pulsar in M82 from 15 years of Chandra Observations and Analysis of the Pulsed Emission Using NuSTAR. *Astrophysical Journal*, **816**(2), 60.

Broekgaarden, F. S., Justham, S., de Mink, S. E. et al. 2019. STROOPWAFEL: Simulating Rare Outcomes from Astrophysical Populations, with Application to Gravitational-Wave Sources. *Monthly Notices of the Royal Astronomical Society*, **490**(4), 5228–5248.

Brorby, M., Kaaret, P., Prestwich, A. et al. 2016. Enhanced X-ray Emission from Lyman Break Analogues and a Possible L_X-SFR Metallicity Plane. *Monthly Notices of the Royal Astronomical Society*, **457**(4), 4081–4088.

Chary, R., Petitjean, P., Robertson, B. et al. 2016. Gamma-Ray Bursts and the Early Star-Formation History. GRBs and z>4 Star-Formation. *Space Science Reviews*, **202**(1–4), 181–194.

Cignoni, M., Sacchi, E., Tosi, M. et al. 2019. Star Formation Histories of the LEGUS Dwarf Galaxies. III. The Nonbursty Nature of 23 Star-Forming Dwarf Galaxies. *Astrophysical Journal*, **887**(2), 112.

de Mink, S. E., and Belczynski, K. 2015. Merger Rates of Double Neutron Stars and Stellar Origin Black Holes: The Impact of Initial Conditions on Binary Evolution Predictions. *Astrophysical Journal*, **814**(1), 58.

Done, C., Gierliński, M., and Kubota, A. 2007. Modelling the Behaviour of Accretion Flows in X-ray Binaries. Everything You Always Wanted to Know about Accretion but Were Afraid to Ask. *Astronomy and Astrophysics Reviews*, **15**(1), 1–66.

Dragicevich, P. M., Blair, D. G., and Burman, R. R. 1999. Why Are Supernovae in Our Galaxy So Frequent? *Monthly Notices of the Royal Astronomical Society*, **302**(4), 693–699.

Dray, L. M. 2006. On the Metallicity Dependence of High-Mass X-ray Binaries. *Monthly Notices of the Royal Astronomical Society*, **370**(4), 2079–2090.

Eggleton, P. 2006. *Evolutionary Processes in Binary and Multiple Stars.*

Eldridge, J. J., and Stanway, E. R. 2016. BPASS Predictions for Binary Black Hole Mergers. *Monthly Notices of the Royal Astronomical Society*, **462**(3), 3302–3313.

Eldridge, J. J., Stanway, E. R., Xiao, L. et al. 2017. Binary Population and Spectral Synthesis Version 2.1: Construction, Observational Verification, and New Results. *Publications of the Astronomical Society of Australia*, **34**(Nov), e058.

Eufrasio, R. T., Lehmer, B. D., Zezas, A. et al. 2017. On the Spatially Resolved Star Formation History in M51. I. Hybrid UV+IR Star Formation Laws and IR Emission from Dust Heated by Old Stars. *Astrophysical Journal*, **851**(1), 10.

Fabbiano, G. 1989. X-rays from Normal Galaxies. *Annual Review of Astronomy and Astrophysics*, **27**(Jan.), 87–138.

Fabbiano, G. 2006. Populations of X-ray Sources in Galaxies. *Annual Review of Astronomy and Astrophysics*, **44**(1), 323–366.

Fabian, A. C., Pringle, J. E., and Rees, M. J. 1975. Tidal Capture Formation of Binary Systems and X-ray Sources in Globular Clusters. *Monthly Notices of the Royal Astronomical Society*, **172**(Aug.), 15.

Ferrand, G., and Safi-Harb, S. 2012. A Census of High-Energy Observations of Galactic Supernova Remnants. *Advances in Space Research*, **49**(9), 1313–1319.

Fields, B. D., Pavlidou, V., and Prodanović, T. 2010. Cosmic Gamma-ray Background from Star-Forming Galaxies. *Astrophysical Journal*, **722**(2), L199–L203.

Fishman, G. J., and Meegan, C. A. 1995. Gamma-Ray Bursts. *Annual Review of Astronomy and Astrophysics*, **33**(Jan), 415–458.

Fornasini, F. M., Civano, F., Fabbiano, G. et al. 2018. Low-Luminosity AGN and X-ray Binary Populations in COSMOS Star-Forming Galaxies. *Astrophysical Journal*, **865**(1), 43.

Fragos, T., Lehmer, B., Tremmel, M. et al. 2013. X-ray Binary Evolution Across Cosmic Time. *Astrophysical Journal*, **764**(1), 41.

Gaensler, B. M., and Slane, P. O. 2006. The Evolution and Structure of Pulsar Wind Nebulae. *Annual Review of Astronomy and Astrophysics*, **44**(1), 17–47.

Galbany, L., Stanishev, V., Mourão, A. M. et al. 2014. Nearby Supernova Host Galaxies from the CALIFA Survey. I. Sample, Data Analysis, and Correlation to Star-Forming Regions. *Astronomy and Astrophysics*, **572**(Dec.), A38.

Garofali, K., Williams, B. F., Hillis, T. et al. 2018. Formation Time-Scales for High-Mass X-ray Binaries in M33. *Monthly Notices of the Royal Astronomical Society*, **479**(3), 3526–3544.

Gehrels, N., Sarazin, C. L., O'Brien, P. T. et al. 2005. A Short γ-ray Burst Apparently Associated with an Elliptical Galaxy at Redshift z = 0.225. *Nature*, **437**(7060), 851–854.

Gilfanov, M. 2004. Low-Mass X-ray Binaries as a Stellar Mass Indicator for the Host Galaxy. *Monthly Notices of the Royal Astronomical Society*, **349**(1), 146–168.

Gilfanov, M., Grimm, H. J., and Sunyaev, R. 2004. Statistical Properties of the Combined Emission of a Population of Discrete Sources: Astrophysical Implications. *Monthly Notices of the Royal Astronomical Society*, **351**(4), 1365–1378.

Graham, M. J., Kulkarni, S. R., Bellm, E. C. et al. 2019. The Zwicky Transient Facility: Science Objectives. *Publications of the Astronomical Society of the Pacific*, **131**(1001), 078001.

Graur, Or, Bianco, F. B., and Modjaz, M. 2015. A Unified Explanation for the Supernova Rate-Galaxy Mass Dependence Based on Supernovae Detected in Sloan Galaxy Spectra. *Monthly Notices of the Royal Astronomical Society*, **450**(1), 905–925.

Green, D. A. 2019. A Revised Catalogue of 294 Galactic Supernova Remnants. *Journal of Astrophysics and Astronomy*, **40**(4), 36.

Greiner, J., Michałowski, M. J., Klose, S. et al. 2016. Probing Dust-Obscured Star Formation in the Most Massive Gamma-Ray Burst Host Galaxies. *Astronomy and Astrophysics*, **593**(Aug.), A17.

Grimes, J. P., Heckman, T., Strickland, D. et al. 2005. A Chandra X-ray Investigation of the Violent Interstellar Medium: From Dwarf Starbursts to Ultraluminous Infrared Galaxies. *Astrophysical Journal*, **628**(1), 187–204.

Grimm, H. J., Gilfanov, M., and Sunyaev, R. 2003. High-Mass X-ray Binaries as a Star Formation Rate Indicator in Distant Galaxies. *Monthly Notices of the Royal Astronomical Society*, **339**(3), 793–809.

Groves, B., Dopita, M. A., Sutherland, R. S. et al. 2008. Modeling the Pan-Spectral Energy Distribution of Starburst Galaxies. IV. The Controlling Parameters of the Starburst SED. *Astrophysical Journal Supplement Series*, **176**(2), 438–456.

Harris, J., and Zaritsky, D. 2004. The Star Formation History of the Small Magellanic Cloud. *Astronomical Journal*, **127**(3), 1531–1544.

Hornschemeier, A. E., Wolter, A., and Kim, D.-W. 2016. What Dominates the X-ray Emission of Normal Galaxies? *IAU Focus Meeting*, **29B**(Jan), 124–135.

Humphrey, P. J., and Buote, D. A. 2008. Low-Mass X-ray Binaries and Globular Clusters in Early-Type Galaxies. I. Chandra Observations. *Astrophysical Journal*, **689**(2), 983–1004.

Hurley, J. R., Tout, C. A., and Pols, O. R. 2002. Evolution of Binary Stars and the Effect of Tides on Binary Populations. *Monthly Notices of the Royal Astronomical Society*, **329**(4), 897–928.

Irwin, J. A. 2005. The Birthplace of Low-Mass X-ray Binaries: Field Versus Globular Cluster Populations. *Astrophysical Journal*, **631**(1), 511–517.

Ivanova, N., and Taam, R. E. 2003. Magnetic Braking Revisited. *Astrophysical Journal*, **599**(1), 516–521.

Ivanova, N., Justham, S., Chen, X. et al. 2013. Common Envelope Evolution: Where We Stand and How We Can Move Forward. *Astronomy and Astrophysics Reviews*, **21**(Feb), 59.

Izzard, R. G., Preece, H., Jofre, P. et al. 2018. Binary Stars in the Galactic Thick Disc. *Monthly Notices of the Royal Astronomical Society*, **473**(3), 2984–2999.

Kaaret, P., Feng, H., and Roberts, T. P. 2017. Ultraluminous X-ray Sources. *Annual Review of Astronomy and Astrophysics*, **55**(1), 303–341.

Kennicutt, R. C., Jr. 1998. The Global Schmidt Law in Star-Forming Galaxies. *Astrophysical Journal*, **498**(2), 541–552.

Kennicutt, R. C., and Evans, N. J. 2012. Star Formation in the Milky Way and Nearby Galaxies. *Annual Review of Astronomy and Astrophysics*, **50**(Sep.), 531–608.

Kim, D.-W., and Fabbiano, G. 2010. X-ray Properties of Young Early-type Galaxies. I. X-ray Luminosity Function of Low-Mass X-ray Binaries. *Astrophysical Journal*, **721**(2), 1523–1530.

Kim, D. W., Fabbiano, G., Kalogera, V. et al. 2006. Probing the Low-Luminosity X-ray Luminosity Function in Normal Elliptical Galaxies. *Astrophysical Journal*, **652**(2), 1090–1096.

Kistler, M. D., Yüksel, H., Beacom, J. F. et al. 2009. The Star Formation Rate in the Reionization Era as Indicated by Gamma-Ray Bursts. *Astrophysical Journal*, **705**(2), L104–L108.

Komis, I., Pavlidou, V., and Zezas, A. 2019. Extragalactic Gamma-ray Background from Star-Forming Galaxies: Will Empirical Scalings Suffice? *Monthly Notices of the Royal Astronomical Society*, **483**(3), 4020–4030.

Kopsacheili, M., Zezas, A., and Leonidaki, I. 2020. A Diagnostic Tool for the Identification of Supernova Remnants. *Monthly Notices of the Royal Astronomical Society*, **491**(1), 889–902.

Kovlakas K., Zezas A., Andrews J. J. et al. A Census of Ultraluminous X-ray Sources in the Local Universe. *Monthly Notices of the Royal Astronomical Society*, **498**(4), 4790.

Kouroumpatzakis, K., Zezas, A., Sell, P. et al. 2020. Sub-Galactic Scaling Relations between X-ray Luminosity, Star-Formation Rate, and Stellar Mass. *Monthly Notices of the Royal Astronomical Society*, **494**(4), 5967.

Krivonos, R., Tsygankov, S., Lutovinov, A. et al. 2012. INTEGRAL/IBIS Nine-Year Galactic Hard X-ray Survey. *Astronomy and Astrophysics*, **545**(Sep.), A27.

Lacki, B. C., and Thompson, T. A. 2013. Diffuse Hard X-ray Emission in Starburst Galaxies as Synchrotron from Very High Energy Electrons. *Astrophysical Journal*, **762**(1), 29.

Lehmer, B. D., Alexander, D. M., Bauer, F. E. et al. 2010. A Chandra Perspective on Galaxy-Wide X-ray Binary Emission and Its Correlation with Star Formation Rate and Stellar Mass: New Results from Luminous Infrared Galaxies. *Astrophysical Journal*, **724**(1), 559–571.

Lehmer, B. D., Berkeley, M., Zezas, A. et al. 2014. The X-ray Luminosity Functions of Field Low-Mass X-ray Binaries in Early-Type Galaxies: Evidence for a Stellar Age Dependence. *Astrophysical Journal*, **789**(1), 52.

Lehmer, B. D., Basu-Zych, A. R., Mineo, S. et al. 2016. The Evolution of Normal Galaxy X-ray Emission through Cosmic History: Constraints from the 6 MS Chandra Deep Field-South. *Astrophysical Journal*, **825**(1), 7.

Lehmer, B. D., Brandt, W. N., Alexander, D. M. et al. 2008. Tracing the Mass-Dependent Star Formation History of Late-Type Galaxies Using X-ray Emission: Results from the Chandra Deep Fields. *Astrophysical Journal*, **681**(2), 1163–1182.

Lehmer, B. D., Eufrasio, R. T., Markwardt, L. et al. 2017. On the Spatially Resolved Star Formation History in M51. II. X-ray Binary Population Evolution. *Astrophysical Journal*, **851**(1), 11.

Lehmer, B. D., Eufrasio, R. T., Tzanavaris, P. et al. 2019. X-ray Binary Luminosity Function Scaling Relations for Local Galaxies Based on Subgalactic Modeling. *Astrophysical Journal Supplement Series*, **243**(1), 3.

Leibler, C. N., and Berger, E. 2010. The Stellar Ages and Masses of Short Gamma-ray Burst Host Galaxies: Investigating the Progenitor Delay Time Distribution and the Role of Mass and Star Formation in the Short Gamma-ray Burst Rate. *Astrophysical Journal*, **725**(1), 1202–1214.

Leonidaki, I., Boumis, P., and Zezas, A. 2013. A Multiwavelength Study of Supernova Remnants in Six Nearby Galaxies – II. New Optically Selected Supernova Remnants. *Monthly Notices of the Royal Astronomical Society*, **429**(1), 189–220.

Lin, D., Irwin, J. A., Wong, K.-W. et al. 2015. The Megasecond Chandra X-ray Visionary Project Observation of NGC 3115. III. Luminosity Functions of LMXBs and Dependence on Stellar Environments. *Astrophysical Journal*, **808**(1), 20.

Linden, T., Kalogera, V., Sepinsky, J. F. et al. 2010. The Effect of Starburst Metallicity on Bright X-ray Binary Formation Pathways. *Astrophysical Journal*, **725**(2), 1984–1994.

Long, K. S. 2017. *Galactic and Extragalactic Samples of Supernova Remnants: How They Are Identified and What They Tell Us*. Page 2005.

Lopez, L. A., Ramirez-Ruiz, E., Huppenkothen, D. et al. 2011. Using the X-ray Morphology of Young Supernova Remnants to Constrain Explosion Type, Ejecta Distribution, and Chemical Mixing. *Astrophysical Journal*, **732**(2), 114.

Madau, P., and Dickinson, M. 2014. Cosmic Star-Formation History. *Annual Review of Astronomy and Astrophysics*, **52**(Aug.), 415–486.

Maoz, D., and Badenes, C. 2010. The Supernova Rate and Delay Time Distribution in the Magellanic Clouds. *Monthly Notices of the Royal Astronomical Society*, **407**(2), 1314–1327.

Mapelli, M., Ripamonti, E., Zampieri, L. et al. 2010. Ultra-luminous X-ray Sources and Remnants of Massive Metal-Poor Stars. *Monthly Notices of the Royal Astronomical Society*, **408**(1), 234–253.

Martínez-Galarza, J. R., Smith, H. A., Lanz, L. et al. 2016. Variations of the ISM Compactness across the Main Sequence of Star Forming Galaxies: Observations and Simulations. *Astrophysical Journal*, **817**(1), 76.

Mathewson, D. S., and Clarke, J. N. 1973. Supernova Remnants in the Large Magellanic Cloud. *Astrophysical Journal*, **180**(Mar.), 725–738.

Matsumoto, H., and Tsuru, T. G. 1999. X-ray Evidence of an AGN in M82. *Publications of the Astronomical Society of Japan*, **51**(June), 321–331.

Metzger, B. D. 2019. Kilonovae. *Living Reviews in Relativity*, **23**(1), 1.

Mészáros, P. 2002. Theories of Gamma-Ray Bursts. *Annual Review of Astronomy and Astrophysics*, **40**(Jan.), 137–169.

McQuinn, K. B. W., Skillman, E. D., Cannon, J. M. et al. 2010. The Nature of Starbursts. I. The Star Formation Histories of Eighteen Nearby Starburst Dwarf Galaxies. *Astrophysical Journal*, **721**(1), 297–317.

Mineo, S., Gilfanov, M., and Sunyaev, R. 2012a. X-ray Emission from Star-Forming Galaxies – I. High-Mass X-ray Binaries. *Monthly Notices of the Royal Astronomical Society*, **419**(3), 2095–2115.

Mineo, S., Gilfanov, M., and Sunyaev, R. 2012b. X-ray Emission from Star-Forming Galaxies – II. Hot Interstellar Medium. *Monthly Notices of the Royal Astronomical Society*, **426**(3), 1870–1883.

Mineo, S., Gilfanov, M., Lehmer, B. D. et al. 2014. X-ray Emission from Star-Forming Galaxies – III. Calibration of the L_X-SFR Relation Up to Redshift $z \approx 1.3$. *Monthly Notices of the Royal Astronomical Society*, **437**(2), 1698–1707.

Olejak, A., Belczynski, K., Bulik, T. et al. 2020. Synthetic Catalog of Black Holes in the Milky Way. *Astronomy and Astrophysics*, **638**(June), A94.

Palmerio, J. T., Vergani, S. D., Salvaterra, R. et al. 2019. Are Long Gamma-Ray Bursts Biased Tracers of Star Formation? Clues from the Host Galaxies of the Swift/BAT6 Complete Sample of Bright LGRBs. III. Stellar Masses, Star Formation Rates, and Metallicities at z ¿ 1. *Astronomy and Astrophysics*, **623**(Mar.), A26.

Pavlidou, V., and Fields, B. D. 2001. Diffuse Gamma Rays from Local Group Galaxies. *Astrophysical Journal*, **558**(1), 63–71.

Peacock, M. B., and Zepf, S. E. 2016. The X-ray Luminosity Function of Low-Mass X-ray Binaries in Early-Type Galaxies, Their Metal-Rich, and Metal-Poor Globular Clusters. *Astrophysical Journal*, **818**(1), 33.

Podsiadlowski, Ph., Langer, N., Poelarends, A. J. T. et al. 2004. The Effects of Binary Evolution on the Dynamics of Core Collapse and Neutron Star Kicks. *Astrophysical Journal*, **612**(2), 1044–1051.

Prestwich, A. H., Tsantaki, M., Zezas, A. et al. 2013. Ultra-Luminous X-ray Sources in the Most Metal Poor Galaxies. *Astrophysical Journal*, **769**(2), 92.

Ptak, A., and Griffiths, R. 1999. Hard X-ray Variability in M82: Evidence for a Nascent Active Galactic Nucleus? *Astrophysical Journal*, **517**(June), L85–L89.

Ranalli, P., Comastri, A., and Setti, G. 2003. The 2–10 keV Luminosity as a Star Formation Rate indicator. *Astronomy and Astrophysics*, **399**(Feb.), 39–50.

Rauw, G., and Nazé, Y. 2016. X-ray Emission from Interacting Wind Massive Binaries: A Review of 15 Years of Progress. *Advances in Space Research*, **58**(5), 761–781.

Remillard, R. A., and McClintock, J. E. 2006. X-ray Properties of Black-Hole Binaries. *Annual Review of Astronomy and Astrophysics*, **44**(1), 49–92.

Reynolds, S. P. 2008. Supernova Remnants at High Energy. *Annual Review of Astronomy and Astrophysics*, **46**(Sep.), 89–126.

Schady, P. 2017. Gamma-ray Bursts and their Use as Cosmic Probes. *Royal Society Open Science*, **4**(7), 170304.

Sell, P. H., Pooley, D., Zezas, A. et al. 2011. Luminosity Functions and Point-Source Properties from Multiple Chandra Observations of M81. *Astrophysical Journal*, **735**(1), 26.

Shtykovskiy, P. E., and Gilfanov, M. 2005. High Mass X-ray Binaries in the LMC: Dependence on the Stellar Population Age and the "Propeller" Effect. *Astronomy and Astrophysics*, **431**(Feb.), 597–614.

Shtykovskiy, P. E., and Gilfanov, M. R. 2007. High-Mass X-ray Binaries and Recent Star Formation History of the Small Magellanic Cloud. *Astronomy Letters*, **33**(7), 437–454.

Smith, B. J., Campbell, K., Struck, C. et al. 2018. Diffuse X-ray-Emitting Gas in Major Mergers. *Astronomical Journal*, **155**(2), 81.

Soria, R., and Wu, K. 2003. Properties of Discrete X-ray Sources in the Starburst Spiral Galaxy M 83. *Astronomy and Astrophysics*, **410**(Oct), 53–74.

Stevens, I. R., and Strickland, D. K. 1998. A ROSAT Survey of Wolf–Rayet Galaxies – II. The Extended Sample. *Monthly Notices of the Royal Astronomical Society*, **301**(1), 215–230.

Strickland, D. K., and Heckman, T. M. 2009. Supernova Feedback Efficiency and Mass Loading in the Starburst and Galactic Superwind Exemplar M82. *Astrophysical Journal*, **697**(2), 2030–2056.

Strickland, D. K., Heckman, T. M., Colbert, E. J. M. et al. 2004. A High Spatial Resolution X-ray and Hα Study of Hot Gas in the Halos of Star-Forming Disk Galaxies. I. Spatial and Spectral Properties of the Diffuse X-ray Emission. *Astrophysical Journal Supplement Series*, **151**(2), 193–236.

Strong, A. W. 2007. Source Population Synthesis and the Galactic Diffuse Gamma-Ray Emission. *Astrophysics and Space Science*, **309**(1–4), 35–41.

Swartz, D. A., Ghosh, K. K., McCollough, M. L. et al. 2003. Chandra X-ray Observations of the Spiral Galaxy M81. *Astrophysical Journal Supplement Series*, **144**(2), 213–242.

Tammann, G. A., Loeffler, W., and Schroeder, A. 1994. The Galactic Supernova Rate. *Astrophysical Journal Supplement Series*, **92**(June), 487.

Tauris, T. M., and van den Heuvel, E. P. J. 2006. *Formation and Evolution of Compact Stellar X-ray Sources*. Page 623.

Tetarenko, B. E., Sivakoff, G. R., Heinke, C. O. et al. 2016. WATCHDOG: A Comprehensive All-Sky Database of Galactic Black Hole X-ray Binaries. *Astrophysical Journal Supplement Series*, **222**(2), 15.

The Fermi-LAT collaboration. 2019. Fermi Large Area Telescope Fourth Source Catalog. *arXiv e-prints*, Feb, arXiv:1902.10045.

Trenti, M., Perna, R., and Tacchella, S. 2013. Gamma-Ray Burst and Star Formation Rates: The Physical Origin for the Redshift Evolution of their Ratio. *Astrophysical Journal*, **773**(2), L22.

Trenti, M., Perna, R., Levesque, E. M. et al. 2012. Gamma-Ray Burst Host Galaxy Surveys at Redshift z >~4: Probes of Star Formation Rate and Cosmic Reionization. *Astrophysical Journal*, **749**(2), L38.

Tzanavaris, P., Fragos, T., Tremmel, M. et al. 2013. Modeling X-ray Binary Evolution in Normal Galaxies: Insights from SINGS. *Astrophysical Journal*, **774**(2), 136.

van Paradijs, J., Kouveliotou, C., and Wijers, R. A. M. J. 2000. Gamma-Ray Burst Afterglows. *Annual Review of Astronomy and Astrophysics*, **38**(Jan.), 379–425.

Veilleux, S., Cecil, G., and Bland-Hawthorn, J. 2005. Galactic Winds. *Annual Review of Astronomy and Astrophysics*, **43**(1), 769–826.

Veitch, J., Raymond, V., Farr, B. et al. 2015. Parameter Estimation for Compact Binaries with Ground-Based Gravitational-Wave Observations Using the LALInference Software Library. *Physical Review D: Particles, Fields, Gravitation and Cosmology*, **91**(4), 042003.

Vink, J. 2012. Supernova Remnants: The X-ray Perspective. *Astronomy and Astrophysics Reviews*, **20**(Dec), 49.

Vitale, S., Farr, W. M., Ng, K. K. Y. et al. 2019. Measuring the Star Formation Rate with Gravitational Waves from Binary Black Holes. *Astrophysical Journal*, **886**(1), L1.

Voss, R., and Gilfanov, M. 2007. The Dynamical Formation of LMXBs in Dense Stellar Environments: Globular Clusters and the Inner Bulge of M31. *Monthly Notices of the Royal Astronomical Society*, **380**(4), 1685–1702.

Vulic, N., Hornschemeier, A. E., Wik, D. R. et al. 2018. Black Holes and Neutron Stars in Nearby Galaxies: Insights from NuSTAR. *Astrophysical Journal*, **864**(2), 150.

Wanderman, D., and Piran, T. 2015. The Rate, Luminosity Function and Time Delay of Non-Collapsar Short GRBs. *Monthly Notices of the Royal Astronomical Society*, **448**(4), 3026–3037.

Wang, S., Qiu, Y., Liu, J. et al. 2016. Chandra ACIS Survey of X-ray Point Sources in Nearby Galaxies. II. X-ray Luminosity Functions and Ultraluminous X-ray Sources. *Astrophysical Journal*, **829**(1), 20.

Wang, X., and Fields, B. D. 2018. Are Starburst Galaxies Proton Calorimeters? *Monthly Notices of the Royal Astronomical Society*, **474**(3), 4073–4088.

Weisz, D. R., Dalcanton, J. J., Williams, B. F. et al. 2011. The ACS Nearby Galaxy Survey Treasury. VIII. The Global Star Formation Histories of 60 Dwarf Galaxies in the Local Volume. *Astrophysical Journal*, **739**(1), 5.

Wik, D. R., Lehmer, B. D., Hornschemeier, A. E. et al. 2014. Spatially Resolving a Starburst Galaxy at Hard X-ray Energies: NuSTAR, Chandra, and VLBA Observations of NGC 253. *Astrophysical Journal*, **797**(2), 79.

Wiktorowicz, G., Sobolewska, M., Lasota, J.-P. et al. 2017. The Origin of the Ultraluminous X-ray Sources. *Astrophysical Journal*, **846**(1), 17.

Williams, B. F., Binder, B. A., Dalcanton, J. J. et al. 2013. The Ages of High-Mass X-ray Binaries in NGC 2403 and NGC 300. *Astrophysical Journal*, **772**(1), 12.

Williams, B. F., Lazzarini, M., Plucinsky, Paul P. et al. 2018. Comparing Chandra and Hubble in the Northern Disk of M31. *Astrophysical Journal Supplement Series*, **239**(1), 13.

Woosley, S. E., and Bloom, J. S. 2006. The Supernova Gamma-Ray Burst Connection. *Annual Review of Astronomy and Astrophysics*, **44**(1), 507–556.

Yüksel, H., Kistler, M. D., Beacom, J. F. et al. 2008. Revealing the High-Redshift Star Formation Rate with Gamma-Ray Bursts. *Astrophysical Journal*, **683**(1), L5.

Zafar, T., Watson, D., Møller, P. et al. 2018. VLT/X-shooter GRBs: Individual Extinction Curves of Star-Forming Regions. *Monthly Notices of the Royal Astronomical Society*, **479**(2), 1542–1554.

Zezas, A., Fabbiano, G., Baldi, A. et al. 2007. Chandra Monitoring Observations of the Antennae Galaxies. II. X-ray Luminosity Functions. *Astrophysical Journal*, **661**(1), 135–148.

Zezas, A. 2008 (May). X-ray Source Populations in Nearby Galaxies. Pages 281–290 of: Bandyopadhyay, R. M., Wachter, S., Gelino, D. et al. (eds), *A Population Explosion: The Nature & Evolution of X-ray Binaries in Diverse Environments*. American Institute of Physics Conference Series, vol. 1010.

Zhang, Z., Gilfanov, M., and Bogdán, Á. 2012. Dependence of the Low-Mass X-ray Binary Population on Stellar Age. *Astronomy and Astrophysics*, **546**(Oct.), A36.

Index

For EU product safety concerns, contact us at Calle de José Abascal, 56–1°, 28003 Madrid, Spain or eugpsr@cambridge.org.

www.ingramcontent.com/pod-product-compliance
Ingram Content Group UK Ltd.
Pitfield, Milton Keynes, MK11 3LW, UK
UKHW050650180926
15242UKWH00007B/1
* 9 7 8 1 1 0 7 1 8 4 1 6 9 *